Deep Learning in Biometrics

Deep Learning in Biometrics

Edited by
Mayank Vatsa
Richa Singh
Angshul Majumdar

CRC Press
Taylor & Francis Group
Boca Raton London New York

CRC Press is an imprint of the
Taylor & Francis Group, an **informa** business

CRC Press
Taylor & Francis Group
6000 Broken Sound Parkway NW, Suite 300
Boca Raton, FL 33487-2742

First issued in paperback 2023

© 2018 by Taylor & Francis Group, LLC
CRC Press is an imprint of Taylor & Francis Group, an Informa business

No claim to original U.S. Government works

**Visit the Taylor & Francis Web site at
http://www.taylorandfrancis.com**

**and the CRC Press Web site at
http://www.crcpress.com**

ISBN 13: 978-1-032-65310-5 (pbk)
ISBN 13: 978-1-138-57823-4 (hbk)
ISBN 13: 978-1-351-26500-3 (ebk)

DOI: 10.1201/b22524

Contents

Editors   vii

Contributors   ix

1   Deep Learning: Fundamentals and Beyond   1
Shruti Nagpal, Maneet Singh, Mayank Vatsa, and Richa Singh

2   Unconstrained Face Identification and Verification Using
Deep Convolutional Features   33
*Jun-Cheng Chen, Rajeev Ranjan, Vishal M. Patel,
Carlos D. Castillo, and Rama Chellappa*

3   Deep Siamese Convolutional Neural Networks
for Identical Twins and Look-Alike Identification   65
Xiaoxia Sun, Amirsina Torfi, and Nasser Nasrabadi

4   Tackling the Optimization and Precision Weakness
of Deep Cascaded Regression for Facial
Key-Point Localization   85
Yuhang Wu, Shishir K. Shah, and Ioannis A. Kakadiaris

5   Learning Deep Metrics for Person Reidentification   109
Hailin Shi, Shengcai Liao, Dong Yi, and Stan Z. Li

6   Deep Face-Representation Learning for
Kinship Verification   127
*Naman Kohli, Daksha Yadav, Mayank Vatsa, Richa Singh,
and Afzel Noore*

7   What's Hiding in My Deep Features?   153
*Ethan M. Rudd, Manuel Günther, Akshay R. Dhamija,
Faris A. Kateb, and Terrance E. Boult*

8   Stacked Correlation Filters   175
*Jonathon M. Smereka, Vishnu Naresh Boddeti,
and B. V. K. Vijaya Kumar*

9   **Learning Representations for Unconstrained Fingerprint Recognition**   **197**
    Aakarsh Malhotra, Anush Sankaran, Mayank Vatsa, and Richa Singh

10 **Person Identification Using Handwriting Dynamics and Convolutional Neural Networks**   **227**
    Gustavo H. Rosa, João P. Papa, and Walter J. Scheirer

11 **Counteracting Presentation Attacks in Face, Fingerprint, and Iris Recognition**   **245**
    Allan Pinto, Helio Pedrini, Michael Krumdick, Benedict Becker, Adam Czajka, Kevin W. Bowyer, and Anderson Rocha

12 **Fingervein Presentation Attack Detection Using Transferable Features from Deep Convolution Neural Networks**   **295**
    Raghavendra Ramachandra, Kiran B. Raja, Sushma Venkatesh, and Christoph Busch

Index   **307**

Editors

Mayank Vatsa received M.S. and Ph.D. degrees in computer science from West Virginia University, Morgantown in 2005 and 2008, respectively. He is currently an Associate Professor with the Indraprastha Institute of Information Technology, Delhi, India, and a Visiting Professor at West Virginia University. He is also the Head for the Infosys Center on Artificial Intelligence at IIIT Delhi. His research has been funded by UIDAI and DeitY, Government of India. He has authored over 200 publications in refereed journals, book chapters, and conferences. His areas of interest are biometrics, image processing, computer vision, and information fusion. He is a recipient of the AR Krishnaswamy Faculty Research Fellowship, the FAST Award by DST, India, and several best paper and best poster awards in international conferences. He is also the Vice President (Publications) of IEEE Biometrics Council, an Associate Editor of the IEEE ACCESS, and an Area Editor of *Information Fusion* (Elsevier). He served as the PC Co-Chair of ICB 2013, IJCB 2014, and ISBA 2017.

Richa Singh received a Ph.D. degree in Computer Science from West Virginia University, Morgantown, in 2008. She is currently an Associate Professor with the IIIT Delhi, India, and an Adjunct Associate Professor at West Virginia University. She is a Senior Member of both IEEE and ACM. Her areas of interest are biometrics, pattern recognition, and machine learning. She is a recipient of the Kusum and Mohandas Pai Faculty Research Fellowship at the IIIT Delhi, the FAST Award by the Department of Science and Technology, India, and several best paper and best poster awards in international conferences. She has published over 200 research papers in journals, conferences, and book chapters. She is also an Editorial Board Member of Information Fusion (Elsevier), Associate Editor of Pattern Recognition, IEEE Access, and the EURASIP Journal on Image and Video Processing (Springer). She has also served as the Program Co-Chair of IEEE BTAS 2016 and General Co-Chair of ISBA 2017. She is currently serving as Program Co-Chair of International Workshop on Biometrics and Forensics, 2018, and International Conference on Automatic Face and Gesture Recognition, 2019.

Angshul Majumdar received his Master's and Ph.D. from the University of British Columbia in 2009 and 2012, respectively. Currently he is an assistant professor at Indraprastha Institute of Information Technology, Delhi. His research interests are broadly in the areas of signal processing and machine learning. He has co-authored over 150 papers in journals and reputed conferences. He is the author of *Compressed Sensing for Magnetic Resonance Image Reconstruction* published by Cambridge University Press and co-editor of *MRI: Physics, Reconstruction, and Analysis* published by CRC Press. He is currently serving as the chair of the IEEE SPS Chapter's committee and the chair of the IEEE SPS Delhi Chapter.

Contributors

Benedict Becker
Department of Computer Science
and Engineering
University of Notre Dame
Notre Dame, Indiana

Vishnu Naresh Boddeti
Department of Computer Science
and Engineering
Michigan State University
East Lansing, Michigan

Terrance E. Boult
Vision and Security Technology Lab
University of Colorado–Colorado
Springs
Colorado Springs, Colorado

Kevin W. Bowyer
Department of Computer Science
and Engineering
University of Notre Dame
Notre Dame, Indiana

Christoph Busch
Norwegian Biometrics Laboratory
Norwegian University of Science
and Technology (NTNU)
Gjøvik, Norway

Carlos D. Castillo
Department of Electrical
and Computer Engineering
University of Maryland
College Park, Maryland

Rama Chellappa
Department of Electrical
and Computer Engineering
University of Maryland
College Park, Maryland

Jun-Cheng Chen
Department of Electrical
and Computer Engineering
University of Maryland
College Park, Maryland

Adam Czajka
Department of Computer Science
and Engineering
University of Notre Dame
Notre Dame, Indiana
Research and Academic Computer
Network (NASK)
and
Warsaw University of Technology
Warsaw, Poland

Akshay R. Dhamija
Vision and Security Technology Lab
University of Colorado–Colorado
Springs
Colorado Springs, Colorado

Manuel Günther
Vision and Security Technology Lab
University of Colorado–Colorado
Springs
Colorado Springs, Colorado

Ioannis A. Kakadiaris
Computational Biomedicine
    Laboratory
University of Houston
Houston, Texas

Faris A. Kateb
Vision and Security Technology Lab
University of Colorado–Colorado
    Springs
Colorado Springs, Colorado

Naman Kohli
Lane Department of
    Computer Science and
    Electrical Engineering
West Virginia University
Morgantown, West Virginia

Michael Krumdick
Department of Computer Science
    and Engineering
University of Notre Dame
Notre Dame, Indiana

Stan Z. Li
Center for Biometrics and Security
    Research, National Laboratory
    of Pattern Recognition
Institute of Automation,
    Chinese Academy of Sciences
Beijing, China

Shengcai Liao
Center for Biometrics and Security
    Research, National Laboratory
    of Pattern Recognition
Institute of Automation,
    Chinese Academy of Sciences
Beijing, China

Aakarsh Malhotra
Department of Computer Science
IIIT–Delhi
New Delhi, India

Shruti Nagpal
Department of Computer Science
IIIT–Delhi
New Delhi, India

Nasser Nasrabadi
Lane Department of
    Computer Science and
    Electrical Engineering
West Virginia University
Morgantown, West Virginia

Afzel Noore
Lane Department of
    Computer Science and
    Electrical Engineering
West Virginia University
Morgantown, West Virginia

João P. Papa
Department of Computing
São Paulo State University
Bauru, Brazil

Vishal M. Patel
Department of Electrical
    and Computer Engineering
Rutgers University
New Brunswick, New Jersey

Helio Pedrini
Institute of Computing
University of Campinas
Campinas, Brazil

Allan Pinto
Institute of Computing
University of Campinas
Campinas, Brazil

Kiran B. Raja
Norwegian Biometrics Laboratory
Norwegian University of Science
    and Technology (NTNU)
Gjøvik, Norway

Raghavendra Ramachandra
Norwegian Biometrics Laboratory
Norwegian University of Science
    and Technology (NTNU)
Gjøvik, Norway

Rajeev Ranjan
Department of Electrical
    and Computer Engineering
University of Maryland
College Park, Maryland

Anderson Rocha
Institute of Computing
University of Campinas
Campinas, Brazil

Gustavo H. Rosa
Department of Computing
São Paulo State University
Bauru, Brazil

Ethan M. Rudd
Vision and Security Technology Lab
University of Colorado–Colorado
    Springs
Colorado Springs, Colorado

Anush Sankaran
Department of Computer Science
IIIT–Delhi
New Delhi, India

Walter J. Scheirer
Department of Computer Science
    and Engineering
University of Notre Dame
Notre Dame, Indiana

Shishir K. Shah
Computational Biomedicine
    Laboratory
University of Houston
Houston, Texas

Hailin Shi
Center for Biometrics and Security
    Research, National Laboratory
    of Pattern Recognition
Institute of Automation,
    Chinese Academy of Sciences
Beijing, China

Maneet Singh
Department of Computer Science
IIIT–Delhi
New Delhi, India

Richa Singh
Department of Computer Science
IIIT–Delhi
New Delhi, India

Jonathon M. Smereka
Department of Electrical
    and Computer Engineering
Carnegie Mellon University
Pittsburgh, Pennsylvania

Xiaoxia Sun
Department of Electrical
    and Computer Engineering
Johns Hopkins University
Baltimore, Maryland

Amirsina Torfi
Lane Department of
    Computer Science and
    Electrical Engineering
West Virginia University
Morgantown, West Virginia

Mayank Vatsa
Department of Computer Science
IIIT–Delhi
New Delhi, India

Sushma Venkatesh
Norwegian Biometrics Laboratory
Norwegian University of Science
    and Technology (NTNU)
Gjøvik, Norway

B. V. K. Vijaya Kumar
Department of Electrical
 and Computer Engineering
Carnegie Mellon University
Pittsburgh, Pennsylvania

Yuhang Wu
Computational Biomedicine
 Laboratory
University of Houston
Houston, Texas

Daksha Yadav
Lane Department of
 Computer Science and
 Electrical Engineering
West Virginia University
Morgantown, West Virginia

Dong Yi
Alibaba Group
Hangzhou, China

1

Deep Learning: Fundamentals and Beyond

Shruti Nagpal, Maneet Singh, Mayank Vatsa, and Richa Singh

CONTENTS

1.1	Introduction	1
1.2	Restricted Boltzmann Machine	5
	1.2.1 Incorporating supervision in RBMs	7
	1.2.2 Other advances in RBMs	7
	1.2.3 RBMs for biometrics	8
1.3	Autoencoder	9
	1.3.1 Incorporating supervision in AEs	11
	1.3.2 Other variations of AEs	12
1.4	Convolutional Neural Networks	14
	1.4.1 Architecture of a traditional CNNs	14
	1.4.2 Existing architectures of CNNs	15
1.5	Other Deep Learning Architectures	20
1.6	Deep Learning: Path Ahead	21
References	....	22

1.1 Introduction

The science of uniquely identifying a person based on his or her physiological or behavioral characteristics is termed *biometrics*. Physiological characteristics include face, iris, fingerprint, and DNA, whereas behavioral modalities include handwriting, gait, and keystroke dynamics. Jain et al. [1] lists seven factors that are essential for any trait (formally termed *modality*) to be used for biometric authentication. These factors are: universality, uniqueness, permanence, measurability, performance, acceptability, and circumvention.

An automated biometric system aims to either correctly predict the identity of the instance of a modality or verify whether the given sample is the same as the existing sample stored in the database. Figure 1.1 presents a traditional pipeline of a biometric authentication system. Input data corresponds to the

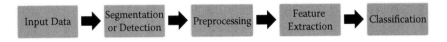

FIGURE 1.1
Illustrating the general biometrics authentication pipeline, which consists of five stages.

raw data obtained directly from the sensor(s). Segmentation, or detection, refers to the process of extracting the region of interest from the given input. Once the required region of interest has been extracted, it is preprocessed to remove the noise, enhance the image, and normalize the data for subsequent ease of processing. After segmentation and preprocessing, the next step in the pipeline is feature extraction. Feature extraction refers to the process of extracting unique and discriminatory information from the given data. These features are then used for performing classification. Classification refers to the process of creating a model, which given a seen/unseen input feature vector is able to provide its correct label. For example, in case of a face recognition pipeline, the aim is to identify the individual in the given input sample. Here, the input data consists of images captured from the camera, containing at least one face image along with background or other objects. Segmentation, or detection, corresponds to detecting the face in the given input image. Several techniques can be applied for this step [2,3]; the most common being the Viola Jones face detector [4]. Once the faces are detected, they are normalized with respect to their geometry and intensity. For feature extraction, hand-crafted features such as Gabor filterbank [5], histogram of oriented gradients [6], and local binary patterns [7] and more recently, representation learning approaches have been used. The extracted features are then provided to a classifier such as a support vector machine [8] or random decision forest [9] for classification.

Automated biometric authentication systems have been used for several real-world applications, ranging from fingerprint sensors on mobile phones to border control applications at airports. One of the large-scale applications of automated biometric authentication is the ongoing project of Unique Identification Authority of India, pronounced "Aadhaar." Initiated by the Indian Government,* the project aims to provide a unique identification number for each resident of India and capture his or her biometric modalities—face, fingerprint, and irises. This is done in an attempt to facilitate digital authentication anytime, anywhere, using the collected biometric data. Currently, the project has enrolled more than 1.1 billion individuals. Such large-scale projects often result in data having large intraclass variations, low interclass variations, and unconstrained environments.

---

*https://uidai.gov.in/

FIGURE 1.2
Sample images showcasing the large intraclass and low interclass variations that can be observed for the problem of face recognition. All images are taken from the Internet: (a) Images belonging to the same subject depicting high intraclass variations and (b) Images belonging to the different subjects showing low interclass variations. (Top, from left to right: https://tinyurl. com/y7hbvwsy, https://tinyurl.com/ydx3mvbf, https://tinyurl.com/y9uryu, https://tinyurl.com/y8lrnvrm; bottom from left to right, https://tinyurl.com/ ybgvst84, https://tinyurl.com/y8762gl3, https://tinyurl.com/y956vrb6.)

Figure 1.2 presents sample face images that illustrate the low interclass and high intraclass variations that can be observed in face recognition. In an attempt to model the challenges of real-world applications, several large-scale data sets, such as MegaFace [10], CelebA [11], and CMU Multi-PIE [12] have been prepared. The availability of large data sets and sophisticated technologies (both hardware and algorithms) provide researchers the resources to model the variations observed in the data. These variations can be modeled in either of the four stages shown in Figure 1.1. Each of the four stages in the biometrics pipeline can also be viewed as separate machine learning tasks, which involve learning of the optimal parameters to enhance the final authentication performance. For instance, in the segmentation stage, each pixel can be classified into modality (foreground) or background [13,14]. Similarly, at the time of preprocessing, based on prior knowledge, different techniques can be applied depending on the type or quality of input [15]. Moreover, because of the progress in machine learning research, feature extraction is now viewed as a learning task.

Traditionally, research in feature extraction focused largely on hand-crafted features such as Gabor and Haralick features [5,16], histogram of oriented gradients [6], and local binary patterns [7]. Many such hand-crafted

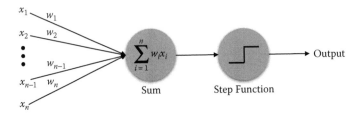

FIGURE 1.3
Pictorial representation of a perceptron.

features encode the pixel variations in the images to generate robust feature vectors for performing classification. Building on these, more complex hand-crafted features are also proposed that encode rotation and scale variations in the feature vectors as well [17,18]. With the availability of training data, researchers have started focusing on learning-based techniques, resulting in several representation learning-based algorithms. Moreover, because the premise is to train the machines for tasks performed with utmost ease by humans, it seemed fitting to understand and imitate the functioning of the human brain. This led researchers to reproduce similar structures to automate complex tasks, which gave rise to the domain of deep learning. Research in deep learning began with the single unit of a perceptron [19], which was able to mimic the behavior of a single brain neuron. Figure 1.3 illustrates a perceptron for an input vector of dimensionality $n \times 1$, that is, $[x_1, x_2, \ldots, x_n]$. The perceptron generates an output based on the input as follows:

$$\text{output} = \begin{cases} 1, \text{ if } \sum_{i=1}^{n} w_i x_i > 0 \\ 0, \text{ if } \sum_{i=1}^{n} w_i x_i \leq 0 \end{cases} \tag{1.1}$$

where w_i corresponds to the weight for the ith element of the input. The behavior of the perceptron is said to be analogous to that of a neuron, since, depending on a fixed threshold, the output would become 1 or 0. Thus, behaving like a neuron receiving electrical signal (input), and using the synapse (weight) to *fire* its output. Treating the perceptron as a building block, several complex architectures have further been proposed. Over the past few years, the domain of deep learning has seen steep development. It is being used to address a multitude of problems with applications in biometrics, object recognition, speech, and natural language processing.

Deep learning architectures can broadly be categorized into three paradigms: restricted Boltzmann machines (RBMs), autoencoders, and convolutional neural networks (CNNs). Restricted Boltzmann machines and

autoencoders are traditionally unsupervised models used for learning meaningful representations of the given data. CNNs, on the other hand, are traditionally supervised models with the objective of improving the overall classification
performance. Each of these architectures are discussed in detail in following
sections.

1.2 Restricted Boltzmann Machine

A restricted Boltzmann machine (RBM) is an unsupervised generative artificial neural network model used for learning representations of a given set
of input. It was first introduced in 1986 with the name Harmonium [20]
and was built on the traditional Boltzmann machine. As can be seen from
Figure 1.4a, a Boltzmann machine is a fully connected graphical model consisting of hidden and visible layers, such that each node (unit) is connected to
all other nodes of the graph. An RBM is created by *restricting* the within-layer
connections of the hidden and visible layers (Figure 1.4b). The visible layer
corresponds to the known input data, and the hidden layer corresponds to the
representation learned by the model. For a given binary vector of visible units
(v) and a binary vector of hidden units (h), the energy function of the model
is written as:

$$E(v,h) = -\sum_{i=1}^{n} a_i v_i - \sum_{j=1}^{r} b_j h_j - \sum_{i=1}^{n}\sum_{j=1}^{r} v_i h_j w_{i,j} \qquad (1.2)$$

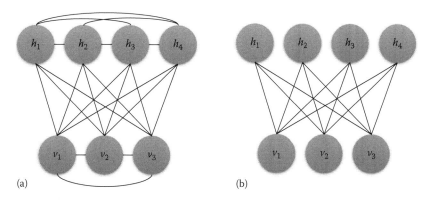

(a)   (b)

FIGURE 1.4
Pictorial representation of a Boltzmann machine and an RBM having a single
visible layer of three nodes (v_1-v_3) and a single hidden layer of four nodes
(h_1-h_4). The RBM does not have within-layer connections for the hidden and
visible layers: (a) Boltzmann Machine and (b) RBM.

where:

   n and r correspond to the number of visible and hidden units in the model, and $v \in \{0, 1\}^n$, $h \in \{0, 1\}^r$

   a and b are the visible and hidden bias vectors, respectively

   $w_{i,j}$ is the weight connection between the visible unit v_i and the hidden unit h_j

Therefore, the energy function consists of three terms, one for the visible (input) data, one for the hidden representation, and the third for modeling the relationship between the hidden and visible vectors. In matrix form, Equation 1.2 can be written as:

$$E(v, h) = -a^T v - b^T h - v^T \mathbf{W} h \qquad (1.3)$$

Being a probabilistic model, the network defines the probability distribution over the visible and hidden vectors as follows:

$$P(v, h) = \frac{1}{Z} e^{-E(v,h)} \qquad (1.4)$$

where, Z is a normalization constant (termed *partition function*), defined as the sum of the energy function over all combinations ($Z = \sum_{v,h} e^{-E(v,h)}$). Building on Equation 1.4, the probability that a network assigns to a particular visible vector can be calculated as follows:

$$P(v) = \frac{1}{Z} \sum_h e^{-E(v,h)} \qquad (1.5)$$

Thus, the loss function of an RBM can be expressed as the negative log-likelihood of the probability that a network assigns to the visible vector and written as:

$$\ell_{\text{RBM}} = -\sum_{i=1}^{n} log(P(v_i)) \qquad (1.6)$$

For a real-valued input vector, the data can be modeled as Gaussian variables, resulting in the modification of the energy function for the RBM as follows:

$$E(v, h) = -\sum_{i=1}^{n} \frac{(v_i - b_i)^2}{2\sigma_i^2} - \sum_{j=1}^{r} b_j h_j - \sum_{i=1}^{n} \sum_{j=1}^{r} \frac{v_i}{\sigma_i} h_j w_{i,j} \qquad (1.7)$$

Using RBMs as the building blocks, deep architectures of deep belief network (DBN) [21] and deep Boltzmann machine (DBM) [22] have also been proposed in the literature. Both the models are created by stacking RBMs such that the input to the nth RBM is the learned representation of the $(n-1)^{th}$ RBM. A DBN has undirected connections between its first two layers (resulting in an RBM) and directed connections between its remaining layers (resulting in a sigmoid belief network). On the other hand, a DBM constitutes of stacked

RBMs with only undirected connections between the layers. RBMs have been used for addressing several challenging problems such as document modeling [23,24], collaborative filtering [25], audio conversion, and person identification [26–28]. Moreover, building on the unsupervised model of RBM, researchers have also proposed supervised architectures to learn discriminative feature representations [29,30].

1.2.1 Incorporating supervision in RBMs

In 2008, Larochelle and Bengio [29] presented the discriminative restricted Boltzmann machine (DRBM), which incorporates supervision in the traditionally unsupervised feature extraction model. DRBM models the joint distribution of the input data and their corresponding target classes, thereby resulting in a model capable of performing classification. Modifying Equation 1.6, the loss function of DRBM can be expressed as follows:

$$\ell_{\text{DRBM}} = -\sum_{i=1}^{n} \log(P(v_i, y_i)) \tag{1.8}$$

where, y_i corresponds to the target class for input sample, v_i. Modeling Equation 1.8 results in a complete model capable of performing feature extraction as well as classification for a given input. This is followed by several models incorporating supervision in the probabilistic model to learn discriminative features [31,32].

In 2016, inspired by the discriminative properties of DRBM, Sankaran et al. presented Class Sparsity Signature-based RBM (cssRBM) [30]. The proposed cssRBM is a semi-supervised model, built on DRBM by incorporating a $l_{2,1}$-norm–based regularizer on the hidden variables. This is done to ensure that samples belonging to a particular class have a similar sparsity signature, thereby reducing the within-class variations. For a k-class problem, the loss function for cssRBM is formulated as follows:

$$\ell_{\text{cssRBM}} = \ell_{\text{DRBM}} + \lambda \sum_{i=1}^{k} \|\mathbf{H}_i\|_{2,1} \tag{1.9}$$

where $\mathbf{H}_i$ corresponds to a matrix containing representations of samples belonging to the ith class, where the jth row corresponds to the hidden layer representation of the jth training sample of the given class. The model is used to learn discriminative feature representations, which are then provided to a classifier for performing classification.

1.2.2 Other advances in RBMs

Over the past several years, progress has also been made with RBMs in the form of incorporating unsupervised regularizers and modifying them for different applications. In 2008, Lee et al. presented sparse DBNs [33], mimicking certain properties of the human brain's visual area, V2. A regularization term is added to the loss function of an RBM to introduce sparsity in the learned representations. Similar to the performance observed with stacked autoencoders, the first layer was seen to learn edge filters (like the Gabor filters), and the second layer encoded correlations of the first-layer responses in the data, along with learning corners and junctions. Following this, a convolutional deep belief network (CDBN) was proposed by Lee et al. [34] for addressing several visual-recognition tasks. The model incorporated a novel probabilistic max-pooling technique for learning hierarchical features from unlabeled data. CDBN is built using the proposed convolutional RBMs, which incorporate convolution in the feature learning process of traditional RBMs. Probabilistic max-pooling is used at the time of stacking convolutional RBMs to create CDBNs for learning hierarchical representations. To eliminate trivial solutions, sparsity has also been enforced on the hidden representations. Inspired by the observation that both coarse and fine details of images may provide discriminative information for image classification, Tang and Mohamad proposed multiresolution DBNs [35]. The model used multiple independent RBMs trained on different levels of the Laplacian pyramid of an image and combined the learned representations to create the input to a final RBM. This entire model is known as multiresolution DBN, and the objective is to extract meaningful representations from different resolutions of the given input image. Coarse and fine details of the input are used for feature extraction, thereby enabling the proposed model to encode multiple variations. Further, in 2014, in an attempt to model the intermodality variations for a multimodal classification task, Srivastava and Salakhutdinov proposed the multimodal DBM [36]. The model aimed to learn a common (joint) representation for samples belonging to two different modalities such that the learned feature is representative of both the samples. The model also ensures that it is able to generate a common representation given only a sample from a single modality. In the proposed model, two DBMs are trained for two modalities, followed by a DBM trained on the combined learned representations from the two previous DBMs. The learned representation from the third DBM corresponds to the joint representation of the two modalities. Recently, Huang et al. proposed an RBM-based model for unconstrained multimodal multilablel learning [37] termed a multilabel conditional RBM. It aims to learn a joint feature representation over multiple modalities and predict multiple labels.

1.2.3 RBMs for biometrics

RBMs have also been used to address several bometrics-related applications, including kinship verification and face and iris recognition. In 2009, Lee et al. built on the CDBN [34] for addressing the task of audio classification [38]. They modified the model to work with single-dimensional input data and thus learn hierarchical representations using the probabilistic max-pooling technique. Another hybrid model of RBM and CNNs, termed ConvNet-RBM, has been proposed for performing face verification in the wild [28]. Multiple deep CNNs are trained using pairs of face images with the aim of extracting visual relational features. Each ConvNet is trained on a separate patch-pair of geometrically normalized face images. The high-level features learned from the deep ConvNets are then provided as input to a discriminative RBM for learning the joint distribution of the samples, labels, and the hidden representations. The entire model is then used for performing classification of face images in the wild. Taking inspiration from the multimodal DBM model [36], Alam et al. presented a joint DBM for the task of person identification using mobile data [26]. A joint model is built on two unimodal DBMs and trained using a novel three-step algorithm. Learned representations from the unimodal DBMs are provided as input to a common RBM, which then learns the shared representation over two different modalities. In 2017, RBMs have also been used to perform kinship verification on face images [39]. A hierarchical kinship verification via representation learning framework is presented by Kohli et al. [39] which uses the proposed filtered contractive (fc) DBN. fc-RBMs are used as the building blocks of the architecture, wherein a contractive term has been added to the loss function of the traditional RBM, to learn representations robust to the local variations in the images. Moreover, a filtering approach has also been incorporated in the RBM, such that the model uses the structural properties of face images and extracts meaningful facial features for representation learning. Multiple independent fc-DBNs are trained for local and global facial features; the learned representations are then combined and provided as input to a final fc-DBN for feature learning.

It can thus be observed that RBMs have widely been used for addressing the task of biometric authentication. As mentioned, models such as ConvNet-RBM [28], joint DBM [26], and fc-DBN [39] have shown to perform well with face images. Teh and Hinton proposed a rate-coded RBM, which is a neurally inspired generative model for performing face recognition [40]. The proposed algorithm creates generative models for pairs of face images belonging to the same individual, which are then used for identifying a given test image. Goswami et al. [41] also proposed a deep learning architecture, which is a combination of a DBM and a stacked denoising autoencoder for performing face verification in videos. One of the key contributions of the work is the inclusion of sparsity and low-rank regularization in the formulation of a traditional DBM. Other than face recognition, RBMs have also been explored for other modalities, such as recognition of fingerprint and periocular images [30,42].

1.3   Autoencoder

Autoencoders (AEs) are unsupervised neural network models aimed at learning meaningful representations of the given data [43]. An AE model consists of two components: the encoder and the decoder. The encoder learns a feature

Deep Learning in Biometrics

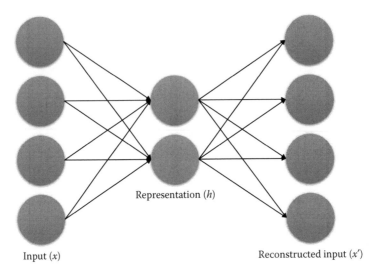

FIGURE 1.5
Diagrammatic representation of a single-layer AE having input as x, learned
representation as h, and the reconstructed sample as x'.

representation of the given input sample, and the decoder reconstructs the
input from the learned feature vector. The model aims to reduce the error
between the input and the reconstructed sample to learn representative features of the input data. Figure 1.5 presents a diagrammatic representation
of a single-layer AE. For a given input vector x, a single-layer AE can be
formulated as follows:

$$\arg\min_{\mathbf{W_e}, \mathbf{W_d}} \|x - \mathbf{W_d}\phi(\mathbf{W_e}x)\|_2^2 \qquad (1.10)$$

where:

  $\mathbf{W_e}$ and $\mathbf{W_d}$ are the encoding and decoding weights respectively

  ϕ is the activation function

Nonlinear functions such as *sigmoid* or *tanh* are often used as the activation
functions. If no activation function is used at the encoding layers, the model
is termed a *linear AE*. Equation 1.10 aims to learn a hidden representation
($h = \phi(\mathbf{W_e}x)$) for the given input x, such that the error between the original
sample and the reconstructed sample ($\mathbf{W_d}h$) is minimized. To create *deeper*
models, stacked AEs are used. Stacked AEs contain multiple AEs, such that
the learned representation of the first AE is provided as input to the second
one. A stacked AE with l layers is formulated as follows:

$$\arg\min_{\mathbf{W_e}, \mathbf{W_d}} \|x - g \circ f(x)\|_2^2 \qquad (1.11)$$

where, $\mathbf{W_e} = \{\mathbf{W_e^1}, \dots, \mathbf{W_e^l}\}$ and $\mathbf{W_d} = \{\mathbf{W_d^1}, \dots, \mathbf{W_d^l}\}$. $f(x) = \phi(\mathbf{W_e^L}\phi(\mathbf{W_e^{L-1}} \dots \phi(\mathbf{W_e^1}(x))))$, such that $\mathbf{W_e^i}$ refers to the encoding weights of the ith layer and $g(x) = \mathbf{W_d^1}(\mathbf{W_d^2} \dots \mathbf{W_d^L}(x))$, where $\mathbf{W_d^i}$ refers to the decoding weights of the ith layer. Since deeper models require learning large number of parameters, Bengio et al. [44] and Hinton and Salakhutdinov [45] proposed greedy layer-by-layer optimization of deep models. This approach aims at learning the weight parameters of one layer at a time, while keeping the remaining fixed, and thus reducing the number of parameters to be optimized simultaneously. AEs have been used to address a variety of tasks such as face verification, object classification, magnetic-resonance image reconstruction, and audio processing. There also exists several architectural variations of AEs, where authors have incorporated different forms of regularizations and even incorporated supervision to learn robust or task-specific features. The following subsections provide details of such models.

1.3.1 Incorporating supervision in AEs

Researchers have proposed incorporating supervision in the unsupervised feature learning models of AEs in order to learn discriminative features. This is done either by directly using the class label at the time of feature learning or incorporating the class information (*same* or *different*) while training the model. Gao et al. [46] proposed supervised AEs for performing single-sample recognition. The model aims to map two different images of the same person on to the same representation to reduce intraclass variations during feature encoding. Equation 1.12 presents the loss function of the proposed model, where x_i corresponds to the gallery image, and x_{ni} refers to the probe image of the training set. The model aims to learn a representation for the probe image, x_{ni}, such that when it is reconstructed via the proposed model, it generates the gallery image x_i. The second term corresponds to a similarity preserving term, which ensures that samples of the same class have a similar representation. The loss function of the proposed supervised AE is formulated as follows:

$$\underset{\mathbf{W_e}, \mathbf{W_d}}{\arg\min} \frac{1}{N} \sum_i \left(\|x_i - g \circ f(x_{ni})\|_2^2 + \lambda \|f(x_i) - f(x_{ni})\|_2^2 \right)$$

$$+ \alpha \left(KL(\rho_x \| \rho_o) + KL(\rho_{x_n} \| \rho_o) \right) \quad (1.12)$$

$$where \ \rho_x = \frac{1}{N} \sum_i \frac{1}{2}\left(f(x_i) + 1\right), \rho_{x_{ni}} = \frac{1}{N} \sum_i \frac{1}{2}\left(f(x_{ni}) + 1\right)$$

The first term corresponds to the reconstruction error, the second is the similarity-preserving term, and the remaining two correspond to the Kullback–Leibler divergence [47]. Following this, Zheng et al. [48] proposed the contrastive AE (CsAE), which aimed at reducing the intraclass variations. The architecture consists of two AEs, which learn representations of an input pair of images. For a given pair of images belonging to the same class, the architecture minimizes the difference between the learned representation at the final layer. For a k-layered architecture, the loss function of the CsAE is modeled as follows:

$$\underset{\mathbf{W_e, W_d}}{\arg\min} \lambda(\|x_1 - g_1 \circ f_1(x_1)\|_2^2 + \|x_2 - g_2 \circ f_2(x_2)\|_2^2)$$

$$+ (1 - \lambda)\left\|O_1^k(x_1) - O_2^k(x_2)\right\|_2^2 \qquad (1.13)$$

where:

x_1 and x_2 refer to two input samples of the same class

$f_j(x)$ and $g_j(x)$ correspond to the encoding and decoding functions of the jth AE

For each AE, $f(\mathbf{x}) = \phi(\mathbf{W_e^k}\phi(\mathbf{W_e^{k-1}} \ldots \phi(\mathbf{W_e^1}(x))))$ and $g(\mathbf{x}) = \mathbf{W_d^1}(\mathbf{W_d^2} \ldots \mathbf{W_d^k}(x))$, where $\mathbf{W_e^i}$ and $\mathbf{W_d^i}$ refer to the encoding and decoding weights of the ith layer for both the AEs, and $O_j^k(\mathbf{x})$ is the output of the kth layer of the jth AE.

Recently, Zhuang et al. [49] proposed a transfer learning-based supervised AE. They modified the AE model to incorporate a layer based on softmax regression for performing classification on the learned-feature vectors. Although the encoding–decoding layers ensure that features are learned such that the reconstruction error is minimized, the label-encoding layer aims to incorporate discrimination in the learned features based on their classification performance. In 2017, Majumdar et al. [50] presented a class sparsity–based supervised encoding algorithm for the task of face verification. Class information is used to modify the loss function of an unsupervised AE, by incorporating a $l_{2,1}$-norm–based regularizer. For input samples $\mathbf{X}$, the proposed architecture is formulated as:

$$\underset{\mathbf{W_e, W_d}}{\arg\min} \|\mathbf{X} - g \circ f(\mathbf{X})\|_2^2 + \lambda \|\mathbf{W_e X_c}\|_{2,1} \qquad (1.14)$$

where, $\mathbf{X_c}$ refers to the samples belonging to class $\mathbf{c}$. The regularization parameter ensures that samples belonging to a particular class have a similar sparsity signature during feature encoding. This helps in reducing the intraclass variations, thereby promoting the utility of the learned features for classification.

1.3.2   Other variations of AEs

Other than incorporating supervision, researchers have also focused on modifying the architecture of the unsupervised AE to learn robust features from the given data. In 2010, Vincent et al. introduced the stacked denoising AE, which learns robust representations of the input data, even in the presence of noise [51]. During feature learning, the authors introduced noise in the input data with the aim of reconstructing the original, clean sample. That is, a noisy sample is provided as input to the model, and the reconstruction error is minimized with respect to the clean, original sample. For a given input sample x, the loss function of stacked denoising AE can be formulated as follows:

$$\underset{\mathbf{W_e}, \mathbf{W_d}}{\arg\min} \|x - g \circ f(x_n)\|_2^2 \tag{1.15}$$

where x_n is the noisy input sample. Experimental evaluation and analysis suggested the model to be learning Grabor-like features, thereby encoding edge information in the feature-learning process. To extract robust features from noisy data, Rifai et al. [52] proposed the contractive AE. The loss function of the model consists of the Frobenius norm of the Jacobian matrix of the encoder activations with respect to the input. This additional term helps in learning robust features, irrespective of noisy input or minute corruptions in the input data. This is followed by higher order contractive AEs [53], where an additional regularization term consisting of the Hessian of the output with respect to the encoder, is added in the loss function. Experimental evaluation and analysis depicts that the model performs well with noisy input, and the learned weight matrices can be used for efficient initialization of deep models.

AEs have also been used for addressing tasks involving multimodal recognition [54,55]. In 2011, Ngiam et al. [55] proposed using AEs for learning a shared representation for a given video and audio input. A bimodal deep AE was proposed that learned a shared representation from a pair of audio and video data, such that the input can be reconstructed back from the shared representation. In 2015, Hong et al. [54] proposed using a multimodal deep AE for performing human-pose recovery. The architecture consists of two AEs and a neural network. The AEs are trained independently on two-dimensional (2D) images and three-dimensional (3D) poses. This is followed by a neural network that learns nonlinear mapping between the hidden representations of both AEs. Once the entire architecture is learned, the model is able to recover the pose from a given input image.

AEs have been explored for performing biometric authentication for different modalities. As mentioned previously, several models have been built on the traditional unsupervised AE model for performing the task of face recognition [46,48,50], as well as for performing face segmentation and alignment [56–58]. Recently, Singh et al. [59] proposed a class representative AE for performing gender classification in low resolution, multispectral face images. The proposed model uses the mean feature vectors of the two classes for learning discriminative feature representations. Dehghan et al. [60] proposed a gated AE-based architecture for determining parent–offspring resemblance from face images. The proposed architecture uses a pair of face images as input and learns patch-wise features for the given pair, which is followed by a neural network for classification. AEs have also been used for analyzing latent fingerprints and periocular images [61–63]. Raghavendra and Busch proposed deeply coupled AEs for performing periocular verification for smartphones [42]. The proposed architecture uses a combination of maximum-response filters and coupled AEs, along with neural networks for classification. Overall, AEs have been used for different tasks related to biometric authentication, such as segmentation, feature extraction, and quality assessment in supervised and unsupervised manners.

1.4   Convolutional Neural Networks

A convolutional neural network (CNN) is a supervised deep learning model used for performing classification. The CNN architecture is inspired from the arrangement of neurons in the visual cortex of animals [64]. CNNs are used to learn an efficient feature representation for a given set of images by performing spatial convolution on a two-dimensional input followed by pooling to ensure translational invariance. During each forward pass of a given CNN, the model learns filters or kernels, which are used for performing convolution. Deep CNNs are hierarchical in nature, that is, they learn low-level features in the shallow layers, such as edges, which are combined to learn higher levels of abstraction in the deeper layers of the network.

1.4.1   Architecture of a traditional CNNs

A CNN is made up of several different types of layers, each performing a specific function. A traditional CNN is made up of convolution and pooling layers alternatively, followed by a fully connected layer to perform classification. An explanation of each layer follows.

Convolutional layer: As the name suggests, this is the building block of a CNN and is of utmost importance. Several filters are used to perform convolutions on the input vector by sliding it over the image. These filters are learned as part of the training process. The feature vector obtained after convolving an image with a filter is referred to as an *activation map* or *filter map*. The number of activation maps obtained is equal to the number of filters learned over the input. This operation encodes the fine details as well as the spatial information of the input in the feature maps. Given an input image of size $n \times n \times d$, the convolutional layer of the model learns m kernels of size $k \times k$, with a stride of size s. Thus, the output of the given layer is of the following dimension:

$$\text{Output Size} = (n-k)/s + 1 \tag{1.16}$$

Rectified linear units layer (ReLU): This is used to introduce nonlinearity in the network to learn more discriminative features. It is usually applied after each convolutional layer. In the past, activation functions such as *sigmoid* and *tanh* have been used to introduce nonlinearity, however it has been observed that ReLU is faster and reduces the training time significantly [65]. It also eliminates the vanishing gradient problem by converting all negative values to 0. The function applied in this layer is given as follows:

$$f(x) = \max(0, x) \tag{1.17}$$

Pooling layer: Pooling layers or downsampling layers are used for dimensionality reduction of the feature maps after the convolution and ReLU layers. Generally, a filter size is chosen and an operation such as max or average is applied on the input space, which results in a single output for the given subregion. For example, if the operation defined is max-pooling for a filter size of 2×2, the max of all values in the subregion is the output of the filter. This is done for the entire feature map by sliding the filter over it. The aim of this operation is to encode the most representative information, while preserving the relative spatial details. This step not only enables dimensionality reduction, but also prevents over-fitting.

Fully connected layer: After the convolutional and pooling layers, fully connected layers are attached in the network. These layers function like a traditional neural network, where each element is considered an independent node of the neural network. The output dimension of the final layer is equal to the number of classes, and each value of the output vector is the probability value associated with a class. This type of layer is used to encode supervision in the feature-learning process of the CNNs because the last layer is used for classification.

In CNNs, there is no fixed order in which its constituent layers are stacked. However, typically a convolutional layer is followed by a pooling layer forming a convolutional-pooling block. This block is repeated, depending on the desired size of the network. These layers are followed by fully connected layers and the final layer is responsible for classification. ReLU is often attached after each convolutional and fully connected layer to incorporate nonlinearity in the feature-learning process. Figure 1.6 is an example of a traditional CNN model consisting of five layers: two convolutional and two pooling, stacked alternatively, and the final layer being the fully connected layer. Owing to the flexibility in the architecture, researchers have developed different models for performing feature extraction and classification tasks. Some recent developments involving CNNs are discussed in the next subsections.

1.4.2 Existing architectures of CNNs

Inspired by the work of Fukushima [66] on artificial neural networks, LeCun et al. [67] proposed the first model of a CNN in 1990 for the application of hand-written zip code recognition. The back-propagation technique was used to train the existing neural networks. The proposed model consists of four hidden layers, the first and third layers being shared-weight feature-extraction layers and the second and fourth are subsampling layers.

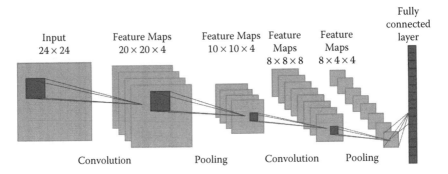

FIGURE 1.6
Diagrammatic representation of a CNN having input as 24×24. The first convolution layer learns four filters of 5×5, resulting in four feature maps of size 20×20. This is followed by pooling with a filter size of 2×2. Pooling layer is again followed by convolution and pooling layers. The final layer is a fully connected layer, the output of which is used to perform classification.

Convolution is used to perform feature extraction. This architecture was further refined and several frameworks were proposed between 1990 and 2000 [68,69]. One such model was **Le-Net5** [69], composed of seven layers. Similar to the first model, the first and third layer are convolutional layers, and the second and fourth layers correspond to the pooling layers. However, in this model, these are followed by three fully connected layers where the output of the final layer provides the output label. Since then, CNNs have undergone major evolution and are being used to achieve state-of-the-art results on several applications such as image classification [70,71], face recognition [72,73], detection [74,75], natural language processing [76], and playing Go [77].

Krizhevsky et al. [65] proposed **AlexNet** to perform image classification, as part of the ImageNet Large-Scale Visual Recognition Challenge (ILSVRC) 2012. The proposed architecture comprises eight layers, out of which the first five are the convolutional layers with overlapping max-pooling and the following three are fully connected layers. The dimension of each fully connected layer is 4096. Softmax is used to predict the class label of the learned representation. AlexNet used 15 million images belonging to 22,000 categories for training, using two GPUs. The model takes RGB images of dimension 224×224 as input, on which 96 kernels are learned in the first layer, followed by 256 kernels in the second convolutional layer. No pooling or normalization is performed between the remaining convolutional layers. The third and fourth layers consisted of learning 384 kernels, and the final layer consisted of learning 256 kernels. It can be observed that the first layer alone required learning 105,705,600 parameters, making it one of the largest networks of that time. ReLU has also been used to introduce nonlinearity between layers because it significantly helps in reducing the training time of the model. Data augmentation is performed on the training images, along with utilization of dropout for preventing over-fitting. The entire model was learned using stochastic gradient descent approach. The authors report a top-5 error rate of 15.3%, displaying an improvement of at least 10% as compared to the next-best result reported in the ImageNet challenge, making CNNs popular in the computer vision community.

In the following year, Zeiler et al. [78] proposed a model built over AlexNet and termed it **ZFNet**. The model used 1.3 million images for training, as opposed to the 15 million used by AlexNet. However, ZFNet achieved state-of-the-art performance on the ImageNet challenge with an error rate of 11.80%, displaying an improvement of about 5% as compared to AlexNet. They achieved this by reducing the filter size of the first convolutional layer to 7×7, while keeping the rest of the architecture consistent with AlexNet. The authors also studied the visualization of weights and layers in depth by building a model to reverse the steps of a traditional CNN, which is termed a *DeconvNet*. They established that CNNs learn features in a hierarchical manner. The initial layers learn low-level features such as edge and color, and as one goes deeper, a higher level of abstraction is learned. DeconvNet is a great tool for not only visualizing the inner workings of a CNN, but also for understanding and improving the architecture based on the learned features.

Lin et al. [79] proposed **Network-in-Network**, a method to learn feature maps using a nonlinear operation, instead of the traditional convolution operation. The authors suggested that using the convolution filter resulted in a linear model capable of learning low-level abstractions. In contrast, adding nonlinearity at this stage results in learning robust features. Therefore, the proposed model uses nonlinearity by learning the feature map using a *micronetwork* structure that consists of multilayers perceptrons resulting in multiple fully connected layers. The features are learned by sliding the network, similar to traditional CNNs. This module of learning the features is termed *mlpconv*. Stacking multiple such modules resulted in the modified network, termed a network-in-network. Alternatively, rather than a fully connected network, global average pooling is performed, followed by a softmax layer for performing classification. Mlpconv layers use nonlinearity in the network to learn discriminative features, and global average pooling is used as a regularizer to prevent over-fitting of the model.

Simonyan et al. [71] reestablished the fact that CNNs need to be deep to learn hierarchical features. The authors proposed VGGNet, a 19-layer CNN, in which very small filters of size 3×3 are learned, as compared to AlexNet's 11×11. The authors observed that learning smaller filters for deeper architectures requires fewer parameters, rather than learning larger filters for fewer layers for the same data. This resulted in lesser training time and also increased the nonlinearity introduced at each layer, learning more discriminative features. In this case, learning a stack of three convolutional layers (without pooling), becomes equivalent to learning a receptive field of 7×7. The convolutional layers are followed by three fully connected layers, in which the last one performs classification. The proposed model, VGGNet reports an error rate of 7.3% on the ImageNet database, being at least 4% better than ZFNet, the best reported results in 2013.

In 2014, Taigman et al. [73] proposed a deep neural network–based system, **DeepFace** to perform face verification. A Siamese network is built, where each unit of the Siamese network consists of a CNN module. Each CNN module consists of a convolution layer with 32 filters resulting in 32 feature maps, which are provided as input to the pooling layer. Max-pooling is applied over 3×3 neighborhoods. This is followed by another convolutional layer that learns 16 filters. This is followed by three locally connected layers, where at each location a different filter is learned for convolution. The output of these is finally given as input to a fully connected layer. ReLU activation is applied after each convolutional layer, locally connected and fully connected layer. The output of the Siamese network is then classified using weighted chi-squared distance. The aim of the model is to minimize the distance between pairs of face images belonging to the same class while maximizing the interclass distance. In the same year, Szegedy et al. [80] proposed **GoogLeNet**, a deep CNN model comprising of 27 layers. However, unlike the traditional CNN models where convolution and pooling layers are stacked one after the other, it performs both the operations parallel at the same level. Based on this idea, the authors introduced the concept of *Inception* in CNNs. The proposed model consists of several inception modules stacked together. Each inception module consists of a 1×1 convolution filter, 1×1 followed by 3×3 convolution filters, 1×1 followed by 5×5 convolution filters, and pooling being performed parallel on the input for the given layer. The output of all these operations is concatenated and used as input for the next layer. In the proposed model, nine such inception modules are stacked one after the other. This architecture enables the use of a small-, medium-, and large-sized filter convolution and pooling at each layer. This leads to learning information about very fine details as well as the spatial details, while pooling reduces the size and prevents over-fitting. The model does not consist of any fully connected layer, instead only softmax is added at the end. Despite being deeper and a more complex architecture, the proposed model learned at least 10 times fewer parameters as compared to AlexNet. This illustrates the computational efficiency of the proposed concept of Inception in CNNs. GoogLeNet used 1.2 million images of 1000 classes from the ImageNet challenge database for its training. It reported state-of-the-art results at the time, with a top-5 error rate of about 6.7%, winning ImageNet Large-Scale Visual Recognition Challenge 2014.

Continuing with the revolution of depth in deep learning architectures, He et al. [70] proposed a deep CNN-based model, **ResNet**. It consists of 152 layers and reports state-of-the-art results on the ImageNet database with a top-5 error rate of 3.6%. The authors proposed the concept of a residual block. In traditional CNN models, the output obtained from the convolution-ReLU layers is learned. However, in a residual block, the difference between the input and the learned representation is learned. This is based on the hypothesis that it is easier to learn a residual term as opposed to a completely new representation for the input. The proposed architecture also prevents the problem of vanishing gradient that arises as the architectures go deeper. The filter sizes are small throughout the network in the convolution layers, followed by a fully connected layer that is used to perform classification.

Schroff et al. [72] developed a CNN-based face-recognition system, **FaceNet**. In the model, face images are mapped to a Euclidean space such that the distance between feature vectors of same identity are minimized and for different identities is maximized. The model consists of the input layer, which is sent to a deep CNN, followed by a triplet loss [81] on the normalized output. The model is trained using triplets of two images belonging to the same identity (genuine), whereas the third one belongs to a different identity (imposter). One of the same identity images is termed the *pivot*. The aim of the model is to make the pivot closer to the genuine image than to the imposter.

Girshick et al. [74] proposed region-based CNNs for object detection, popularly known as **R-CNNs**. The proposed model is used for object detection and reported an improvement of more than 30% as compared to the existing methods, achieving a mean average precision of 53.3% on the Pascal VOC data set [82]. It predicts the presence of an object in the image and also computes the exact location of the object in the image. The pipeline consists of two main steps: selecting possible regions and classifying the objects in the regions. An existing approach of selective search [83] is used to deduce the possible regions of the image. Irrespective of the dimension of the image, it is warped into 224×224. AlexNet is used for feature extraction from these region specific images. The features are used as input for a set of support vector machines, which are trained for classification, to obtain the required result.

CNN-based models such as DeepFace, FaceNet, and VGG-Face have performed well for the task of face recognition [72,73,84]. Moreover, different architectures have also been proposed for performing face detection in challenging environments [85–87]. Farfade et al. [85] proposed Deep Dense Face Detector for performing face detection on a large range of pose variations, without requiring landmark annotation. The proposed model is built over AlexNet [65] and uses a sliding window technique to generate a heat map where each point corresponds to the probability of it being a face. Research also been performed on attribute classification of face images and spoofing detection for iris, face, and fingerprint modalities [88–90]. Levi and Hassner [88] proposed a CNN architecture consisting of three convolutional layers and two fully connected layers for performing age and gender classification on face images. As compared to other existing architectures, the authors used a relatively smaller architecture because the classification task consisted of only eight classes for age and of only two classes for gender. Moreover, CNN models have been explored for the modalities of fingerprint, pupil, and other soft-biometric traits as well [91–93].

1.5   Other Deep Learning Architectures

Although there has been substantial research in the mentioned paradigms of RBM, AE, and CNN, researchers have also focused on other deep learning–based models for addressing different tasks. Most of these models are built on the fundamental building block of a neuron; however, their architectures vary significantly from those mentioned here. For example, recurrent neural networks (RNNs) are neural networks with loops (as shown in Figure 1.7), allowing information to remain in the network by introducing the concept of *memory*. In traditional neural networks, information learned previously cannot be used to classify or predict new information at a later stage. The ability of RNNs to hold on to the information via the additional loops enables the use of sequential information. These networks can be viewed as multiple neural networks connected to each other, having the same architecture and the same parameters with different inputs. RNNs have been used extensively for several applications such as language modeling, text generation, machine translation, image generation, and speech recognition [94–97] and have achieved promising results.

To provide more flexibility to an RNN, Hochreiter and Schmidhuber developed a special kind of RNN, termed a long short-term memory, popularly referred to as LSTM [98]. Such networks provide the ability to control whether the learned information needs to be in the memory for *long* or *short* durations. This is done by incorporating a linear activation function between the recurring loops. LSTMs usually consist of three different types of gates: input, forget, and output. *Input* controls the amount of information flow from the previous iteration, *forget* controls the amount of information to be retained in the memory, and the *output* gate restricts the information that is used to compute the value for the next block. LSTMs have provided state-of-the-art results for various applications and are being used extensively for tasks like sentiment analysis, speech recognition, and language modeling [99–102].

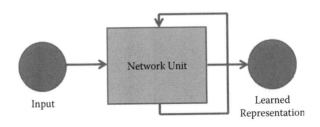

FIGURE 1.7
Diagrammatic representation of a single RNN unit.

Recently, Goodfellow et al. [103] proposed the generative adversarial network (GAN), which is built using two models being trained simultaneously. A GAN framework consists of a generative model, which works on the data distribution of the training data and generates new samples, and a discriminative model, which estimates whether the sample generated from the generative model actually came from the training data or not. The aim of the entire framework is to increase the error of the discriminative model, that is, facilitate generation of samples similar to the training data. Tran et al. [104] proposed a disentangled representation learning GAN for performing pose invariant face recognition. Given one or multiple input faces of varying pose, the ultimate aim of disentangled representation GAN is to synthesize a face image at a predefined target pose, along with the learned representations. Xin et al. [105] also proposed a face frontalization GAN aimed at performing pose invariant face recognition. The model incorporates a three-dimensional morphable model into the GAN structure, along with a recognition engine for optimizing the classification performance.

Apart from traditional models, researchers have also proposed hybrid models such as Deep Dictionary [106]. Deep Dictionary aims to learn sparse representations of data while using the capabilities of deep learning models to learn hierarchical features. In the proposed model, learned sparse codes are treated as the learned features from the given layer, which are used as input to the next layer of dictionary. Such architectures combine the advantages of representation learning techniques such as dictionaries and deep learning models into a single model and achieve promising results [107].

1.6   Deep Learning: Path Ahead

Deep learning has achieved state-of-the-art results for several applications; however, it has a few limitations as well. Most of the models based on deep learning require large amount of training data, thereby being computationally intensive. For example, the AlexNet model is trained on 15 million images and learned 105,705,600 parameters in the first layer alone. Therefore, the requirement of large amount of training data leads to limited deep learning based research in problems with less data. For example, in the application of face-sketch recognition, even on combining all publicly available data sets belonging to viewed, forensic, semi-forensic, and composite sketch data sets, the total number of images is still less than 2000. In particular, real-world forensic images are further less than 500, and thus restrict the applicability of deep learning-based models. Therefore, focused attention is required to develop algorithms that solve *small sample size* problems.

Another major challenge associated with the paradigm of deep learning is the lack of *theoretical bounds* for most of its models, along with the lack of in-depth understanding of their functioning. Although there has been some progress in understanding the theory behind these architectures, most of the models are still treated as black boxes [108,109]. Biometric modalities are often used as evidence for associating an identity with a crime scene. Fingerprints have been used in the court of law to identify suspects and charge them based on the confidence level of match provided by a forensic artist. However, without the knowledge of theoretical bounds of deep learning-based systems, in court of law, it is challenging to associate confidence levels to prosecute suspects. This defeats the purpose of matching biometrics in surveillance or criminal investigation. Therefore, it is essential to work toward understanding these models better and deriving theoretical comprehension.

Homo sapiens are capable of performing challenging tasks such as object and speech recognition, face recognition, and analysis with utmost ease. Motivated by the human brain's functioning, deep learning models and approaches are designed to mimic it. However, for applications such as face recognition with age variations or variations caused by plastic surgery, the data is limited, and the task at hand is further challenging in nature because of the added variations. To facilitate research and further enhance the performance in such scenarios, dedicated attention is required to understand brain functioning and create automated bio-inspired algorithms. Recently Nagpal et al. [110] proposed understanding the human brain via functional magnetic resonance imaging studies to develop algorithms that imitate the human brain. Fong et al. [111] have also proposed *neurally weighted* machine learning, in which the weights of deep learning architectures are learned from functional magnetic resonance imaging responses collected to measure brain activity while volunteers are viewing images. Focused research to develop brain-inspired algorithms can lead to advancements in deep learning and the development of sophisticated architectures to improve performance of automated systems for these challenging tasks.

References

1. A. K. Jain, A. A. Ross, and K. Nandakumar. *Introduction to Biometrics*. Springer Publishing Company, Gewerbestrasse, Switzerland, 2011.

2. E. Hjelms and B. K. Low. Face detection: A survey. *Computer Vision and Image Understanding*, 83(3):236–274, 2001.

3. S. Zafeiriou, C. Zhang, and Z. Zhang. A survey on face detection in the wild. *Computer Vision and Image Understanding*, 138:1–24, 2015.

4. P. Viola and M. J. Jones. Robust real-time face detection. *International Journal of Computer Vision*, 57(2):137–154, 2004.

5. J. G. Daugman. Complete discrete 2-d gabor transforms by neural networks for image analysis and compression. *IEEE Transactions on Acoustics, Speech, and Signal Processing*, 36(7):1169–1179, 1988.

6. N. Dalal and B. Triggs. Histograms of oriented gradients for human detection. In *IEEE Conference on Computer Vision and Pattern Recognition*, pp. 886–893, IEEE Computer Society, Washington, DC, 2005.

7. T. Ojala, M. Pietikinen, and D. Harwood. A comparative study of texture measures with classification based on featured distributions. *Pattern Recognition*, 29(1):51–59, 1996.

8. C. Cortes and V. Vapnik. Support-vector networks. *Machine Learning*, 20(3):273–297, 1995.

9. T. K. Ho. Random decision forests. In *International Conference on Document Analysis and Recognition*, volume 1, pp. 278–282, IEEE Computer Society, Washington, DC, 1995.

10. I. Kemelmacher-Shlizerman, S. M. Seitz, D. Miller, and E. Brossard. The megaface benchmark: 1 million faces for recognition at scale. In *IEEE Conference on Computer Vision and Pattern Recognition*, pp. 4873–4882, IEEE Computer Society, Washington, DC, 2016.

11. Z. Liu, P. Luo, X. Wang, and X. Tang. Deep learning face attributes in the wild. In *International Conference on Computer Vision*, Santiago, Chile, 2015.

12. R. Gross, I. Matthews, J. Cohn, T. Kanade, and S. Baker. Multi-PIE. *Image and Vision Computing*, 28(5):807–813, 2010.

13. H. A. Rowley, S. Baluja, and T. Kanade. Neural network-based face detection. *IEEE Transactions on Pattern Analysis and Machine Intelligence*, 20(1):23–38, 1998.

14. K. K. Sung and T. Poggio. Example-based learning for view-based human face detection. *IEEE Transactions on Pattern Analysis and Machine Intelligence*, 20(1):39–51, 1998.

15. A. Sankaran, M. Vatsa, and R. Singh. Automated clarity and quality assessment for latent fingerprints. In *IEEE International Conference on Biometrics: Theory, Applications and Systems*, pp. 1–6, IEEE Computer Society, Washington, DC, 2013.

16. R. M. Haralick. Statistical and structural approaches to texture. *Proceedings of the IEEE*, 67(5):786–804, 1979.

17. Z. Guo, L. Zhang, and D. Zhang. Rotation invariant texture classification using LBP variance (LBPV) with global matching. *Pattern Recognition*, 43(3):706–719, 2010.

18. D. G. Lowe. Object recognition from local scale-invariant features. In *International Conference on Computer Vision*, Kerkyra, Greece, 1999.

19. F. Rosenblatt. The perceptron: A probabilistic model for information storage and organization in the brain. *Psychological Review*, 65:386–408, 1958.

20. P. Smolensky. Information processing in dynamical systems: Foundations of harmony theory. In D.E. Rumelhart and J.L. McClelland (Eds.), *Parallel Distributed Processing: Volume 1: Foundations*, pp. 194–281, Cambridge, MA: MIT Press, 1987.

21. G. E. Hinton, S. Osindero, and Y.-W. Teh. A fast learning algorithm for deep belief nets. *Neural Computation*, 18(7):1527–1554, 2006.

22. R. Salakhutdinov and G. E. Hinton. Deep boltzmann machines. In *International Conference on Artificial Intelligence and Statistics*, pp. 448–455, PMLR, 2009.

23. N. Srivastava, R. Salakhutdinov, and G. Hinton. Modeling documents with a deep boltzmann machine. In *Conference on Uncertainty in Artificial Intelligence*, pp. 616–624, AUAI Press, Arlington, Virginia, 2013.

24. P. Xie, Y. Deng, and E. Xing. Diversifying restricted Boltzmann machine for document modeling. In *ACM International Conference on Knowledge Discovery and Data Mining*, pp. 1315–1324, ACM, New York, 2015.

25. R. Salakhutdinov, A. Mnih, and G. Hinton. Restricted boltzmann machines for collaborative filtering. In *International Conference on Machine Learning*, pp. 791–798, ACM, New York, 2007.

26. M. R. Alam, M. Bennamoun, R. Togneri, and F. Sohel. A joint deep boltzmann machine (jDBM) model for person identification using mobile phone data. *IEEE Transactions on Multimedia*, 19(2):317–326, 2017.

27. G. Huo, Y. Liu, X. Zhu, and J. Wu. An efficient iris recognition method based on restricted boltzmann machine. In *Chinese Conference on Biometrics Recognition*, pp. 349–356, Springer International Publishing, Cham, 2015.

28. Y. Sun, X. Wang, and X. Tang. Hybrid deep learning for face verification. In *IEEE International Conference on Computer Vision*, Sydney, NSW, Australia, 2013.

29. H. Larochelle and Y. Bengio. Classification using discriminative restricted boltzmann machines. In *International Conference on Machine Learning*, pp. 536–543, ACM, New York, 2008.

30. A. Sankaran, G. Goswami, M. Vatsa, R. Singh, and A. Majumdar. Class sparsity signature based restricted boltzmann machine. *Pattern Recognition*, 61:674–685, 2017.

31. R. Mittelman, H. Lee, B. Kuipers, and S. Savarese. Weakly supervised learning of mid-level features with beta-bernoulli process restricted Boltzmann machines. In *IEEE Conference on Computer Vision and Pattern Recognition*, Portland, OR, 2013.

32. G. S. Xie, X. Y. Zhang, Y. M. Zhang, and C. L. Liu. Integrating supervised subspace criteria with restricted Boltzmann machine for feature extraction. In *International Joint Conference on Neural Networks*, pp. 1622–1629, IEEE, 2014.

33. H. Lee, C. Ekanadham, and A. Y. Ng. Sparse deep belief net model for visual area V2. In *Advances in Neural Information Processing Systems*, pp. 873–880, Curran Associates, Red Hook, NY, 2008.

34. H. Lee, R. Grosse, R. Ranganath, and A. Y. Ng. Convolutional deep belief networks for scalable unsupervised learning of hierarchical representations. In *International Conference on Machine Learning*, pp. 609–616, ACM, New York, 2009.

35. Y. Tang and A. R. Mohamed. Multiresolution deep belief networks. *International Conference on Artificial Intelligence and Statistics*, 22: 1203–1211, 2012.

36. N. Srivastava and R. Salakhutdinov. Multimodal learning with deep Boltzmann machines. *Journal of Machine Learning Research*, 15:2949–2980, 2014.

37. Y. Huang, W. Wang, and L. Wang. Unconstrained multimodal multi-label learning. *IEEE Transactions on Multimedia*, 17(11):1923–1935, 2015.

38. H. Lee, P. Pham, Y. Largman, and A. Y. Ng. Unsupervised feature learning for audio classification using convolutional deep belief networks. In *Advances in Neural Information Processing Systems*, pp. 1096–1104, Curran Associates, Red Hook, NY, 2009.

39. N. Kohli, M. Vatsa, R. Singh, A. Noore, and A. Majumdar. Hierarchical representation learning for kinship verification. *IEEE Transactions on Image Processing*, 26(1):289–302, 2017.

40. Y. W. Teh and G. E. Hinton. Rate-coded restricted boltzmann machines for face recognition. In *Advances in Neural Information Processing Systems*, pp. 908–914, MIT Press, Cambridge, MA, 2001.

41. G. Goswami, M. Vatsa, and R. Singh. Face verification via learned representation on feature-rich video frames. *IEEE Transactions on Information Forensics and Security*, 12(7):1686–1698, 2017.

42. L. Nie, A. Kumar, and S. Zhan. Periocular recognition using unsupervised convolutional RBM feature learning. In *International Conference on Pattern Recognition*, pp. 399–404, IEEE Computer Society, Los Alamitos, CA, 2014.

43. D. E. Rumelhart, G. E. Hinton, and R. J. Williams. Learning internal representations by error propagation. In *Parallel Distributed Processing: Explorations in the Microstructure of Cognition*, pp. 318–362, MIT Press, Cambridge, MA, 1986.

44. Y. Bengio, P. Lamblin, D. Popovici, and H. Larochelle. Greedy layerwise training of deep networks. In *Advances in Neural Information Processing Systems*, pp. 153–160, MIT Press, Cambridge, MA, 2007.

45. G. E. Hinton and R. R. Salakhutdinov. Reducing the dimensionality of data with neural networks. *Science*, 313(5786):504–507, 2006.

46. S. Gao, Y. Zhang, K. Jia, J. Lu, and Y. Zhang. Single sample face recognition via learning deep supervised autoencoders. *IEEE Transactions on Information Forensics and Security*, 10(10):2108–2118, 2015.

47. S. Kullback and R. A. Leibler. On information and sufficiency. *Annals of Mathematical Statistics*, 22(1):79–86, 1951.

48. X. Zheng, Z. Wu, H. Meng, and L. Cai. Contrastive auto-encoder for phoneme recognition. In *International Conference on Acoustics, Speech and Signal Processing*, pp. 2529–2533, IEEE, 2014.

49. F. Zhuang, X. Cheng, P. Luo, S. J. Pan, and Q. He. Supervised representation learning: Transfer learning with deep autoencoders. In *International Joint Conference on Artificial Intelligence*, pp. 4119–4125, AAAI Press, Palo Alto, CA, 2015.

50. A. Majumdar, R. Singh, and M. Vatsa. Face verification via class sparsity based supervised encoding. *IEEE Transactions on Pattern Analysis and Machine Intelligence*, 39(6):1273–1280, 2017.

51. P. Vincent, H. Larochelle, I. Lajoie, Y. Bengio, and P. Antoine Manzagol. Stacked denoising autoencoders: Learning useful representations in a deep network with a local denoising criterion. *Journal of Machine Learning Research*, 11:3371–3408, ACM, New York, NY, 2010.

52. S. Rifai, P. Vincent, X. Muller, X. Glorot, and Y. Bengio. Contractive auto-encoders: Explicit invariance during feature extraction. In *International Conference on Machine Learning*, pp. 833–840, 2011.

53. S. Rifai, G. Mesnil, P. Vincent, X. Muller, Y. Bengio, Y. Dauphin, and X. Glorot. Higher order contractive auto-encoder. In *European Conference on Machine Learning and Knowledge Discovery in Databases*, pp. 645–660, Springer Berlin Heidelberg, Berlin, Heidelberg, 2011.

54. C. Hong, J. Yu, J. Wan, D. Tao, and M. Wang. Multimodal deep autoencoder for human pose recovery. *IEEE Transactions on Image Processing*, 24(12):5659–5670, 2015.

55. J. Ngiam, A. Khosla, M. Kim, J. Nam, H. Lee, and A. Y. Ng. Multimodal deep learning. In *International Conference on Machine Learning*, pp. 689–696, ACM, New York, NY, 2011.

56. P. Luo, X. Wang, and X. Tang. Hierarchical face parsing via deep learning. In *IEEE Conference on Computer Vision and Pattern Recognition*, pp. 2480–2487, 2012.

57. M. Kan, S. Shan, H. Chang, and X. Chen. Stacked progressive auto-encoders (spae) for face recognition across poses. In *IEEE Conference on Computer Vision and Pattern Recognition*, Columbus, OH, 2014.

58. J. Zhang, S. Shan, M. Kan, and X. Chen. Coarse-to-fine auto-encoder networks (cfan) for real-time face alignment. In *European Conference on Computer Vision*, pp. 1–16, Springer International Publishing, Cham, 2014.

59. M. Singh, S. Nagpal, R. Singh, and M. Vatsa. Class representative autoencoder for low resolution multi-spectral gender classification. In *International Joint Conference on Neural Networks*, Anchorage, AK, 2017.

60. A. Dehghan, E. G. Ortiz, R. Villegas, and M. Shah. Who do I look like? Determining parent-offspring resemblance via gated autoencoders. In *IEEE Conference on Computer Vision and Pattern Recognition*, Columbus, OH, 2014.

61. R. Raghavendra and C. Busch. Learning deeply coupled autoencoders for smartphone based robust periocular verification. In *IEEE International Conference on Image Processing*, pp. 325–329, IEEE, 2016.

62. A. Sankaran, P. Pandey, M. Vatsa, and R. Singh. On latent fingerprint minutiae extraction using stacked denoising sparse autoencoders. In *International Joint Conference on Biometrics*, pp. 1–7, IEEE, 2014.

63. A. Sankaran, M. Vatsa, R. Singh, and A. Majumdar. Group sparse autoencoder. *Image and Vision Computing*, 60:64–74, 2017.

64. D. H. Hubel and T. N. Wiesel. Receptive fields and functional architecture of monkey striate cortex. *The Journal of Physiology*, 195(1):215–243, 1968.

65. A. Krizhevsky, I. Sutskever, and G. E. Hinton. Imagenet classification with deep convolutional neural networks. In *Advances in Neural Information Processing Systems*, pp. 1097–1105, ACM, New York, NY, 2012.

66. K. Fukushima. Neocognitron: A self-organizing neural network model for a mechanism of pattern recognition unaffected by shift in position. *Biological Cybernetics*, 36(4):193–202, 1980.

67. Y. L. Cun, O. Matan, B. Boser, J. S. Denker, D. Henderson, R. E. Howard, W. Hubbard, L. D. Jacket, and H. S. Baird. Handwritten zip code recognition with multilayer networks. In *IEEE International Conference on Pattern Recognition*, volume 2, pp. 35–40, IEEE, 1990.

68. Y. Bengio, Y. LeCun, C. Nohl, and C. Burges. LeRec: A NN/HMM hybrid for on-line handwriting recognition. *Neural Computation*, 7(6):1289–1303, 1995.

69. Y. Lecun, L. Bottou, Y. Bengio, and P. Haffner. Gradient-based learning applied to document recognition. *Proceedings of the IEEE*, 86(11): 2278–2324, 1998.

70. K. He, X. Zhang, S. Ren, and J. Sun. Deep residual learning for image recognition. In *IEEE Conference on Computer Vision and Pattern Recognition*, pp. 770–778, IEEE Computer Society, Washington, DC, 2016.

71. K. Simonyan and A. Zisserman. Very deep convolutional networks for large-scale image recognition. *CoRR*, abs/1409.1556, 2014.

72. F. Schroff, D. Kalenichenko, and J. Philbin. Facenet: A unified embedding for face recognition and clustering. In *IEEE Conference on Computer Vision and Pattern Recognition*. Boston, MA, June 7–12, 2015.

73. Y. Taigman, M. Yang, M. Ranzato, and L. Wolf. Deepface: Closing the gap to human-level performance in face verification. In *IEEE Conference on Computer Vision and Pattern Recognition*, pp. 1701–1708, IEEE Computer Society, Washington, DC, 2014.

74. R. Girshick, J. Donahue, T. Darrell, and J. Malik. Rich feature hierarchies for accurate object detection and semantic segmentation. In *IEEE Conference on Computer Vision and Pattern Recognition*, pp. 580–587, IEEE Computer Society, Washington, DC, 2014.

75. S. Ren, K. He, R. Girshick, and J. Sun. Faster R-CNN: Towards real-time object detection with region proposal networks. In *Advances in neural information processing systems*, pp. 91–99, Curran Associates, Red Hook, NY, 2015.

76. Y. Shen, X. He, J. Gao, L. Deng, and G. Mesnil. Learning semantic representations using convolutional neural networks for web search. In *International Conference on World Wide Web*, pp. 373–374, ACM, New York, NY, 2014.

77. C. Clark and A. J. Storkey. Teaching deep convolutional neural networks to play Go. *CoRR*, abs/1412.3409, 2014.

78. M. D. Zeiler and R. Fergus. Visualizing and understanding convolutional networks. In *European Conference on Computer Vision*, pp. 818–833. Springer, Zurich, Switzerland, 2014.

79. M. Lin, Q. Chen, and S. Yan. Network in network. *CoRR*, abs/1312.4400, 2013.

80. C. Szegedy, W. Liu, Y. Jia, P. Sermanet, S. Reed, D. Anguelov, D. Erhan, V. Vanhoucke, and A. Rabinovich. Going deeper with convolutions. In *IEEE Conference on Computer Vision and Pattern Recognition*, pp. 1–9, IEEE Computer Society, Los Alamitos, CA, 2015.

81. K. Q. Weinberger, J. Blitzer, and L. Saul. Distance metric learning for large margin nearest neighbor classification. *Advances in Neural Information Processing Systems*, 18:1473, 2006.

82. M. Everingham, L. Van Gool, C. K. I. Williams, J. Winn, and A. Zisserman. The PASCAL visual object classes (VOC) challenge. *International Journal of Computer Vision*, 88(2):303–338, 2010.

83. J. R. R. Uijlings, K. E. A. Van De Sande, T. Gevers, and A. W. M. Smeulders. Selective search for object recognition. *International Journal of Computer Vision*, 104(2):154–171, 2013.

84. O. M. Parkhi, A. Vedaldi, and A. Zisserman. Deep face recognition. In *British Machine Vision Conference*, Swansea, UK, 2015.

85. S. S. Farfade, M. J. Saberian, and L.-J. Li. Multi-view face detection using deep convolutional neural networks. In *International Conference on Multimedia Retrieval*, pp. 643–650, ACM, New York, 2015.

86. C. Garcia and M. Delakis. Convolutional face finder: A neural architecture for fast and robust face detection. *IEEE Transactions on Pattern Analysis and Machine Intelligence*, 26(11):1408–1423, 2004.

87. H. Li, Z. Lin, X. Shen, J. Brandt, and G. Hua. A convolutional neural network cascade for face detection. In *IEEE Conference on Computer Vision and Pattern Recognition*, pp. 5325–5334, IEEE Computer Society, Los Alamitos, CA, 2015.

88. G. Levi and T. Hassner. Age and gender classification using convolutional neural networks. In *IEEE Conference on Computer Vision and Pattern Recognition Workshops*, pp. 34–42, IEEE Computer Society, Los Alamitos, CA, 2015.

89. D. Menotti, G. Chiachia, A. Pinto, W. R. Schwartz, H. Pedrini, A. X. Falco, and A. Rocha. Deep representations for iris, face, and fingerprint spoofing detection. *IEEE Transactions on Information Forensics and Security*, 10(4):864–879, 2015.

90. H. A. Perlin and H. S. Lopes. Extracting human attributes using a convolutional neural network approach. *Pattern Recognition Letters*, 68: 250–259, 2015.

91. W. Fuhl, T. Santini, G. Kasneci, and E. Kasneci. Pupilnet: Convolutional neural networks for robust pupil detection. *CoRR*, abs/1601. 04902, 2016.

92. R. F. Nogueira, R. de Alencar Lotufo, and R. Campos Machado. Fingerprint liveness detection using convolutional neural networks. *IEEE Transactions on Information Forensics and Security*, 11(6):1206–1213, 2016.

93. J. Zhu, S. Liao, D. Yi, Z. Lei, and S. Z. Li. Multi-label cnn based pedestrian attribute learning for soft biometrics. In *International Conference on Biometrics*, pp. 535–540, 2015.

94. A. Graves and N. Jaitly. Towards end-to-end speech recognition with recurrent neural networks. In *International Conference on Machine Learning*, pp. 1764–1772, JMLR, 2014.

95. A. Karpathy and L. Fei-Fei. Deep visual-semantic alignments for generating image descriptions. In *IEEE Conference on Computer Vision and Pattern Recognition*, pp. 3128–3137, IEEE Computer Society, Washington, DC, 2015.

96. T. Mikolov, M. Karafiát, L. Burget, J. Cernockỳ, and S. Khudanpur. Recurrent neural network based language model. *Interspeech*, 2:3, 2010.

97. I. Sutskever, J. Martens, and G. E. Hinton. Generating text with recurrent neural networks. In *International Conference on Machine Learning*, pp. 1017–1024, ACM, New York, 2011.

98. S. Hochreiter and J. Schmidhuber. Long short-term memory. *Neural computation*, 9(8):1735–1780, 1997.

99. T. Mikolov, A. Joulin, S. Chopra, M. Mathieu, and M. A. Ranzato. Learning longer memory in recurrent neural networks. *CoRR*, abs/1412.7753, 2014.

100. H. Sak, A. Senior, and F. Beaufays. Long short-term memory recurrent neural network architectures for large scale acoustic modeling. In *International Speech Communication Association*, Singapore, 2014.

101. I. Sutskever, O. Vinyals, and Q. V. Le. Sequence to sequence learning with neural networks. In *Neural Information Processing Systems*, Montreal, Canada, 2014.

102. H. Zen and H. Sak. Unidirectional long short-term memory recurrent neural network with recurrent output layer for low-latency speech synthesis. In *International Conference on Acoustics, Speech, and Signal Processing*, pp. 4470–4474, IEEE Computer Society, Washington, DC, 2015.

103. I. Goodfellow, J. Pouget-Abadie, M. Mirza, B. Xu, D. Warde-Farley, S. Ozair, A. Courville, and Y. Bengio. Generative adversarial nets. In *Advances in Neural Information Processing Systems*, pp. 2672–2680, Curran Associates, Red Hook, NY, 2014.

104. L. Tran, X. Yin, and X. Liu. Disentangled representation learning GAN for pose-invariant face recognition. In *Conference on Computer Vision and Pattern Recognition*, Vol. 4, p. 7, 2017.

105. X. Yin, X. Yu, K. Sohn, X. Liu, and M. Chandraker. Towards large-pose face frontalization in the wild. *CoRR*, abs/1704.06244, 2017.

106. S. Tariyal, A. Majumdar, R. Singh, and M. Vatsa. Deep dictionary learning. *IEEE Access*, 4:10096–10109, 2016.

107. I. Manjani, S. Tariyal, M. Vatsa, R. Singh, and A. Majumdar. Detecting silicone mask based presentation attack via deep dictionary learning. *IEEE Transactions on Information Forensics and Security*, 12(7): 1713–1723, 2017.

108. S. Arora, A. Bhaskara, R. Ge, and T. Ma. Provable bounds for learning some deep representations. In *International Conference on Machine Learning*, pp. 584–592, JMLR, 2014.

109. R. Shwartz-Ziv and N. Tishby. Opening the black box of deep neural networks via information. *CoRR*, abs/1703.00810, 2017.

110. S. Nagpal, M. Vatsa, and R. Singh. Sketch recognition: What lies ahead? *Image and Vision Computing*, 55:9–13, 2016.

111. R. Fong, W. J. Scheirer, and D. D. Cox. Using human brain activity to guide machine learning. *CoRR*, abs/1703.05463, 2017.

2

Unconstrained Face Identification and Verification Using Deep Convolutional Features

Jun-Cheng Chen⋆, Rajeev Ranjan⋆, Vishal M. Patel,
Carlos D. Castillo, and Rama Chellappa

CONTENTS

2.1	Introduction	....	34
2.2	Related Work	....	35
	2.2.1	Face detection	35
	2.2.2	Facial landmark detection	36
	2.2.3	Feature representation for face recognition	37
	2.2.4	Metric learning	38
2.3	Proposed System	....	38
	2.3.1	HyperFace: A multitask face detector	40
	2.3.2	Deep convolutional face representation	44
	2.3.3	Joint Bayesian metric learning	46
2.4	Experimental Results	....	47
	2.4.1	Face detection	48
	2.4.2	Facial-landmark detection on IJB-A	48
	2.4.3	IJB-A and JANUS CS2 for face identification/verification	49
	2.4.4	Performance evaluations of face identification/ verification on IJB-A and JANUS CS2	51
	2.4.5	Labeled faces in the wild	53
	2.4.6	Comparison with methods based on annotated metadata	54
	2.4.7	Run time	54
2.5	Open Issues	....	56
2.6	Conclusion	....	56
Acknowledgments		....	57
References		....	57

⋆The first two authors equally contributed to this book chapter.

2.1   Introduction

Face recognition is a challenging problem in computer vision and has been actively researched for more than two decades [1]. There are two major tasks of strong interests: face identification and face verification. Face identification aims to identify the subject identity of a query image or video from a set of enrolled persons in the database. On the other hand, face verification, given two images or videos, determines whether they belong to the same person. Since the early 1990s, numerous algorithms have been shown to work well on images and videos that are collected in controlled settings. However, the performance of these algorithms often degrades significantly on images that have large variations in pose, illumination, expression, aging, and occlusion. In addition, for an automated face recognition system to be effective, it also needs to handle errors that are introduced by algorithms for automatic face detection and facial landmark detection.

Existing methods have focused on learning robust and discriminative representations from face images and videos. One approach is to extract an overcomplete and high-dimensional feature representation followed by a learned metric to project the feature vector onto low-dimensional space and then compute the similarity scores. For example, high-dimensional multi-scale local binary pattern (LBP) [2] features extracted from local patches around facial landmarks and Fisher vector (FV) [3,4] features have been shown to be effective for face recognition. Despite significant progress, the performance of these systems has not been adequate for deployment. However, with the availability of millions of annotated data, faster GPUs and a better understanding of the nonlinearities, deep convolutional neural networks (DCNNs) yield much better performance on tasks such as object recognition [5,6], object/face detection [7,8], and face recognition [9,10]. It has been shown that DCNN models can not only characterize large data variations, but also learn a compact and discriminative representation when the size of the training data is sufficiently large. In addition, it can be generalized to other vision tasks by fine-tuning the pretrained model on the new task [11].

In this chapter, we present an automated face-recognition system. Because of the robustness of DCNNs, we build each component of our system based on separate DCNN models. The module for face detection and face alignment tasks uses the AlexNet proposed in [5] as the base DCNN architecture coupled with multitask learning to perform both tasks simultaneously. For face identification and verification, we train a DCNN model using the CASIA-WebFace [12] data set. Finally, we compare the performance of our approach with many face matchers on the Intelligence Advanced Research Projects Activity (IARPA) Janus Benchmark A (IJB-A) data set, which are being carried out or have been recently reported by the National Institute of Standards and Technology [13]. The proposed system is automatic and yields comparable or better performance than other existing algorithms when evaluated on IJB-A and JANUS CS2 data sets. Although the IJB-A data set contains significant variations in pose, illumination, expression, resolution, and occlusion, which are much harder than the Labeled Faces in the Wild (LFW) data set, we present verification results for the LFW data set, too.

In addition, the system presented in this chapter differs from its predecessor by Chen et al. [14] in the following ways: (1) instead of separate components for face detection and facial landmark detection, we adopt HyperFace [15] as the preprocessing module, which not only simultaneously performs face detection and facial landmark localization, but also achieves better results than its predecessor [14]. Meanwhile, because the quality of the detected facial landmarks has improved, the proposed system can yield improved results without applying any fine-tuning for the pretrained DCNN model on the target training data as its predecessor [14] did. In the experimental section, we also demonstrate the improvement as a result of media-sensitive pooling where we first average the features separately according to their media types and then perform the second average.

In the rest of the chapter, we briefly review closely related works in Section 2.2; we present the design details of a deep learning system for unconstrained face recognition, including face detection, face alignment, and face identification/verification in Section 2.3; experimental results using IJB-A, JANUS CS2, and LFW data sets are presented in Section 2.4; some open issues regarding the use of DCNNs for face recognition problems are discussed in Section 2.5; and we conclude with a brief summary in Section 2.6.

2.2 Related Work

An automatic face-recognition system consists of the following components: (1) face detection, (2) facial landmark detection to align faces, and (3) face identification/verification to identify a subject's identity or to determine two faces from the same identity or not. As a result of the large number of published papers, we briefly review some relevant works for each component.

2.2.1 Face detection

The face detection method introduced by Viola and Jones [16] is based on cascaded classifiers built using the Haar wavelet features. Since then, a variety of sophisticated cascade-based face detectors such as Joint Cascade [17], SURF Cascade [18], and CascadeCNN [19] have demonstrated improved performance. Zhu and Ramanan [20] improved the performance of the face-detection algorithm using the deformable part model (DPM) approach, which treats each facial landmark as a part and uses the histogram of oriented gradient features to simultaneously perform face detection, pose estimation, and landmark localization. A recent face detector, HeadHunter [21], shows competitive performance using a simple DPM. However, the key challenge in unconstrained face detection is that features like Haar wavelets and histogram of gradient do not capture the salient facial information at different poses and illumination conditions. To overcome these limitations, few DCNN-based face-detection methods have been proposed in the literature such as Faceness [22], DDFD [23], and CascadeCNN [19]. It has been shown in [11] that a DCNN pretrained with the Imagenet data set can be used as a meaningful feature extractor for various vision tasks. The method based on Regions with CNN (R-CNN) [24] computes region-based deep features and attains state-of-the-art face detection performance. Additionally, because the deep pyramid [25] removes the fixed-scale input dependency in DCNNs, it is attractive to be integrated with the DPM approach to further improve the detection accuracy across scale [8]. A deep feature-based face detector for mobile devices was proposed in [26]. Jiang and Learned-Miller [27] adapted the Faster R-CNN [28] for face detection using the new face-detection data set, WIDER FACE [29], for training and achieved top performance on various face-detection benchmarks. Hu and Ramanan [30] showed that contextual information is useful for finding tiny faces, and it models both appearance and context of faces by using the features of different layers of a convolutional neural network (CNN) to detect tiny faces. (i.e., the lower level features are for semantics and higher level for context.) In particular, Ranjan et al. [15,31] developed a single multi-task deep network for various facial analytic tasks, and the results showed that multi-task learning helps improve the performance for face detection, but also for other tasks, including fiducial point detection and face recognition.

2.2.2 Facial landmark detection

Facial landmark detection is an important component for a face identification/verification system to align the faces into canonical coordinates and to improve the performance of verification algorithms. Pioneering works such as Active Appearance Models [32] and Active Shape Models [33] are built using the PCA constraints on appearance and shape. Cristinacce and Cootes [34] generalized the Active Shape Model to a Constrained Local Model, in which every landmark has a shape-constrained descriptor to capture the appearance. Zhu and Ramanan [20] used a part-based model for face detection, pose estimation, and landmark localization assuming the face shape to be a tree structure. Asthana et al. [35] combined the discriminative response map fitting with a Constrained Local Model. In addition, Cao et al. [36] follows the procedure as cascaded pose regression proposed by Dollár et al. [37], which is feature extraction followed by a regression stage. However, unlike cascaded pose regression which uses pixel difference as features, it trains a random forest based on LBPs. In general, these methods learn a model that directly maps the image appearance to the target output. Nevertheless, the performance of these methods depends on the robustness of local descriptors. The deep features are shown to be robust to different challenging variations. Sun et al. [38] proposed a cascade of carefully designed CNNs, in which at each level, outputs of multiple networks are fused for landmark estimation and achieve good performance. Unlike [38], the multi-task detectors [15,31] fuse the features of different layers and perform the face detection and facial landmark detection at the same time, which not only resolves the scale issue, but also uses the rich information from different layers to achieve better performance. Kumar et al. [39] proposed an iterative conditional deep regression framework for fiducial point detection, which uses the input face along with the predicted points of the previous iteration together to improve the results. Kumar and Chellappa [40] proposed a novel deep regression framework that consists of several branches of subnetworks from a main deep network to take the spatial relationship of fiducial points into consideration.

2.2.3 Feature representation for face recognition

Learning invariant and discriminative feature representations is a critical step in a face-verification system. Ahonen et al. [41] showed that the LBP is effective for face recognition. Chen et al. [2] demonstrated good results for face verification using the high-dimensional multi-scale LBP features extracted from patches obtained around facial landmarks. However, recent advances in deep learning methods have shown that compact and discriminative representations can be learned using a DCNN trained with very large data sets. Taigman et al. [42] learned a DCNN model on the frontalized faces generated with a general three-dimensional (3D) shape model from a large-scale face data set and achieved better performance than many traditional methods. Sun et al. [43] achieved results that surpass human performance for face verification on the LFW data set using an ensemble of 25 simple DCNNs with fewer layers trained on weakly aligned face images from a much smaller data set than [42]. Schroff et al. [9] adapted a state-of-the-art object recognition network to face recognition and trained it using a large-scale unaligned private face data set with the triplet loss. Parkhi et al. [10] trained a very deep convolutional network based on VGGNet for face verification and demonstrated impressive results. Abd-Almageed et al. [44] handles pose variations by learning separate DCNN models for frontal, half-profile, and full-profile poses to improve face recognition performance in the real world. Furthermore, Masi et al. [45] used 3D morphable models to augment the CASIA-WebFace data set with large amounts of synthetic faces to improve the recognition performance instead of collecting more data through crowdsourcing the annotation tasks. Ding and Tao [46] proposed to fuse the deep features around facial landmarks from different layers followed by a new triplet-loss function, which achieves state-of-the-art performance for video-based face recognition. A neural aggregated network [47] has been proposed to aggregate the multiple face images in the video for a succinct and robust representation for video face recognition. Bodla et al. [48] proposed a fusion network to combine the face representations from two different DCNN models to improve the recognition performance. These studies essentially demonstrate the effectiveness of the DCNN model for feature learning and detection, recognition, and verification problems.

2.2.4   Metric learning

Learning a similarity measure from data is the other key component for improving the performance of a face verification system. Many approaches have been proposed that essentially exploit the label information from face images or face pairs. For instance, Weinberger et al. [49] proposed the Large Margin Nearest Neighbor metric, which enforces the large margin constraint among all triplets of labeled training data. Taigman et al. [50] learned the Mahalanobis distance using the Information Theoretic Metric Learning method [51]. Chen et al. [52] proposed a joint Bayesian approach for face verification that models the joint distribution of a pair of face images and uses the ratio of between-class and within-class probabilities as the similarity measure. Hu et al. [53] learned a discriminative metric within the deep neural network framework. Schroff et al. [9] and Parkhi et al. [10] optimized the DCNN parameters based on the triplet loss, which directly embeds the DCNN features into a discriminative subspace and presented promising results for face verification. In addition, Wen et al. [54] proposed a new loss that takes the centroid for each class into consideration and uses it as a regularization to the softmax loss in addition to a residual neural network for learning more discriminative face representation. Liu et al. [55] proposed a modified softmax loss by imposing discriminative constraints on a hypersphere manifold to learn angularly discriminative features, which greatly improves the recognition performance. Ranjan et al. [56] also proposed a modified softmax loss regularized with the scale L_2-norm constraint, which helps to learn angularly discriminative features as well and improves the recognition performance significantly.

2.3   Proposed System

The proposed system includes the whole pipeline for performing automatic face identification and verification. We first perform face detection and facial-landmark detection simultaneously with the proposed multitask face detector for each image and video frame. Then, we align the faces into canonical coordinates using the detected landmarks. Finally, we perform face recognition by computing the similarity scores between the query and the enrolled gallery for face identification or between a pair of images or video frames for face verification. The system is illustrated in Figure 2.1. The details of each component are presented in the following subsections.

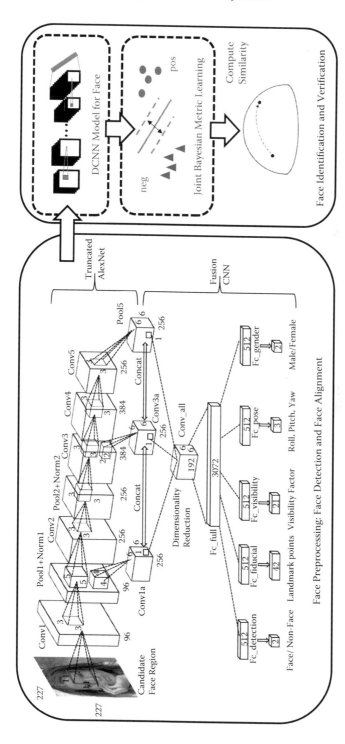

FIGURE 2.1

An overview of the proposed automated DCNN-based face identification/verification system.

2.3.1   HyperFace: A multitask face detector

HyperFace [15] is a single DCNN model for simultaneous face detection, landmark localization, pose estimation, and gender classification. The network architecture is deep in both vertical and horizontal directions as shown in the face preprocessing part of Figure 2.1. HyperFace consists of three modules. The first one generates class-independent region proposals from the given image and scales them to the resolution of 227×227 pixels. The second module is a DCNN model that takes in the resized candidate regions and classifies them as face or nonface. If a region gets classified as a face, the network additionally predicts facial-landmark locations, 3D head pose, and gender information. The third module is a postprocessing step, which involves iterative region proposals and landmarks-based non-maximum suppression to boost the face-detection score and improve the performance of individual tasks (Figure 2.2).

HyperFace uses the well-known AlexNet [57] for image classification as the base network. The network consists of five convolutional layers along with three fully connected layers. The network is created by the following two observations: (1) The features in CNN are distributed hierarchically in the network. The lower-layer features are informative for landmark localization and pose

FIGURE 2.2
Sample results of HyperFace with detected face bounding boxes, fiducial points, and 3D head pose in which magenta boxes refer the gender as female and blue boxes as male.

estimation, and the higher-layer features are suitable for more semantic tasks such as detection or classification [58]. (2) Learning multiple correlated tasks simultaneously builds a synergy and improves the performance of individual tasks. Hence, to simultaneously learn face detection, landmarks, pose, and gender, HyperFace fuses the features from the intermediate layers of the network and learns multiple tasks on top of it. Because the adjacent layers are highly correlated, we skip some layers and fuse the max1, conv3, and pool5 layers of AlexNet. The feature maps for these layers have different dimensions, $27 \times 27 \times 96$, $13 \times 13 \times 384$, $6 \times 6 \times 256$, respectively, so we add conv1a and conv3a convolutional layers to pool1 and conv3 layers to obtain consistent feature maps of dimensions $6 \times 6 \times 256$ at the output. We then concatenate the output of these layers along with pool5 to form a $6 \times 6 \times 768$ dimensional feature maps followed by a 1×1 kernel convolution layer (*conv_all*) to reduce the dimensions to $6 \times 6 \times 192$. We add a fully connected layer (*fc_all*) to *conv_all*, which outputs a 3072 dimensional feature vector. Then, we split the network into five separate branches corresponding to the different tasks. We add *fc_detection*, *fc_landmarks*, *fc_visibility*, *fc_pose*, and *fc_gender* fully connected layers, each of dimension 512, to *fc_all*. Finally, a fully connected layer is added to each of the branches to predict the labels for each task. After every convolution or a fully connected layer, we deploy the Rectified Layer Unit. In addition, we did not include any pooling operation in the fusion network because it provides local invariance, which is not desired for the facial landmark localization task. Task-specific loss functions are then used to learn the weights of the network.

For HyperFace, we use a AFLW [59] data set for training. It contains 25,993 faces in 21,997 real-world images with full pose, expression, ethnicity, age, and gender variations. It provides annotations for 21 landmark points per face, along with the face bounding box, face pose (yaw, pitch, and roll), and gender information. We randomly selected 1000 images for testing and keep the rest for training the network. Different objective functions are used for training the tasks of face detection, facial-landmark localization, facial-landmark visibility, 3D head-pose estimation, and gender-classification tasks.

Face Detection

We use the Selective Search [60] algorithm to generate region proposals for faces in an image. A region having an overlap of more than 0.5 with the ground-truth bounding box is considered a positive sample ($l = 1$). The candidate regions with overlap less than 0.35 are treated as negative instance ($l = 0$). All the other regions are ignored. We use the binary cross entropy loss function given by (2.1) for training the face-detection task.

$$loss_D = -(1-\ell)\log(1-p) - \ell\log(p) \qquad (2.1)$$

where p is the probability that the candidate region is a face.

Landmark Localization

We use the 21 point annotations of facial-landmark locations as provided in the AFLW data set [59]. Besides the facial-landmark annotations, the data set also provides the annotations of visibility for each landmark because some of the landmark points are invisible under different head poses. We consider regions with overlap greater than 0.35 with the ground truth for learning this task, while ignoring the rest. A region is represented as $\{x, y, w, h\}$, where (x, y) are the coordinates of the center of the region and (w, h) the width and height of the region, respectively. Each visible landmark point is shifted with respect to the region center (x, y) and normalized by (w, h) as given by $(a_i, b_i) = (\frac{x_i - x}{w}, \frac{y_i - y}{h})$, where (x_i, y_i)'s are the given ground-truth fiducial coordinates. The (a_i, b_i)'s are treated as labels for training the landmark-localization task using the Euclidean loss weighted by the visibility factor. The labels for landmarks, which are not visible, are taken to be $(0, 0)$. The loss in predicting the landmark location is computed from (2.2) as follows:

$$loss_L = \frac{1}{2N} \sum_{i=1}^{N} v_i((\hat{x}_i - a_i)^2 + ((\hat{y}_i - b_i)^2)) \tag{2.2}$$

where: (x_i, y_i) is the ith landmark location predicted by the network, relative to a given region, N is the total number of landmark points, which is 21 for the AFLW data set. The visibility factor, v_i, is 1 if the ith landmark is visible in the candidate region and is 0 otherwise. This implies that there is no loss corresponding to invisible points, and hence they do not take part during back-propagation.

Learning Visibility

We also learn the visibility factor to test the presence of the predicted landmark. For a given region with overlap higher than 0.35, we use the Euclidean loss to train the visibility as shown in (2.3):

$$loss_V = \frac{1}{N} \sum_{i=1}^{N} N(\hat{v}_i - v_i)^2 \tag{2.3}$$

where $\hat{v}_i$ is the predicted visibility of ith landmark. The visibility label, v_i, is 1 if the ith landmark is visible in the candidate region and is 0 otherwise.

Pose Estimation

We use the Euclidean loss to train the head pose estimates of roll (p_1), pitch (p_2), and yaw (p_3). We compute the loss for candidate regions having an overlap more than 0.5 with the ground truth by using (2.4):

$$loss_P = \frac{(\hat{p}_1 - p_1)^2 + (\hat{p}_2 - p_2)^2 + (\hat{p}_3 - p_3)^2}{3} \tag{2.4}$$

where $(\hat{p}_1, \hat{p}_2, \hat{p}_3)$ are the estimates for roll, pitch, and yaw, respectively.

Gender Recognition

Gender classification is a two-class problem. For a candidate region with overlap of 0.5 with the ground truth, we compute the loss by using (2.5):

$$loss_G = -(1-g) \cdot \log(1-g_0) - g \cdot \log(g_1), \qquad (2.5)$$

where $g = 0$ if the gender is male, and $g = 1$ otherwise. In addition, (g_0, g_1) are the probabilities for male and female, respectively, computed from the network. The total loss is computed as the weighted sum of the five individual losses as shown in (2.6):

$$loss_{all} = \sum_{t=1}^{t=5} \lambda_t loss_t, \qquad (2.6)$$

where $loss_t$ is the individual loss corresponding to t-th task. We choose $(\lambda_D = 1, \lambda_L = 5, \lambda_V = 0.5, \lambda_P = 5, \lambda_G = 2)$ for our experiments. Higher weights are assigned to landmark localization and pose-estimation tasks because they need spatial accuracy.

For a given a test image, we first extract the candidate region proposals using selective search. For each of the regions, we predict the task labels by a forward pass through the HyperFace network. Only regions with detection scores above a certain threshold are classified as face and processed for subsequent tasks. Sample results are illustrated in Figure 2.3. The results demonstrate

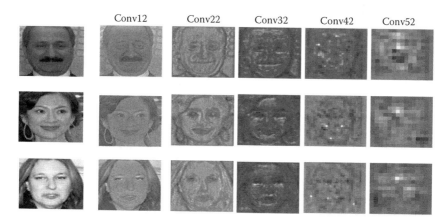

FIGURE 2.3
An illustration of some feature maps of conv12, conv22, conv32, conv42, and conv52 layers of the DCNN trained for the face-identification task. At upper layers, the feature maps capture more global shape features, which are also more robust to illumination changes than conv12. The feature maps are rescaled to the same size for visualization purpose. The green pixels represent high activation values, and the blue pixels represent low activation values as compared to the green.

that HyperFace can reliably detect the faces, localize fiducial points, classify the gender, and estimate 3D head pose for faces in large pose variations. For this chapter, we focus on discussing face detection and facial-landmark detection. Other details and results including training part of HyperFace, we refer interested readers to [15].

2.3.2   Deep convolutional face representation

In this work, we train the deep convolutional networks, which are trained using tight face bounding boxes. The architecture of the network is summarized in Table 2.1.

Stacking small filters to approximate large filters and building very deep convolutional networks reduce the number of parameters, but also increase the nonlinearity of the network in [6,61]. In addition, the resulting feature representation is compact and discriminative. Therefore, we use the same network architecture presented in [62] and train it using the CASIA-WebFace data set [12]. The dimensionality of the input layer is $100 \times 100 \times 3$ for RGB images. The network includes 10 convolutional layers, 5 pooling layers, and 1 fully connected layer. Each convolutional layer is followed by a parametric rectified linear unit [63], except the last one, conv52. Moreover, two local normalization layers are added after conv12 and conv22, respectively, to mitigate the effect of illumination variations. The kernel size of all filters is 3×3. The first four

TABLE 2.1
The architectures of the DCNN for face identification

Name	Type	Filter size/stride	#Params
conv11	convolution	$3\times3/1$	0.84K
conv12	convolution	$3\times3/1$	18K
pool1	max pooling	$2\times2/2$	
conv21	convolution	$3\times3/1$	36K
conv22	convolution	$3\times3/1$	72K
pool2	max pooling	$2\times2/2$	
conv31	convolution	$3\times3/1$	108K
conv32	convolution	$3\times3/1$	162K
pool3	max pooling	$2\times2/2$	
conv41	convolution	$3\times3/1$	216K
conv42	convolution	$3\times3/1$	288K
pool4	max pooling	$2\times2/2$	
conv51	convolution	$3\times3/1$	360K
conv52	convolution	$3\times3/1$	450K
pool5	avg pooling	$7\times7/1$	
dropout	dropout (40%)		
fc6	fully connected	10,548	3296K
loss	softmax	10,548	
total			5M

pooling layers use the max operator, and pool5 uses average pooling. The feature dimensionality of pool5 is thus equal to the number of channels of conv52, which is 320. The dropout ratio is set as 0.4 to regularize fc6 because of the large number of parameters (i.e., $320 \times 10{,}548$). The pool5 feature is used for face representation. The extracted features are further L_2-normalized to unit length before the metric learning stage. If there are multiple images and frames available for the subject template, we use the average of pool5 features as the overall feature representation.

In Figure 2.4, we show some feature-activation maps of the DCNN model. At upper layers, the feature maps capture more global shape features, which are also more robust to illumination changes than conv12 in which the green pixels represent high-activation values, and the blue pixels represent low-activation values compared to the green.

For face recognition, the DCNN is implemented using caffe [64] and trained on the CASIA-WebFace data set. The CASIA-WebFace data set contains 494,414 face images of 10,575 subjects downloaded from the IMDB website. After removing 27 overlapping subjects with the IJB-A data set, there are 10,548 subjects and 490,356 face images. For each subject, there still exists several false images with wrong identity labels and few duplicate images. All images are scaled into $[0, 1]$ and subtracted from the mean. The data are augmented with horizontal flipped face images. We use the standard batch size 128 for the training phase. Because it only contains sparse positive and negative pairs per batch in addition to the false-image problems, we do not take the verification cost into consideration as is done in [43]. The initial negative slope for a parametric rectified linear unit is set to 0.25 as suggested in [63]. The weight decay of all convolutional layers are set to 0, and the weight decay of the final fully connected layer to 5e-4. In addition, the learning rate is set to 1e-2 initially and reduced by half every 100,000 iterations.

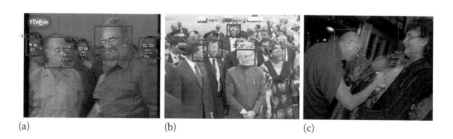

(a)   (b)   (c)

FIGURE 2.4

Faces not detected or failed for other tasks by HyperFace usually in very extreme pose, low-resolution, occlusion, and other conditions contain limited information for face recognition. Although the gender for Dalai Lama (c) is wrongly classified, the face bounding boxes and fiducial points are still detected with reasonable quality. This also demonstrates the robustness of HyperFace.

The momentum is set to 0.9. Finally, we use the snapshot of 1,000,000th iteration for all our experiments.

2.3.3   Joint Bayesian metric learning

To use the positive and negative label information available from the training data set, we learn a joint Bayesian metric, which has achieved good performances on face-verification problems [52,65]. Instead of modeling the difference vector between two faces, this approach directly models the joint distribution of feature vectors of both ith and jth images, $\{\mathbf{x}_i, \mathbf{x}_j\}$, as a Gaussian. Let $P(\mathbf{x}_i, \mathbf{x}_j | H_I) \sim N(0, \mathbf{\Sigma}_I)$ when $\mathbf{x}_i$ and $\mathbf{x}_j$ belong to the same class, and $P(\mathbf{x}_i, \mathbf{x}_j | H_E) \sim N(0, \mathbf{\Sigma}_E)$ when they are from different classes. In addition, each face vector can be modeled as, $\mathbf{x} = \mathbf{\mu} + \mathbf{\epsilon}$, where $\mathbf{\mu}$ stands for the identity and $\mathbf{\epsilon}$ for pose, illumination, and other variations. Both $\mathbf{\mu}$ and $\mathbf{\epsilon}$ are assumed to be independent zero-mean Gaussian distributions, $N(0, \mathbf{S}_\mu)$ and $N(0, \mathbf{S}_\epsilon)$, respectively.

The log likelihood ratio of intra- and inter-classes, $r(\mathbf{x}_i, \mathbf{x}_j)$, can be computed as follows:

$$r(\mathbf{x}_i, \mathbf{x}_j) = \log \frac{P(\mathbf{x}_i, \mathbf{x}_j | H_I)}{P(\mathbf{x}_i, \mathbf{x}_j | H_E)} = \mathbf{x}_i^T \mathbf{M} \mathbf{x}_i + \mathbf{x}_j^T \mathbf{M} \mathbf{x}_j - 2\mathbf{x}_i^T \mathbf{R} \mathbf{x}_j \qquad (2.7)$$

where $\mathbf{M}$ and $\mathbf{R}$ are both negative semi-definite matrices. Equation 2.7 can be rewritten as $(\mathbf{x}_i - \mathbf{x}_j)^T \mathbf{M} (\mathbf{x}_i - \mathbf{x}_j) - 2\mathbf{x}_i^T \mathbf{B} \mathbf{x}_j$, where $\mathbf{B} = \mathbf{R} - \mathbf{M}$. More details can be found in [52]. Instead of using the EM algorithm to estimate $\mathbf{S}_\mu$ and $\mathbf{S}_\epsilon$, we optimize the distance in a large-margin framework as follows:

$$\arg\min_{\mathbf{M},\mathbf{B},b} \sum_{i,j} max[1 - y_{ij}(b - (\mathbf{x}_i - \mathbf{x}_j)^T \mathbf{M} (\mathbf{x}_i - \mathbf{x}_j) + 2\mathbf{x}_i^T \mathbf{B} \mathbf{x}_j), 0], \quad (2.8)$$

where:

$b \in \mathbb{R}$ is the threshold

y_{ij} is the label of a pair: $y_{ij} = 1$ if person i and j are the same and $y_{ij} = -1$, otherwise

For simplicity, we denote $(\mathbf{x}_i - \mathbf{x}_j)^T \mathbf{M} (\mathbf{x}_i - \mathbf{x}_j) - 2\mathbf{x}_i^T \mathbf{B} \mathbf{x}_j$ as $d_{\mathbf{M},\mathbf{B}}(\mathbf{x}_i, \mathbf{x}_j)$. $\mathbf{M}$ and $\mathbf{B}$ are updated using stochastic gradient descent as follows and are equally trained on positive and negative pairs in turn:

$$\mathbf{M}_{t+1} = \begin{cases} \mathbf{M}_t, & \text{if } y_{ij}(b_t - d_{\mathbf{M},\mathbf{B}}(\mathbf{x}_i, \mathbf{x}_j)) > 1 \\ \mathbf{M}_t - \gamma y_{ij} \mathbf{\Gamma}_{ij}, & \text{otherwise,} \end{cases}$$

$$\mathbf{B}_{t+1} = \begin{cases} \mathbf{B}_t, & \text{if } y_{ij}(b_t - d_{\mathbf{M},\mathbf{B}}(\mathbf{x}_i, \mathbf{x}_j)) > 1 \\ \mathbf{B}_t + 2\gamma y_{ij} \mathbf{x}_i \mathbf{x}_j^T, & \text{otherwise,} \end{cases} \qquad (2.9)$$

$$b_{t+1} = \begin{cases} b_t, & \text{if } y_{ij}(b_t - d_{\mathbf{M},\mathbf{B}}(\mathbf{x}_i, \mathbf{x}_j)) > 1 \\ b_t + \gamma_b y_{ij}, & \text{otherwise,} \end{cases}$$

where:

$$\mathbf{\Gamma}_{ij} = (\mathbf{x}_i - \mathbf{x}_j)(\mathbf{x}_i - \mathbf{x}_j)^T$$

γ is the learning rate for $\mathbf{M}$ and $\mathbf{B}$, and γ_b for the bias b

We use random semi-definite matrices to initialize both $\mathbf{M} = \mathbf{V}\mathbf{V}^T$ and $\mathbf{B} = \mathbf{W}\mathbf{W}^T$ where both $\mathbf{V}$ and $\mathbf{W} \in \mathbb{R}^{d \times d}$, and v_{ij} and $w_{ij} \sim N(0, 1)$. Note that $\mathbf{M}$ and $\mathbf{B}$ are updated only when the constraints are violated. In our implementation, the ratio of the positive and negative pairs that we generate based on the identity information of the training set is 1 verus 40. In addition, the other reason to train the metric instead of using traditional EM is that for IJB-A training and test data, some templates only contain a single image. More details about the IJB-A data set are given in Section 2.4.

In general, to learn a reasonable distance measure directly using pairwise or triplet metric learning approach requires a huge amount of data (i.e., the state-of-the-art approach [9] uses 200 million images). In addition, the proposed approach decouples the DCNN feature learning and metric learning because of memory constraints. To learn a reasonable distance measure requires generating informative pairs or triplets. The batch size used for stochastic gradient descent is limited by the memory size of the graphics card. If the model is trained end to end, then only a small batch size is available for use. Thus, in this work, we perform DCNN model training and metric learning independently. In addition, for the publicly available deep model [10], it is also trained first with softmax loss and followed by fine-tuning the model with verification loss with freezing the convolutional and fully connected layers except the last one to learn the transformation, which is equivalent to the proposed approach.

2.4 Experimental Results

In this section, we present the results of the proposed automatic system for face detection on the FDDB data set [66], facial-landmark localization on the AFW data set [20], and face verification tasks on the challenging IJB-A [67], its extended version Janus Challenging set 2 (JANUS CS2) data set, and the LFW data set. The JANUS CS2 data set contains not only the sampled frames and images in the IJB-A, but also the original videos. In addition, the JANUS CS2 data set* includes considerably more test data for identification and verification problems in the defined protocols than the IJB-A data set. The receiver operating characteristic (ROC) curves and the cumulative match characteristic (CMC) scores are used to evaluate the performance of different algorithms for face verification. The ROC curve measures the performance in the verification scenarios, and the CMC score measures the accuracy in closed set identification scenarios.

---

*The JANUS CS2 data set is not publicly available yet.

2.4.1   Face detection

To demonstrate the effectiveness of HyperFace for face detection, we evaluate it using the challenging FDDB benchmark [66]. HyperFace achieves good performance with mAP of 90.1% in the FDDB data set with a large performance margin compared to most algorithms. Some of the recent published methods compared in the FDDB evaluation include Faceness [22], HeadHunter [21], JointCascade [17], CCF [68], Squares-ChnFtrs-5 [21], CascadeCNN [19], Structured Models [69], DDFD [23], NDPFace [70], PEP-Adapt [71], DP2MFD [8], and TSM [72]. More comparison results with other face-detection data sets are available in [8,15]. In addition, we illustrate typical faces in the IJB-A data set that are not detected by HyperFace, and we can find the faces to be usually in extreme conditions that contain limited information for face identification/verification (Figure 2.5).

2.4.2   Facial-landmark detection on IJB-A

We evaluate the performance of different landmark localization algorithms on the AFW data set [20]. The data set contains faces with full-pose variations. Some of the methods compared include Multiview Active Appearance Model-based method [20], Constrained Local Model [73], Oxford facial landmark detector [74], Zhu [20], FaceDPL [75], Multitask Face, which performs multi-task learning as HyperFace, but without fusing features from intermediate layers, and R-CNN Fiducial, which is trained in the single-task setting. Although the data sets provide ground-truth bounding boxes, we do not use them for evaluating on HyperFace, Multitask Face, and R-CNN Fiducial. Instead we use the

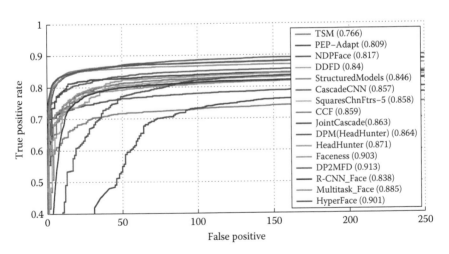

FIGURE 2.5
Face detection performance evaluation on the FDDB data set.

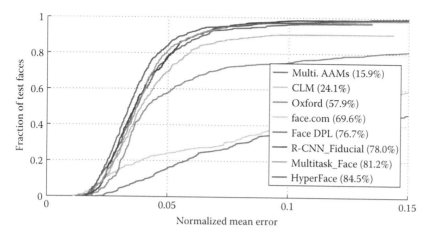

FIGURE 2.6
Facial-landmark detection performance evaluation on the AFW data set. CLM, Constrained Local Model; Multi. AAMs, Multiview Active Appearance Model.

respective algorithms to detect both the face and its fiducial points. Because R-CNN Fiducial cannot detect faces, we provide it with the detections from the HyperFace. Figure 2.6 compares the performance of different landmark localization methods on the AFW data set using the protocol defined in [75]. The data set provides six key points for each face, which are: left eye center, right eye center, nose tip, mouth left, mouth center, and mouth right. We compute the error as the mean distance between the predicted and ground-truth key points, normalized by the face size. The plots for comparison were obtained from [75]. As can be seen from the figure, HyperFace outperforms many recent state-of-the-art landmark-localization methods. The HyperFace has an advantage over them because it uses the intermediate layers for fusion. The local information is contained well in the lower layers of CNN and becomes invariant as depth increases. Fusing the layers brings out that hidden information, which boosts the performance for the landmark-localization task.

2.4.3 IJB-A and JANUS CS2 for face identification/verification

For the face identification/verification task, both IJB-A and JANUS CS2 data sets contain 500 subjects with 5397 images and 2042 videos split into 20,412 frames, which is 11.4 images and 4.2 videos per subject. Sample images and video frames from the data sets are shown in Figure 2.7. The IJB-A evaluation protocol consists of verification (1:1 matching) over 10 splits. Each split contains around 11,748 pairs of templates (1756 positive and 9992 negative pairs) on average. Similarly, the identification (1:N search) protocol also consists of

FIGURE 2.7
Sample images and frames from the IJB-A (*top*) and JANUS CS2 data sets
(*bottom*). Challenging variations resulting from pose, illumination, resolution,
occlusion, and image quality are present in these images.

10 splits, which are used to evaluate the search performance. In each search
split, there are about 112 gallery templates and 1763 probe templates (i.e.,
1187 genuine probe templates and 576 impostor probe templates). On the
other hand, for the JANUS CS2, there are about 167 gallery templates and
1763 probe templates, and all of them are used for both identification and
verification. The training set for both data sets contains 333 subjects, and the
test set contains 167 subjects without any overlapping subjects. Ten random
splits of training and testing are provided by each benchmark, respectively.
The main differences between IJB-A and JANUS CS2 evaluation protocols are
that (1) IJB-A considers the open-set identification problem and the JANUS
CS2 considers the closed-set identification and (2) IJB-A considers the more
difficult pairs, which are the subsets from the JANUS CS2 data set.

Unlike LFW and YTF data sets, which only use a sparse set of negative
pairs to evaluate the verification performance, the IJB-A and JANUS CS2
both divide the images and video frames into gallery and probe sets so that
all the available positive and negative pairs are used for the evaluation. Also,
each gallery and probe set consist of multiple templates. Each template contains a combination of images or frames sampled from multiple image sets or
videos of a subject. For example, the size of the similarity matrix for JANUS
CS2 split1 is 167×1806 where 167 are for the gallery set and 1806 for the
probe set (i.e., the same subject reappears multiple times in different probe
templates). Moreover, some templates contain only one profile face with a
challenging pose with low-quality imagery. In contrast to LFW and YTF data
sets, which only include faces detected by the Viola Jones face detector [16],
the images in the IJB-A and JANUS CS2 contain extreme-pose, illumination, and expression variations. These factors essentially make the IJB-A and
JANUS CS2 challenging face-recognition data sets [67].

2.4.4 Performance evaluations of face identification/verification on IJB-A and JANUS CS2

To evaluate the performance of the proposed approach, we present the identification/verification results of the proposed approach for the IJB-A data set in Table 2.2. For each face, we slightly shift it with a small offset (i.e., three to five pixels) and take 25 crops followed by average pooling. Besides using the average feature representation, we also perform media averaging, which is to first average the features coming from the same media (image or video) and then further average the media average features to generate the final feature representation. We show the results before and after media averaging for the IJB-A data set in Table 2.2. It is clear that media averaging significantly improves the performance.

Tables 2.3 and 2.4 summarize the scores (i.e., both ROC and CMC numbers) produced by different face identification/verification methods on the IJB-A and JANUS CS2 data sets, respectively. For the IJB-A data set, we compare our results with DCNN$_{bl}$ (bilinear CNN [76]), DCNN$_{pose}$ (multipose DCNN models [44]), NAN [47], DCNN$_{3d}$ [45], template adaptation (TP) [77], DCNN$_{tpe}$ [78], DCNN$_{l2+tpe}$ [56], and the one [13] reported recently by NIST where JanusB-092015 achieved the best verification results. For the JANUS CS2 data set, Table 2.4 includes, a DCNN-based method [79], Fisher vector-based method [3], DCNN$_{pose}$ [44], DCNN$_{3d}$ [45], and two commercial

TABLE 2.2

Results on the IJB-A data set. The true acceptance rate (TAR) of all the approaches at false acceptance rate (FAR) = 0.1 and 0.01 for the ROC curves (IJB-A 1:1 verification). The Rank-1, Rank-5, and Rank-10 retrieval accuracies of the CMC curves and TPIR at FPIR = 0.01 and 0.1 (IJB-A 1:N identfication). We also show the results before, DCNN$_{cos}$, and after media averaging, DCNN$_{cos}^{m}$, where m means media averaging. DCNN$_{jb}^{m}$ refers to the results after joint Bayesian metric and media averaging

IJB-A-Verif	DCNN$_{cos}$	DCNN$_{cos}^{m}$	DCNN$_{jb}^{m}$
FAR=1e-3	0.644	0.739	**0.81**
FAR=1e-2	0.821	0.865	**0.912**
FAR=1e-1	0.943	0.949	**0.974**
IJB-A-Ident	DCNN$_{cos}$	DCNN$_{cos}^{m}$	DCNN$_{jb}^{m}$
Rank-1	0.887	0.921	**0.935**
Rank-5	0.955	0.968	**0.973**
Rank-10	0.97	0.979	**0.982**
IJB-A-Ident	DCNN$_{cos}$	DCNN$_{cos}^{m}$	DCNN$_{jb}^{m}$
FPIR=0.01	0.547	0.641	**0.739**
FPIR=0.1	0.73	0.794	**0.883**

TABLE 2.3

Results on the IJB-A data set. The true acceptance rate (TAR) of all the approaches at false acceptance rate (FAR)=0.1, 0.01, and 0.001 for the ROC curves (IJB-A 1:1 verification). The Rank-1, Rank-5, and Rank-10 retrieval accuracies of the CMC curves (IJB-A 1:N identification). We report average and standard deviation of the 10 splits. All the performance results reported in Janus B (JanusB-092015) [13], $DCNN_{bl}$ [76], $DCNN_{fusion}$ [80], $DCNN_{3d}$ [45], NAN [47], $DCNN_{pose}$ [44], $DCNN_{tpe}$ [78], TP [77], and $DCNN_{l2+tpe}$ [56]

IJB-A-Verif	[79]	JanusB [13]	$DCNN_{pose}$ [44]	$DCNN_{bl}$ [76]	NAN [47]	$DCNN_{3d}$ [45]
FAR=1e-3	0.514	0.65	–	–	0.881	–
FAR=1e-2	0.732	0.826	0.787	–	0.941	0.725
FAR=1e-1	0.895	0.932	0.911	–	0.978	0.886

IJB-A-Ident	[79]	JanusB [13]	$DCNN_{pose}$ [44]	$DCNN_{bl}$ [76]	NAN [47]	$DCNN_{3d}$ [45]
Rank-1	0.820	0.87	0.846	0.895	0.958	0.906
Rank-5	0.929	–	–	0.927	**0.980**	0.962
Rank-10	–	0.95	0.947	–	0.986	0.977

IJB-A-Verif	$DCNN_{fusion}$ [80]	$DCNN^m_{cos}$	$DCNN^m_{jb}$	$DCNN_{tpe}$ [78]	TP [77]	$DCNN_{l2+tpe}$ [56]
FAR=1e-3	0.76	0.739	0.81	0.813	–	**0.910**
FAR=1e-2	0.889	0.865	0.912	0.9	0.939	**0.951**
FAR=1e-1	0.968	0.949	0.974	0.964	–	**0.979**

IJB-A-Ident	$DCNN_{fusion}$ [80]	$DCNN^m_{cos}$	$DCNN^m_{jb}$	$DCNN_{tpe}$ [78]	TP [77]	$DCNN_{l2+tpe}$ [56]
Rank-1	0.942	0.921	0.935	0.932	0.928	**0.961**
Rank-5	**0.980**	0.968	0.973	–	–	–
Rank-10	**0.988**	0.979	0.982	0.977	0.986	0.983

TABLE 2.4

Results on the JANUS CS2 data set. The total acceptance rate (TAR) of all the approaches at false acceptance rate (FAR) = 0.1, 0.01, and 0.001 for the ROC curves. The Rank-1, Rank-5, and Rank-10 retrieval accuracies of the CMC curves. We report average and standard deviation of the 10 splits. The performance results of $DCNN_{pose}$ have produced results for setup 1 only

CS2-Verif	COTS	GOTS	FV[3]	$DCNN_{pose}$[44]
FAR=1e-3	—	—	—	—
FAR=1e-2	0.581	0.467	0.411	0.897
FAR=1e-1	0.767	0.675	0.704	0.959

CS2-Ident	COTS	GOTS	FV[3]	$DCNN_{pose}$[44]
Rank-1	0.551	0.413	0.381	0.865
Rank-5	0.694	0.571	0.559	0.934
Rank-10	0.741	0.624	0.637	0.949

CS2-Verif	$DCNN_{3d}$[45]	$DCNN_{fusion}$[80]	$DCNN_{cos}^{m}$	$DCNN_{jb}^{m}$
FAR=1e-3	0.824	0.83	0.790	**0.881**
FAR=1e-2	0.926	0.935	0.901	**0.949**
FAR=1e-1	-	0.986	0.971	**0.988**

CS2-Ident	$DCNN_{3d}$[45]	$DCNN_{fusion}$[80]	$DCNN_{cos}^{m}$	$DCNN_{jb}^{m}$
Rank-1	0.898	**0.931**	0.917	0.93
Rank-5	0.956	**0.976**	0.962	0.968
Rank-10	0.969	**0.985**	0.973	0.977

off-the-shelf matchers, COTS and GOTS [67]. From the ROC and CMC scores, we see that the proposed approach achieves good performances for face identification/verification tasks. This can be attributed to the fact that the DCNN model does capture face variations over a large data set and generalizes well to a new small data set. In addition, with better preprocessing modules, HyperFace, the proposed approach achieves better and comparable face identification/verification than [14] without applying any fine-tuning procedures using training data set as Chen et al. [14,80] did to boost their performances. We conjecture that with better detected face bounding boxes and fiducial points from HyperFace, we can reduce the false alarms caused by face detection and perform better face alignment to mitigate the domain shift between the training and test set.

In addition, the performance results of Janus B (JanusB-092015), $DCNN_{bl}$, and $DCNN_{pose}$, systems are computed based on landmarks provided along with the IJB-A data set.

2.4.5 Labeled faces in the wild

We also evaluate our approach on the well-known LFW data set [81] using the standard protocol that defines 3000 positive pairs and 3000 negative pairs in total and further splits them into 10 disjoint subsets for cross validation.

Each subset contains 300 positive and 300 negative pairs. It contains 7701 images of 4281 subjects. All the faces are preprocessed using HyperFace. We compare the mean accuracy of the proposed deep model with other state-of-the-art deep learning–based methods: DeepFace [42], DeepID2 [43], DeepID3 [82], FaceNet [9], Yi et al. [12], Wang et al. [79], Ding and Tao [83], Parkhi et al. [10], and human performance on the "funneled" LFW images. The results are summarized in Table 2.5. It can be seen that our approach performs comparable to other deep learning–based methods. Note that some of the deep learning–based methods compared in Table 2.5 use millions of data samples for training the model. However, we use only the CASIA data set for training our model, which has less than 500,000 images.

2.4.6 Comparison with methods based on annotated metadata

Most systems compared in this chapter produced the results based on landmarks provided along with the data set. For $DCNN_{3d}$ [45], the number of face images is augmented along with the original CASIA-WebFace data set by around 2 million using 3D morphable models. On the other hand, NAN [47] and TP [77] used data sets with more than 2 million face images to train the model. However, the proposed network was trained with the original CASIA-WebFace data set which contains around 500,000 images. In addition, TP adapted the one-shot similarity framework [84] with a linear support vector machine for set-based face verification and trained the metric on the fly with the help of a preselected negative set during testing. Although TP achieved significantly better results than other approaches, it takes more time during testing than the proposed method because our metric is trained offline and requires much less time for testing than TP. We expect the performance of the proposed approach can also be improved by using the one-shot similarity framework. As shown in Table 2.3, the proposed approach achieves comparable results to other methods and strikes a balance between testing time and performance. Similar to our work, $DCNN_{tpe}$ [78] adopted a probabilistic embedding for similarity computation and HyperFace [15] for improved face detection and fiducial point localization. In addition, we could also adopt the novel softmax loss regularized with the scale L_2-norm constraint used in $DCNN_{l2+tpe}$ [56] to improve the performance of the proposed method.

2.4.7 Run time

The networks for face recognition and HyperFace were both trained and tested with NVidia Titan-X GPU and caffe. The DCNN model for face verification is trained on the CASIA-WebFace data set from scratch for about 2 days. In addition, the overall time taken to perform all the four tasks of HyperFace was 3 seconds per image. The limitation was not because of CNN, but because of selective search, which takes approximately 2 seconds to generate candidate

TABLE 2.5
The accuracies of different methods on the LFW data set

Method	#Net	Training set	Metric	Mean accuracy
DeepFace [42]	1	4.4 million images of 4,030 subjects, private	cosine	95.92%
DeepFace	7	4.4 million images of 4,030 subjects, private	unrestricted, SVM	97.35%
DeepID2 [43]	1	202,595 images of 10,117 subjects, private	unrestricted, Joint-Bayes	95.43%
DeepID2	25	202,595 images of 10,117 subjects, private	unrestricted, Joint-Bayes	99.15%
DeepID3 [82]	50	202,595 images of 10,117 subjects, private	unrestricted, Joint-Bayes	99.53%
FaceNet [9]	1	260 million images of 8 million subjects, private	L2	99.63%
Yi et al. [12]	1	494,414 images of 10,575 subjects, public	cosine	96.13%
Yi et al.	1	494,414 images of 10,575 subjects, public	unrestricted, Joint-Bayes	97.73%
Wang et al. [79]	1	494,414 images of 10,575 subjects, public	cosine	96.95%
Wang et al.	7	494,414 images of 10,575 subjects, public	cosine	97.52%
Wang et al.	1	494,414 images of 10,575 subjects, public	unrestricted, Joint-Bayes	97.45%
Wang et al.	7	494,414 images of 10,575 subjects, public	unrestricted, Joint-Bayes	98.23%
Ding and Tao [83]	8	471,592 images of 9,000 subjects, public	unrestricted, Joint-Bayes	99.02%
Parkhi et al. [10]	1	2.6 million images of 2,622 subjects, public	unrestricted, TDE	98.95%
Human, funneled [79]	N/A	N/A	N/A	99.20%
Our DCNN	1	490,356 images of 10,548 subjects, public	cosine	97.7%
Our DCNN	1	490,356 images of 10,548 subjects, public	unrestricted, JB	98.2%

region proposals. One forward pass through the HyperFace network takes only 0.2 seconds and through the recognition network takes 0.01 seconds.

2.5   Open Issues

Given sufficient number of annotated data and GPUs, DCNNs have been shown to yield impressive performance improvements. Still many issues remain to be addressed to make the DCNN-based recognition systems robust and practical. We briefly discuss design considerations for each component of a automated face identification/verification system, including:

- *Face detection:* Face detection is challenging because of the wide range of variations in the appearance of faces. The variability is caused mainly by changes in illumination, facial expression, viewpoints, occlusions, and so on. Other factors such as blur and low resolution challenge the face detection task.

- *Fiducial detection:* Most of the data sets only contain a few thousand images. A large-scale annotated and unconstrained data set will make the face-alignment system more robust to the challenges, including extreme-pose, low-illumination, small and blurry face images. Researchers have hypothesized that deeper layers of DCNNs can encode more abstract information such as identity, pose, and attributes; however, it has not yet been thoroughly studied which layers exactly correspond to local features for fiducial detection.

- *Face identification/verification:* For face identification/verification, the performance can be improved by learning a discriminative distance measure. However, because of memory constraints limited by graphics cards, how to choose informative pairs or triplets and train the network end to end using online methods (e.g., stochastic gradient descent) on large-scale data sets are still open problems.

2.6   Conclusion

The design and performance of our automatic face identification/verification system was presented, which automatically detects faces, localizes fiducial points, and performs identification/verification on newly released challenging face verification data sets, IJB-A and its extended version, JANUS CS2. It was shown that the proposed DCNN-based system can not only accurately locate the faces from images and video frames, but also learn a robust model for face identification/verification. Experimental results demonstrate that the performance of the proposed system on the IJB-A data set is comparable to other state-of-the-art approaches and much better than COTS and GOTS matchers.

Acknowledgments

This research is based on work supported by the Office of the Director of National Intelligence, IARPA, via IARPA R&D Contract No. 2014-14071600012. The views and conclusions contained herein are those of the authors and should not be interpreted as necessarily representing the official policies or endorsements, either expressed or implied, of the ODNI, IARPA, or the U.S. government. The U.S. Government is authorized to reproduce and distribute reprints for governmental purposes notwithstanding any copyright annotation thereon.

References

1. W. Y. Zhao, R. Chellappa, P. J. Phillips, and A. Rosenfeld. Face recognition: A literature survey. *ACM Computing Surveys*, 35(4):399–458, 2003.

2. D. Chen, X. D. Cao, F. Wen, and J. Sun. Blessing of dimensionality: High-dimensional feature and its efficient compression for face verification. In *IEEE Conference on Computer Vision and Pattern Recognition*, 2013.

3. K. Simonyan, O. M. Parkhi, A. Vedaldi, and A. Zisserman. Fisher vector faces in the wild. *British Machine Vision Conference*, 1:7, 2013.

4. J.-C. Chen, S. Sankaranarayanan, V. M. Patel, and R. Chellappa. Unconstrained face verification using Fisher vectors computed from frontalized faces. In *IEEE International Conference on Biometrics: Theory, Applications and Systems*, 2015.

5. A. Krizhevsky, I. Sutskever, and G. E. Hinton. Imagenet classification with deep convolutional neural networks. In *Advances in Neural Information Processing Systems*, pp. 1097–1105, Lake Tahoe, NV, Curran Associates, 2012.

6. C. Szegedy, W. Liu, Y. Jia, P. Sermanet, S. Reed, D. Anguelov, D. Erhan, V. Vanhoucke, and A. Rabinovich. Going deeper with convolutions. *arXiv preprint arXiv:1409.4842*, 2014.

7. R. Girshick, J. Donahue, T. Darrell, and J. Malik. Rich feature hierarchies for accurate object detection and semantic segmentation. In *IEEE Conference on Computer Vision and Pattern Recognition*, pp. 580–587, 2014.

8. R. Ranjan, V. M. Patel, and R. Chellappa. A deep pyramid deformable part model for face detection. In *IEEE International Conference on Biometrics: Theory, Applications and Systems*, 2015.

9. F. Schroff, D. Kalenichenko, and J. Philbin. Facenet: A unified embedding for face recognition and clustering. *arXiv preprint arXiv:1503.03832*, 2015.

10. O. M. Parkhi, A. Vedaldi, and A. Zisserman. Deep face recognition. In *British Machine Vision Conference*, Swansea, UK, BMVA, 2015.

11. J. Donahue, Y. Jia, O. Vinyals, J. Hoffman, N. Zhang, E. Tzeng, and T. Darrell. Decaf: A deep convolutional activation feature for generic visual recognition. *arXiv preprint arXiv:1310.1531*, 2013.

12. D. Yi, Z. Lei, S. Liao, and S. Z. Li. Learning face representation from scratch. *arXiv preprint arXiv:1411.7923*, 2014.

13. National Institute of Standards and Technology (NIST): IARPA Janus benchmark—a performance report. http://biometrics.nist.gov/cs_links/face/face_challenges/IJBA_reports.zip.

14. J. C. Chen, R. Ranjan, S. Sankaranarayanan, A. Kumar, C. H. Chen, V. M. Patel, C. D. Castillo, and R. Chellappa. Unconstrained still/video-based face verification with deep convolutional neural networks. *International Journal of Computer Vision*, pp. 1–20, 2017.

15. R. Ranjan, V. M Patel, and R. Chellappa. Hyperface: A deep multi-task learning framework for face detection, landmark localization, pose estimation, and gender recognition. In *IEEE Transactions on Pattern Analysis and Machine Intelligence, arXiv preprint arXiv:1603.01249*, 2016.

16. P. Viola and M. J. Jones. Robust real-time face detection. *International Journal of Computer Vision*, 57(2):137–154, 2004.

17. Y. Wei X. Cao D. Chen, S. Ren and J. Sun. Joint cascade face detection and alignment. In D. Fleet, T. Pajdla, B. Schiele, and T. Tuytelaars (Eds.), *European Conference on Computer Vision*, Vol. 8694, pp. 109–122, Zurich, Switzerland: Springer, 2014.

18. J. Li and Y. Zhang. Learning SURF cascade for fast and accurate object detection. In *Computer Vision and Pattern Recognition (CVPR), 2013 IEEE Conference on*, pp. 3468–3475, June 2013.

19. H. Li, Z. Lin, X. Shen, J. Brandt, and G. Hua. A convolutional neural network cascade for face detection. In *IEEE Conference on Computer Vision and Pattern Recognition*, pp. 5325–5334, June 2015.

20. X. G. Zhu and D. Ramanan. Face detection, pose estimation, and landmark localization in the wild. In *IEEE Conference on Computer Vision and Pattern Recognition*, pp. 2879–2886. IEEE, 2012.

21. M. Mathias, R. Benenson, M. Pedersoli, and L. Van Gool. Face detection without bells and whistles. *European Conference on Computer Vision*, 8692:720–735, 2014.

22. S. Yang, P. Luo, C. C. Loy, and X. Tang. From facial parts responses to face detection: A deep learning approach. In *IEEE International Conference on Computer Vision*, 2015.

23. S. S. Farfade, M. J. Saberian, and L.-J. Li. Multi-view face detection using deep convolutional neural networks. In *International Conference on Multimedia Retrieval*, Shanghai, China, ACM, 2015.

24. G. Ross. Fast R-CNN. In *IEEE International Conference on Computer Vision*, pp. 1440–1448, 2015.

25. R. Girshick, F. Iandola, T. Darrell, and J. Malik. Deformable part models are convolutional neural networks. *IEEE Conference on Computer Vision and Pattern Recognition*, 2014.

26. S. Sarkar, V. M. Patel, and R. Chellappa. Deep feature-based face detection on mobile devices. In *IEEE International Conference on Identity, Security and Behavior Analysis*, pp. 1–8, February 2016.

27. H. Jiang and E. Learned-Miller. Face detection with the faster R-CNN. *arXiv preprint arXiv:1606.03473*, 2016.

28. S. Ren, K. He, R. Girshick, and J. Sun. Faster R-CNN: Towards real-time object detection with region proposal networks. In *Advances in Neural Information Processing Systems*, pp. 91–99, Montreal, Canada, Curran Associates, 2015.

29. S. Yang, P. Luo, C.-C. Loy, and X. Tang. Wider face: A face detection benchmark. In *IEEE Conference on Computer Vision and Pattern Recognition*, pp. 5525–5533, 2016.

30. P. Hu and D. Ramanan. Finding tiny faces. In *IEEE Conference on Computer Vision and Pattern Recognition*, 2017.

31. R. Ranjan, S. Sankaranarayanan, C. D Castillo, and R. Chellappa. An all-in-one convolutional neural network for face analysis. In *IEEE International Conference on Automatic Face and Gesture Recognition*, 2017.

32. T. F. Cootes, G. J. Edwards, and C. J. Taylor. Active appearance models. *IEEE Transactions on Pattern Analysis and Machine Intelligence*, 6:681–685, 2001.

33. T. F. Cootes, C. J. Taylor, D. H. Cooper, and J. Graham. Active shape models—their training and application. *Computer vision and image understanding*, 61(1):38–59, 1995.

34. D. Cristinacce and T. F. Cootes. Feature detection and tracking with constrained local models. *British Machine Vision Conference*, 1:3, 2006.

35. A. Asthana, S. Zafeiriou, S. Y. Cheng, and M. Pantic. Robust discriminative response map fitting with constrained local models. In *IEEE Conference on Computer Vision and Pattern Recognition*, pp. 3444–3451, 2013.

36. X. Cao, Y. Wei, F. Wen, and J. Sun. Face alignment by explicit shape regression, July 3, 2014. US Patent App. 13/728,584.

37. P. Dollár, P. Welinder, and P. Perona. Cascaded pose regression. In *IEEE Conference on Computer Vision and Pattern Recognition*, pp. 1078–1085. IEEE, 2010.

38. Y. Sun, X. Wang, and X. Tang. Deep convolutional network cascade for facial point detection. In *IEEE Conference on Computer Vision and Pattern Recognition*, pp. 3476–3483, 2013.

39. A. Kumar, A. Alavi, and R. Chellappa. Kepler: Keypoint and pose estimation of unconstrained faces by learning efficient H-CNN regressors. *IEEE International Conference on Automatic Face and Gesture Recognition*, 2017.

40. A. Kumar and R. Chellappa. A convolution tree with deconvolution branches: Exploiting geometric relationships for single shot keypoint detection. *arXiv preprint arXiv:1704.01880*, 2017.

41. T. Ahonen, A. Hadid, and M. Pietikainen. Face description with local binary patterns: Application to face recognition. *IEEE Transactions on Pattern Analysis and Machine Intelligence*, 28(12):2037–2041, 2006.

42. Y. Taigman, M. Yang, M. A. Ranzato, and L. Wolf. Deepface: Closing the gap to human-level performance in face verification. In *IEEE Conference on Computer Vision and Pattern Recognition*, pp. 1701–1708, 2014.

43. Y. Sun, X. Wang, and X. Tang. Deeply learned face representations are sparse, selective, and robust. *arXiv preprint arXiv:1412.1265*, 2014.

44. W. Abd-Almageed, Y. Wu, S. Rawls, S. Harel, T. Hassne, I. Masi, J. Choi, J. Lekust, J. Kim, P. Natarajana, R. Nevatia, and G. Medioni. Face recognition using deep multi-pose representations. In *IEEE Winter Conference on Applications of Computer Vision (WACV)*, 2016.

45. I. Masi, A. T. Tran, J. T. Leksut, T. Hassner, and G. Medioni. Do we really need to collect millions of faces for effective face recognition? *arXiv preprint arXiv:1603.07057*, 2016.

46. C. Ding and D. Tao. Trunk-branch ensemble convolutional neural networks for video-based face recognition. *arXiv preprint arXiv:1607.05427*, 2016.

47. J. Yang, P. Ren, D. Chen, F. Wen, H. Li, and G. Hua. Neural aggregation network for video face recognition. In *IEEE Conference on Computer Vision and Pattern Recognition*, 2017.

48. N. Bodla, J. Zheng, H. Xu, J.-C. Chen, C. Castillo, and R. Chellappa. Deep heterogeneous feature fusion for template-based face recognition. *IEEE Winter Conference on Applications of Computer Vision (WACV)*, 2017.

49. K. Q. Weinberger, J. Blitzer, and L. K. Saul. Distance metric learning for large margin nearest neighbor classification. In *Advances in Neural Information Processing Systems*, pp. 1473–1480, Vancouver, Canada, MIT Press, 2005.

50. Y. Taigman, L. Wolf, and T. Hassner. Multiple one-shots for utilizing class label information. In *British Machine Vision Conference*, pp. 1–12, London, UK, BMVA, 2009.

51. J. V. Davis, B. Kulis, P. Jain, S. Sra, and I. S. Dhillon. Information-theoretic metric learning. In *International Conference on Machine Learning*, pp. 209–216, Corvalis, OR, ACM, 2007.

52. D. Chen, X. D. Cao, L. W. Wang, F. Wen, and J. Sun. Bayesian face revisited: A joint formulation. In *European Conference on Computer Vision*, pp. 566–579, Firenze, Italy, Springer, 2012.

53. J. Hu, J. Lu, and Y.-P. Tan. Discriminative deep metric learning for face verification in the wild. In *IEEE Conference on Computer Vision and Pattern Recognition*, pp. 1875–1882, 2014.

54. Y. Wen, K. Zhang, Z. Li, and Y. Qiao. A discriminative feature learning approach for deep face recognition. In *European Conference on Computer Vision (ECCV)*, pp. 499–515, Amsterdam, the Netherlands, Springer, 2016.

55. W. Liu, Y. Wen, Z. Yu, M. Li, B. Raj, and L. Song. Sphereface: Deep hypersphere embedding for face recognition. *IEEE Conference on Computer Vision and Pattern Recognition*, 2017.

56. R. Ranjan, C. D Castillo, and R. Chellappa. L2-constrained softmax loss for discriminative face verification. *arXiv preprint arXiv:1703.09507*, 2017.

57. A. Krizhevsky, I. Sutskever, and G. E. Hinton. Imagenet classification with deep convolutional neural networks. In F. Pereira, C.J.C. Burges, L. Bottou, and K.Q. Weinberger (Eds.), *Advances in Neural Information Processing Systems 25*, pp. 1097–1105. Curran Associates, Red Hook, NY, 2012.

58. B. Hariharan, P. Arbeláez, R. Girshick, and J. Malik. Hypercolumns for object segmentation and fine-grained localization. In *IEEE Conference on Computer Vision and Pattern Recognition*, pp. 447–456, 2015.

59. M. Koestinger, P. Wohlhart, P. M. Roth, and H. Bischof. Annotated facial landmarks in the wild: A large-scale, real-world database for facial landmark localization. In *First IEEE International Workshop on Benchmarking Facial Image Analysis Technologies*, 2011.

60. J. R. Uijlings, K. E. van de Sande, T. Gevers, and A. W. Smeulders. Selective search for object recognition. *International Journal of Computer Vision*, 104(2):154–171, 2013.

61. K. Simonyan and A. Zisserman. Very deep convolutional networks for large-scale image recognition. *arXiv preprint arXiv:1409.1556*, 2014.

62. J.-C. Chen, V. M. Patel, and R. Chellappa. Unconstrained face verification using deep CNN features. *arXiv preprint arXiv:1508.01722*, 2015.

63. K. He, X. Zhang, S. Ren, and J. Sun. Delving deep into rectifiers: Surpassing human-level performance on imagenet classification. *arXiv preprint arXiv:1502.01852*, 2015.

64. Y. Jia, E. Shelhamer, J. Donahue, S. Karayev, J. Long, R. Girshick, S. Guadarrama, and T. Darrell. Caffe: Convolutional architecture for fast feature embedding. In *ACM International Conference on Multimedia*, pp. 675–678, Orlando, FL, ACM, 2014.

65. X. D. Cao, D. Wipf, F. Wen, G. Q. Duan, and J. Sun. A practical transfer learning algorithm for face verification. In *IEEE International Conference on Computer Vision*, pp. 3208–3215. IEEE, 2013.

66. V. Jain and E. Learned-Miller. FDDB: A benchmark for face detection in unconstrained settings. Number UM-CS-2010-009, 2010.

67. B. F. Klare, B. Klein, E. Taborsky, A. Blanton, J. Cheney, K. Allen, P. Grother, A. Mah, M. Burge, and A. K. Jain. Pushing the frontiers of unconstrained face detection and recognition: IARPA Janus Benchmark A. In *IEEE Conference on Computer Vision and Pattern Recognition*, 2015.

68. B. Yang, J. Yan, Z. Lei, and S. Z. Li. Convolutional channel features. In *IEEE International Conference on Computer Vision*, 2015.

69. J. Yan, X. Zhang, Z. Lei, and S. Z. Li. Face detection by structural models. *Image and Vision Computing*, 32(10):790 – 799, 2014. Best of Automatic Face and Gesture Recognition 2013.

70. S. Liao, A. Jain, and S. Li. A fast and accurate unconstrained face detector. In *IEEE Transactions on Pattern Analysis and Machine Intelligence*, 2015.

71. H. Li, G. Hua, Z. Lin, J. Brandt, and J. Yang. Probabilistic elastic part model for unsupervised face detector adaptation. In *IEEE International Conference on Computer Vision*, pp. 793–800, December 2013.

72. X. Zhu and D. Ramanan. Face detection, pose estimation, and landmark localization in the wild. In *IEEE Conference on Computer Vision and Pattern Recognition (CVPR)*, pp. 2879–2886. IEEE, 2012.

73. J. M. Saragih, S. Lucey, and J. F. Cohn. Deformable model fitting by regularized landmark mean-shift. *International Journal of Computer Vision*, 91(2):200–215, 2011.

74. M. Everingham, J. Sivic, and A. Zisserman. "Hello! my name is... buffy" – automatic naming of characters in tv video. *British Machine Vision Conference*, 2:6, 2006.

75. X. Zhu and D. Ramanan. Facedpl: Detection, pose estimation, and landmark localization in the wild. *preprint*, 2015.

76. A. RoyChowdhury, T.-Y. Lin, S. Maji, and E. Learned-Miller. One-to-many face recognition with bilinear CNNS. In *IEEE Winter Conference on Applications of Computer Vision (WACV)*, 2016.

77. N. Crosswhite, J. Byrne, O. M. Parkhi, C. Stauffer, Q. Cao, and A. Zisserman. Template adaptation for face verification and identification. *IEEE International Conference on Automatic Face and Gesture Recognition (FG)*, 2017. arXiv preprint arXiv:1603.03958, 2016.

78. S. Sankaranarayanan, A. Alavi, C. Castillo, and R. Chellappa. Triplet probabilistic embedding for face verification and clustering. *arXiv preprint arXiv:1604.05417*, 2016.

79. D. Wang, C. Otto, and A. K. Jain. Face search at scale: 80 million gallery. *arXiv preprint arXiv:1507.07242*, 2015.

80. J.-C. Chen, R. Ranjan, S. Sankaranarayanan, A. Kumar, C.-H. Chen, V. M. Patel, C. D. Castillo, and R. Chellappa. An end-to-end system for unconstrained face verification with deep convolutional neural networks. *CoRR*, abs/1605.02686, 2016.

81. G. B. Huang, M. Ramesh, T. Berg, and E. Learned-Miller. Labeled faces in the wild: A database forstudying face recognition in unconstrained environments. In *Workshop on Faces in Real-Life Images: Detection, Alignment, and Recognition*, Amhurst, University of Massachusetts, 2008.

82. Y. Sun, D. Liang, X. Wang, and X. Tang. Deepid3: Face recognition with very deep neural networks. *arXiv preprint arXiv:1502.00873*, 2015.

83. C. Ding and D. Tao. Robust face recognition via multimodal deep face representation. *arXiv preprint arXiv:1509.00244*, 2015.

84. L. Wolf, T. Hassner, and Y. Taigman. The one-shot similarity kernel. In *International Conference on Computer Vision*, pp. 897–902. IEEE, 2009.

3

Deep Siamese Convolutional Neural
Networks for Identical Twins and
Look-Alike Identification

Xiaoxia Sun, Amirsina Torfi, and Nasser Nasrabadi

CONTENTS

3.1	Introduction	66
3.2	Related Work	68
	3.2.1 Identical twins identification	68
	3.2.2 Siamese architecture	69
3.3	Our Twin Siamese Discriminative Model	69
	3.3.1 Siamese architecture for distinguishing between identical twins	69
	3.3.2 Contrastive loss	70
	3.3.3 Siamese convolutional neural networks architecture	71
3.4	Evaluation and Verification Metric	72
3.5	Experimental Setup	73
	3.5.1 Data sets	73
	3.5.2 Preprocessing	74
3.6	Experiments	74
	3.6.1 Pretraining the model	74
	3.6.2 Twin verification	76
	3.6.2.1 Twin verification with restricted year	76
	3.6.2.2 Twin verification with age progression and unrestricted year	79
	3.6.2.3 Twin verification with age progression and unrestricted year using merged subjects	79
	3.6.3 Face verification with transfer learning on Look-alike data set	80
3.7	Conclusion	80
References	....	81

3.1   Introduction

In recent years, a great amount of research in the computer vision and pattern recognition area has been dedicated to biometrics whose assumption is the uniqueness of people's identity. An important application of biometrics is the problem of face identification. One of the applications, for instance, is whether the designed system can recognize a pair of face images belonging to the same person. Distinguishing between identical twins is classified as one of the most difficult scenarios, which is reported to be complicated even for humans [1].

The major motives behind face identification lie in forensics applications, specifically when other modalities (i.e., iris or fingerprint) are absent. Wrong subject identification of identical twins has been reported as an important security challenge, which may even cause financial issues [2]. As for other modalities, matching of identical twins' fingerprint has also been investigated [3]. The task of face identification for identical twins usually yields exceedingly poor results, as anticipated, because the two subjects are significantly similar, and it is a daunting challenge to develop a concrete framework to perform the job. The visual understanding of the difficulty of the identical twins recognition is depicted in Figure 3.1. All pairs belong to different identities (i.e., identical twins).

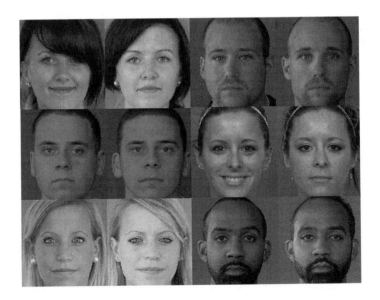

FIGURE 3.1
Visual understanding of the complexity of distinguishing between identical twins.

Different studies have addressed this issue. Some researchers [2,4] use either small data sets or their face-identification evaluations are based on using existing commercial matchers. In [4] provided a comprehensive study on comparing matching algorithms and their performance for face-identification tasks. While using available off-the-shelf algorithms, this study [4] is based on an extensive data set. Nevertheless, it does not offer any new approach optimized to exploit the twin-data characteristics. Some other works like [5] investigate facial features for identical twins by implementing hand-crafted feature descriptors like multiscale local binary patterns, Scale Invariant Feature Transform (SIFT), and facial marks for local feature extraction. However, these feature descriptors do not exploit the subtle characteristics of the twin data. So there is a need to investigate a reliable framework for extracting features from the data itself, which suggests the use of deep learning as an interesting solution for extracting a common feature space, applicable to the characteristics of the identical twins data set. So the general idea is to transform the input feature space to another domain whose features allow distinction between identical twins.

The availability of abundant data and a reliable structure are prerequisites for deep learning. The recognition of identical twins is an example of the case where there might be numerous subjects of twins with only a few of their image pairs per subject available for training. Because in the twin-face identification problem, the focus is image verification and not classification, there is a need to develop new architectures to be trained only for distinguishing between genuine and impostor pairs regardless of their identities. A genuine pair is a pair of images in which the images belong to the same subject and an impostor pair is otherwise.

The main idea in this chapter is to implement an architecture named "Siamese Network" [6] for face verification of identical twins. The goal is to learn a nonlinear mapping that understands a similarity metric from the data itself for identifying the identical twins, which suggests a more data-driven solution compared to the traditional facial-texture analysis methods. This architecture can be learned on pairs of data and later used for distinguishing between subjects that have never been seen in the training phase. Another advantage is that because the similarity metric will be learned using this architecture, as long as there is enough training data (genuine and impostor pairs), the number of training samples per subject does not affect the identification performance, which is an interesting advantage.

Our approach is to implement and train a Siamese architecture using two identical deep convolutional neural networks (CNNs) to find a nonlinear mapping where in the mapped target subspace a simple distance metric can be used for performing the face-verification task. So in the target subspace, the genuine pairs should be as close as possible and the impostor pairs should be as far as possible using simple distance metrics (i.e., Euclidean distance). Recent research in computer vision has demonstrated the superior performance of deep models with global features compared to local features.

Therefore, we have used the whole face as the input feature space (using global features) rather than using different hand-crafted features or facial landmarks for training the deep model.

The contributions of this work are as follows:

- Siamese architecture is used to find a nonlinear mapping to learn a dedicated similarity metric in a target subspace with a significant reduction in dimension compared to the input space.

- A deep CNN model is trained using a large data set, Twin Days Festival Collection (TDFC) to learn a nonlinear mapping function optimized for distinguishing between identical twins. This model, which uses the Siamese architecture to learn the similarity metric consists of two parallel identical CNNs.

- The learned architecture and similarity metric are leveraged to distinguish between the genuine and impostor pairs of data.

To the best of our knowledge, this is the first work using Siamese architecture for distinguishing between the identical twins for the face-verification task.

3.2   Related Work

This section includes two subsections, Identical Twins Identification and Siamese Networks, and treats these subjects separately and then addresses a thorough comparison of our work with previous research works.

3.2.1   Identical twins identification

Various face-recognition works in this area [3,7,8] use a limited number of twin siblings as a data set for their research works. The research was done in [3] is merely based on fingerprint recognition. The work in [7] is dedicated to measuring the discriminating ability of hybrid facial features by the statistical separability between genuine and impostor feature distances. Use of demographic information* and facial marks† for face matching and retrieval has been proposed in [8]. Comprehensive work has been done in [2] in which they perform matching using face, iris, and fingerprint modalities in addition to the fusion of the aforementioned modes. [2] concludes that for distinguishing between identical twins, using face mode is a more complex task compared to using fingerprint and iris modalities. Some other studies have been done for different modalities including iris [9] and speech recognition [10]. Some other

---

*Gender, ethnicity, age, etc.
†Scars, moles, freckles, etc.

works [11,12] were done on the Twins Days data set for investigating the face illumination and expression on identification of identical twins. Another more comprehensive work [4] has been done (as the expansion of [11] and [12]), which takes age progression and gender into consideration as well. All the experiments performed in [4], [11], and [12] use off-the-shelf algorithms on the twin data set for evaluation and do not develop any algorithm or architecture for face identification. Using group classification and a facial-aging features approach was proposed in [13] for recognizing identical twins.

3.2.2 Siamese architecture

As deep learning techniques have attracted a great deal of attention, a powerful model is the Siamese architecture, which can be built by two identical CNNs called the Siamese Convolutional Neural Networks (SCNNs). SCNNs were proposed in Reference 14 as a framework for the face-verification task and also in [15] as a dimensionality reduction framework. More recent work has been done in [16], which shows robustness to geometry distortion. Moreover the designed architecture in [16] shows decent generalization ability by using the same idea of contrastive loss function implementation for similar and dissimilar pairs. In [17], an approach was proposed for person reidentification in a multicamera setup implementing a hierarchical architecture with SCNNs. However, a small-scale data set was used because of computational cost.

3.3 Our Twin Siamese Discriminative Model

The general model that we will use here is a simple distance-based model, which is equivalent to the category of energy-based models proposed in [18]. The following subsections are dedicated to explanation of further details of our proposed twin discriminative model.

3.3.1 Siamese architecture for distinguishing between identical twins

The twin-discriminative model uses a Siamese architecture, which consists of two identical CNNs. The goal is to create a target feature subspace for discrimination between similar and dissimilar pairs based on a simple distance metric. The model is depicted in Figure 3.2. The general idea is that when two images belong to a genuine pair, their distance in the target feature subspace should be as close as possible, and for impostor images, it should be as far as possible. Let X_{p_1} and X_{p_2} be the pair of images as the inputs of the system whether in training or testing mode. The distance between a pair of images in

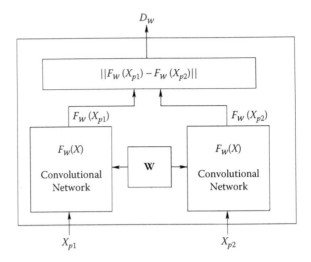

FIGURE 3.2
Siamese model framework.

the target subspace defined as $D_W(X_{p_1}, X_{p_2})$, that is, the ℓ_2-norm between two vectors, in which W is the parameters of the whole network (weights). In a simpler way, $D_W(X_{p_1}, X_{p_2})$ should be low for genuine pairs and high for impostor pairs, which defines the contrastive loss function. Consider Y as the label that is considered to be 1 if both images are genuine and 0 otherwise. F is the network function that maps the input to the target feature subspace, the outputs of the Siamese CNNs are denoted by $F_W(X_{p_1})$ and $F_W(X_{p_2})$, and W is the same because both CNNs share the same weights. The distance is computed as follows:

$$D_W(X_{p_1}, X_{p_1}) = ||F_W(X_{p_1}) - F_W(X_{p_2})||_2 \tag{3.1}$$

3.3.2 Contrastive loss

The goal of the loss function $L_W(X, Y)$ is to minimize the loss in both scenarios of encountering genuine and impostor pairs, so the definition should satisfy both conditions as given by:

$$L_W(X, Y) = \frac{1}{N} \sum_{i=1}^{N} L_W(Y_i, (X_{p_1}, X_{p_2})_i) \tag{3.2}$$

where:
     N is the number of training samples
     i is the index of each sample,

$L_W(Y_i, (X_{p_1}, X_{p_2})_i)$ is defined as follows:

$$L_W(Y_i, (X_{p_1}, X_{p_2})_i) = Y * L_{gen}(D_W(X_{p_1}, X_{p_2})_i)$$
$$+ (1-Y) * L_{imp}(D_W(X_{p_1}, X_{p_2})_i) + \lambda ||W||_2 \qquad (3.3)$$

in which the last term is for regularization, and λ is the regularization parameter. Finally L_{gen} and L_{imp} are defined as the functions of $D_W(X_{p_1}, X_{p_2})$ by Equation 3.4:

$$\begin{cases} L_{gen}(D_W(X_{p_1}, X_{p_2})) = \frac{1}{2} D_W(X_{p_1}, X_{p_2})^2 \\ L_{imp}(D_W(X_{p_1}, X_{p_2})) = \frac{1}{2} max\{0, (M - D_W(X_{p_1}, X_{p_2}))\}^2 \end{cases} \qquad (3.4)$$

where M is a margin, which is obtained by cross-validation. Moreover the *max* argument declares that in case of an impostor pair if the distance in the target feature subspace is larger than the threshold M, there would be no loss.

3.3.3 Siamese convolutional neural networks architecture

The Siamese architecture consists of two identical parallel CNNs as depicted in Figure 3.3. When a pair of images (X_{p_1}, X_{p_2}) is fed as the input to the network, splitting will be done in the net forward procedure in both training and testing modes, and each of the X_{p_1} and X_{p_2} will be forwarded to one of the two parallel CNNs as shown in Figure 3.3. Finally, the outputs of two CNNs will be compared using a contrastive loss function for minimization of the total loss. This procedure feeds images of genuine and impostor pairs to the two identical CNNs structures to learn the similarity metric. In the training phase, the network should learn the similarity of genuine pairs and dissimilarity of the impostor pairs, which is reflected by the outputs CNNs. In the testing phase, the similarity between the two images of any input pair can be calculated based on the outputs of the two CNNs.

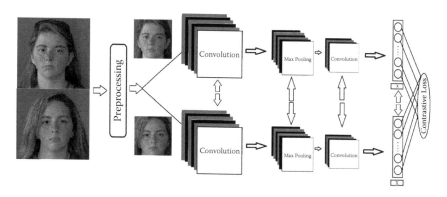

FIGURE 3.3
Twin discriminative model. The Siamese architecture: Two parallel identical CNNs with parameter sharing.

3.4 Evaluation and Verification Metric

In this chapter, we evaluate all the experimental results using the Receiver Operating Characteristic (ROC) curve. The ROC curve consists of the Validation Rate (VR) and False Acceptance Rate (FAR). Because in identical twins' matching, the task is to determine whether the two persons are the same subjects or not, the ROC curve comes to play as an illustration of the binary classifier performance. A genuine pair (also referred to as *match pairs*) is a pair of two images belonging to the same person and the impostor pair (referred to as *nonmatch pairs*) is a pair of images belonging to identical twins or two look-alike persons. All face pairs (X_{P_1}, X_{P_2}) of the same identity are denoted with $\mathcal{P}_{gen}$, whereas all pairs belonging to different identities are denoted as $\mathcal{P}_{imp}$. We define true positive and false acceptance as:

$$TP(thresh) = \{(X_{P_1}, (X_{P_2}) \in \mathcal{P}_{gen}, D_W \leq thresh\} \qquad (3.5)$$

$$FA(thresh) = \{(X_{P_1}, (X_{P_2}) \in \mathcal{P}_{imp}, D_W \leq thresh\} \qquad (3.6)$$

So $TP(thresh)$ is the test samples, which classified as match pairs, whereas $FA(thresh)$ are nonmatch pairs, which incorrectly classified as match pairs. Both the calculations are done using a single predefined threshold, which the output distance will be compared with as a metric for prediction. The validation rate, VAL(thresh), and, FAR(thresh), are calculated as:

$$VR(thresh) = \frac{TP(thresh)}{\mathcal{P}_{gen}}, FAR(thresh) = \frac{FA(thresh)}{\mathcal{P}_{imp}} \qquad (3.7)$$

The VR is the percentage of genuine pairs that are correctly classified as the matching pairs. On the other hand, the FAR is the percentage of nonmatch pairs (impostor pairs) that are incorrectly classified as images belonging to the same subject. According to the definitions, FAR and VR can be computed with regard to impostor and genuine pairs, respectively.

The main metric that has been used for performance evaluation is the Equal Error Rate (EER), which is the point when FAR and False Rejection Rate (FRR) are equal. Because the equality $FRR = 1 - VR$ always holds, the EER is simply equal to $FRR = 1 - VR = FAR$. So by drawing the EER line using equation $VR = 1 - FAR$ on the ROC curve and finding the intersection with the curve, the EER point can be calculated. For the verification, the metric is simply a ℓ_2-norm calculation between the outputs of the two fully connected layers from the two parallel CNNs and a final comparison with a given threshold (predetermined margin). The ROC curve is plotted based on using different thresholds and calculating the corresponding VR and FAR. For having a better understanding of the performance, the test samples are divided into five disjoint parts and the evaluation is done across the five splits.

3.5   Experimental Setup

3.5.1   Data sets

Four data sets have been used in this work. One is a large data set, TDFC, which is used for training and testing. The TDFC data set used in our experiments was gathered from 2010 to 2015 and consists of 1357 distinctive identities. The data set is restricted and provided by West Virginia University. The twin images are taken with a high-resolution camera, Canon 5D Mark III DSLR, under controlled lighting conditions, and each image size is 4288×2848. During data collection, three cameras are used to capture the face image under different viewing angles. In our experiments, we only use the frontal view image for training and testing to alleviate the misalignment problem.

CASIA-WebFace [19] and *LFW* [20] data sets are used for pretraining the model. *LFW* is the de-facto test set for the face-verification task. Although the *CASIA* data set is used to pretrain the network as classier and the *LFW* data set is used to fine-tune the network in verification mode, both are employed to supervise the verification task as the ultimate goal for twin verification.

The fourth data set is *The Look-alike face database* [21] that consists of color images pertaining to 50 well-known personalities from either western, eastern, or Asian origin. Each image is size 220×187. For each subject, its look-alike counterparts are gathered carefully to create the impostor pairs. Each selected celebrity has five genuine images and five look-alike images. These images are captured in the wild, which is similar to that of the LFW data set, making it difficult to recognize because of an enormous registration error. In addition, the small number of samples also prohibits most algorithms from reaching a high verification rate. According to the aforementioned characteristics of the Look-alike data set, it will be only used for fine-tuning our pretrained Siamese network trained on twin data set and tested for performance evaluation.

In real-field scenarios and in the case of face verification, it is impractical to assume the expression of the subjects faces are the same in both images of a pair. As an example, in forensics applications, we may have a natural face in an image and one wants to compare this image to a query image in which the subject has a smiling expression. So to make a more realistic scenario, the pairs are generated in mixed order of smiling and natural expressions, which downgrades the testing performance, but is prone to provide fairness in experiments.

Another assumption is that there might not be any iris information in one or both images of a pair. So the pair creation is based on the assumption of closed eyes verses opened eyes of closed eyes versus closed eyes for a decent portion of the testing and training pairs.

3.5.2   Preprocessing

We have implemented several preprocessing techniques to improve the performance of the proposed model. All face images in both TDFC and Look-alike data sets are first cropped using Dlib C++ library [22] and transformed to color images. Because of misalignment in the data set, we have used the face-frontalization technique [23], also referred to as three-dimensional alignment, to adjust the face poses. The frontalized-face is able to largely improve the network performance especially for the Look-alike data set, where the face images are mostly captured in the uncontrolled wild environment. The frontalized face images are then resized to 224×224 as the standard input to VGG-16 and VGG-19 architecture. The data can be normalized to the range of $[0, 1]$ using Equation 3.8:

$$I_i \leftarrow \left[\frac{I_i - min(I)}{max(I) - min(I)} \right], \tag{3.8}$$

where I_i is the ith pixel of the image I, and $[\cdot]$ is the rounding operator. As the standard preprocessing procedure, each image can be subtracted from the mean face image of the data set, which is usually used to reduce the discrepancy between the distributions of training and testing sets. We did not find it helpful regarding performance to normalize the data to the range of $[0, 1]$. The range kept untouched as is in the range of $[0, 255]$.

3.6   Experiments

The proposed experiments consist of three phases. At first, the VGGNet [24] is used to pretrain the structure on the CASIA data set as a classifier to supervise the learning of deep learning architecture. Then the network is fine-tuned on the twin data set for evaluation of the accuracy and discrimination ability of the proposed method.

All experiments presented here are implemented on a workstation with three Nvidia Geforce Titan X GPUs. The Siamese architecture is trained and evaluated using Caffe as the deep learning framework [25].

3.6.1   Pretraining the model

In generic object recognition or face classification, the categories of the possible samples are within the training set. Therefore, the existing labels determine the performance and softmax or any class-oriented loss would be able to directly address the classification problems. In this manner, the label predictor layer acts as a classifier and the learned features are separable as the loss goes to zero. In face recognition, the learned features should be discriminative even if not be separable from class; it is not practical to assume all the sample categories in testing phase are available in the training. On the contrary, the fair assumption is to assume the training and testing samples classes are mutually exclusive as prompted in forensics application. The learned data-driven features are required to be discriminative and generalized for being able to recognize between unseen classes without category-related label prediction. However, it is proven that discriminative power of features in both the intra- and interclass variations can supervise the learning of deep learning architecture in a more effective way [26].

As noted in [26], training as a classifier makes training significantly easier and faster. Then in the next phase, the VGGNet will be fine-tuned on the LFW data set using the Siamese architecture, which uses two identical VGGNet as the CNNs. The reason behind doing the latter part is to supervise the discriminative features without optimizing the class-separability characteristics. As a technical problem, the learning rate for this fine-tuning must be set to a small value for not disturbing the category separability. Then the fine-tuned network will be used as an initialization for the twin-verification network.

It is worth recalling that in the first stage, which is training on the CASIA data set, the layer SoftmaxWithLoss is used in Caffe implementation, which is numerically more stable than using a cross-entropy loss followed by a softmax operation. The reason is that of the presence of the log. If the output of the network provides a bad prediction and by normalizing that prediction we get zero for the y value, then the loss goes to infinity, and this is unstable. Another issue can be the existence of the exponential in the softmax operator. If the output of any of the neurons is large because in the softmax we do the exponentiation, then the numerator and denominator of the softmax operation can be quite large. So a trick can be adding a number to all of the unscaled outputs.

Before fine-tuning the network on the twin data set, it is worth it to show how good the performance is on the twin test set using only the weights pretrained on the CASIA and LFW data sets. The results performing on both LFW and twin test sets using the CASIA-LFW-Model are reported in Table 3.1. It provides a visual understanding of the discriminative ability of the pretrained model. The demonstrated results in Table 3.1 prove the ability of the model in the task of discrimination of images collected in the wild as a baseline.

TABLE 3.1

Using the pretrained model on CAISA and LFW data sets for evaluation of the performance on LFW test sets

Data set	EER ($\mu \pm \sigma$)	AUC ($\mu \pm \sigma$)
LFW	7.1% $\pm$ 0.005	98% $\pm$ 0.001

TABLE 3.2
Twin verification solely by using the pretrained model on CASIA data set

Test year	No. test subjects	EER (%) ($\mu \pm \sigma$)	AUC (%) ($\mu \pm \sigma$)
2010	159	13.37 ± 0.01	94 ± 0.009
2011	190	11.2 ± 0.016	95.1 ± 0.008
2012	243	14.8 ± 0.01	93 ± 0.004
2013	157	17.5 ± 0.014	90.8 ± 0.013
2014	256	15.5 ± 0.011	91.8 ± 0.011

It is good to have an idea of how good the pretrained model would work on the twin test set without any fine-tuning on twin data set. The *restricted* year refers to considering both images of a test pair belonging to the same year. All the unique subjects presented in the year of X in range (2010–2014), which are not present in any other year, will be used as the test set. Basically, the test set completely separates the subjects that are present by year. The results are shown in Table 3.2 in which the weights pretrained by CASIA are used.

3.6.2 Twin verification

The experiments in this part are solely on the twin data set, which reflects the accuracy on the challenging task of recognizing between identical twins.

The architecture used in this implementation is shown in Table 3.3, which is VGG-16. Activation layers (rectified linear units) representation have been eliminated for simplicity. It is worth mentioning that compared to the VGG architecture, the last 1000-output layer is replaced by an *X-output* FC-layer, which shows the level of compression that is mainly aimed to reduce the overfitting if X is chosen to be less than 1000. X has been chosen to be 500 as empirically proved to be the maximum threshold for feature compression without downgrading the performance specifically for this architecture and experimental setup. If the attribute "Trainable Parameter" is set to "False," the weights of the associated layer will be fixed, which are learned in training on CASIA and LFW datasets. The attributes called "Trainable Parameter," and the margin, M, are chosen concurrently using cross-validation on five folds of the data, which have been divided training set into five folds of disjoint pairs. The margin M for the contrastive loss function is set to 1000 for all experiments, which renders a plausible performance.

3.6.2.1 Twin verification with restricted year

This part in which the network is fine-tuned on the twin data set, consider the year-separation effect. The *restricted* year refers to considering both images of a test pair belonging to the same year. All the subjects presented in year of

TABLE 3.3

Architecture used in twin-verification task

Name	Type	Filter size/stride	Output depth	Trainable parameter
conv1_1	Convolution	$3 \times 3/1$	64	True
conv1_2	Convolution	$3 \times 3/1$	64	False
pool1	Pooling	$2 \times 2/2$	64	—
conv2_1	Convolution	$3 \times 3/1$	128	True
conv2_2	Convolution	$3 \times 3/1$	128	False
pool2	Pooling	$2 \times 2/2$	128	—
conv3_1	Convolution	$3 \times 3/1$	256	True
conv3_2	Convolution	$3 \times 3/1$	256	True
conv3_3	Convolution	$3 \times 3/1$	256	False
pool3	Pooling	$2 \times 2/2$	256	—
conv4_1	Convolution	$3 \times 3/1$	512	True
conv4_2	Convolution	$3 \times 3/1$	512	False
conv4_3	Convolution	$3 \times 3/1$	512	False
pool4	Pooling	$2 \times 2/2$	512	—
conv5_1	Convolution	$3 \times 3/1$	512	True
conv5_2	Convolution	$3 \times 3/1$	512	False
conv5_3	Convolution	$3 \times 3/1$	512	False
pool5	Pooling	$2 \times 2/2$	512	—
FC_1	InnerProduct	4096		False
FC_2	InnerProduct	4096		True
FC_3	InnerProduct	**X**		True

X, which uniquely belong to that year, and not any other year, will be used as the test set, whereas the subjects who are presented in the other years, that is, all the years in range (2010−2015) except for X, will be used for training. X, can be varied in the range of (2010−2014) and the year of 2015 is not considered for testing in this setup, which further investigation will be done on that in combination with 2014 in subsequent experiments. Basically, the test set is completely separated by the year that the subjects are present. Age progression has not been considered in this setup because the testing year can be in the middle of the training set range. A representation of the described identity selection is demonstrated in the Venn diagram depicted in Figure 3.4.

The crucial thing is to make sure that the training and testing sets are mutually exclusive; that is, none of the test identities are present in the training phase. It is worth mentioning that this setup reduces the number of training samples because the repeated subjects that are jointly present in year X and any other year are not considered in the training set.

After fine-tuning the network using the twin data set, the outputs of the FC_3 layer will be used for feature representations of images of each pair. Then by using the Euclidean distance, the similarity score will be calculated in the sense that if two images are closer by the distance metric, their similarity score

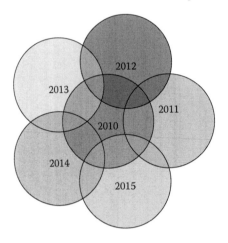

FIGURE 3.4
The Venn diagram for identity splitting.

TABLE 3.4
Twin verification with restricted year

Test year	Test subjects (pairs)	Train subjects (pairs)	EER (%) ($\mu \pm \sigma$)	AUC (%) ($\mu \pm \sigma$)
2010	159(3522)	1198(102,768)	13.3% ± 0.013	94.6% ± 0.009
2011	190(1055)	1167(110,161)	9.5% ± 0.017	96% ± 0.011
2012	243(19,445)	1114(41,537)	10.9% ± 0.003	96.4% ± 0.002
2013	187(1878)	1170(98,745)	13.8% ± 0.018	93.2% ± 0.011
2014	131(2026)	1226(118,238)	9.4% ± 0.019	95.8% ± 0.010

is higher. In the end, by simply using the calculated scores and comparing them by employing different thresholds, the ROC curve for the binary classification task will be demonstrated. The results are shown in Table 3.4.

The base learning rate is set to be 10^{-8}, and the step size of the learning rate drop is initially set to 2000. The batch size is chosen to be 48, which is the maximum that our hardware architecture can tolerate. Total iteration for weights is limited by the moment that the test performance starts to decrease. Each experiment depends on the number of training samples; the process of the weights update continues for at least 10 epochs unless the test performance starts to decrease. Technically, after this step we stop the update process and save the model and then initialize the model again with the saved weights, but start at a lower learning rate by a factor of 0.1. We also enforce an ℓ_2-norm regularizer, also referred to as *weight decay*, on the network weights with a regularization parameter set to 10^{-3}.

TABLE 3.5

Twin verification with age progression and unrestricted year

Test year	No. test subjects	No. train subjects	EER ($\mu \pm \sigma$)	AUC ($\mu \pm \sigma$)
2014–2015	32	1024	13.4% ± 0.02	92.3% ± 0.015

3.6.2.2 Twin verification with age progression and unrestricted year

This part considers the age-progression effect as the previous part. The unique subjects, which are only present in the years between 2010–2013 and not any other year, are used as the training set, whereas the subjects that are present in both years of 2014 and 2015 and absent in years of 2010–2013 are used as test subjects. The results are demonstrated in Table 3.5. In this experiment both subjects in a pair are present in both years, but the images themselves might be taken in different years.

3.6.2.3 Twin verification with age progression and unrestricted year using merged subjects

In this experiment, the subjects that are present in any of the years of 2014 and 2015 and absent in the years of 2010–2013 are used as test subjects, and the training set remains the same as the previous setting. The results are demonstrated in Table 3.6. So the difference between this experiment with the previous one is using merged identities; that is, in nonmatch pairs the subjects are not necessarily present in both years, and for the match pairs, the identity may appear in both years. A total number of 5144 pairs are generated and evaluated as the test set in this setting.

To increase the variation of the test set, the test images are chosen from the years of 2014 and 2015. This has some practical advantages. At first, the effect of age progress is considered while testing on the unseen subjects and moreover, it increases the number of test pairs, which provides a fair experiment. Additionally, choosing the test images from different years has specific forensics applications because one may realize if two images that are

TABLE 3.6

Twin verification with age progression and unrestricted year using merged subjects

Test year	No. test subjects	No. train subjects	EER ($\mu \pm \sigma$)	AUC ($\mu \pm \sigma$)
2014–2015	333	1024	12.64% ± 0.005	94.2% ± 0.007

TABLE 3.7

Test on Look-alike data set using pretrained
model on twins

No. pairs	EER ($\mu \pm \sigma$)	AUC ($\mu \pm \sigma$)
1718	$27.6\% \pm 0.02$	$79.1\% \pm 0.027$

taken in different years belong to the same person. It is worth noting that in
this experiment, the number of test subjects and training subjects is the same
as the previous experiment.

3.6.3 Face verification with transfer learning on Look-alike data set

In our last experiment, we evaluate the proposed model performance on the
Look-alike data set. Because the Look-alike data set suffers from the lack
of data samples, which approximately only contains a total number of 50
identities, we choose to pretrain the Siamese network with the whole TDFC
data set, and we evaluate the pretrained model using the Look-alike data set.

Before testing, the Look-alike data set, the images in the data set should be
aligned and resized to the standard size, 224×224. One of the problems with
using the Dlib face detector is that it misses some of the hard examples (e.g.,
partial occlusion, etc). This makes the data even smaller and hardly reliable
for evaluation. To solve this, other facial-landmark detectors have been used.
One facial-landmark detector that has empirically proven to perform well in
this setting is the Multi-task CNN [27], which has been used for aligning the
Look-alike data set.

For the Look-alike data set, we have gathered 1718 pairs for evaluation
of the model fine-tuned on twin data set. The performance on the Look-alike
data set is shown in Table 3.7 in which the EER achieves 28%.

It is worth noting that the EER does not achieve the best results gained
using the twin data set, which shows the difficulty of the transfer of the learning procedure in this challenging application. However, the results of this work
still beats the state-of-the-art results of [21] on the Look-alike data set.

3.7 Conclusion

This chapter proposed a discriminative model based on Siamese CNNs to be
implemented in the face-verification application. The proposed model was
trained to learn a discriminative similarity metric based on two identical
CNNs. Experimental results on TDFC and Look-alike data sets show plausible performance, verifying the efficiency of the proposed framework. The
effects of age progression and identification splitting have been illustrated in

performance evaluations by using two aforementioned databases. The performance was not changed significantly by altering the dynamic range of years used for training. The reason is that for the nonrestricted scenario, the number of subjects that have been repeated in different years is relatively low because of the specific characteristics of the TDFC data set. By having the same subjects in different years (different ages) for training, improvement in the performance has been observed. Moreover, training on the TDFC data set and fine-tuning on the Look-alike data set resulted in acceptable performance for testing on that data set.

References

1. S. Biswas, K. W. Bowyer, and P. J. Flynn. A study of face recognition of identical twins by humans. *Information Forensics and Security, IEEE International Workshop on*, 1–6, 2011.

2. Z. Sun, A. A. Paulino, J. Feng, Z. Chai, T. Tan, and A. K. Jain. A study of multibiometric traits of identical twins. *Proceedings of the SPIE, Biometric Technology for Human Identification VII*, 7:1–12, 2010.

3. A. K. Jain, S. Prabhakar, and S. Pankanti. On the similarity of identical twin fingerprints. *Pattern Recognition*, 35(11):2653–2663, 2002.

4. J. R. Paone, P. J. Flynn, P. J. Philips, K. W. Bowyer, R. W. V. Bruegge, P. J. Grother, G. W. Quinn, M. T. Pruitt, and J. M. Grant. Double trouble: Differentiating identical twins by face recognition. *IEEE Transactions on Information Forensics and Security*, 9(2):285–295, 2014.

5. B. Klare, A. A. Paulino, and A. K. Jain. Analysis of facial features in identical twins. In *Biometrics (IJCB), 2011 International Joint Conference on*, pp. 1–8, Washington, DC, October 2011.

6. J. Bromley, I. Guyon, Y. LeCun, E. Säckinger, and R. Shah. Signature verification using a "siamese" time delay neural network. In J. D. Cowan, G. Tesauro, and J. Alspector (Eds.), *Advances in Neural Information Processing Systems 6*, pp. 737–744. Holmdel, NJ: Morgan-Kaufmann, 1994.

7. N. Ye and T. Sim. Combining facial appearance and dynamics for face recognition. In *Computer Analysis of Images and Patterns: 13th International Conference, CAIP 2009, Münster, Germany, September 2–4, 2009. Proceedings*, pp. 133–140. Springer Berlin, Germany, 2009.

8. U. Park and A. K. Jain. Face matching and retrieval using soft biometrics. *IEEE Transactions on Information Forensics and Security*, 5(3):406–415, 2010.

9. K. Hollingsworth, K. W. Bowyer, and P. J. Flynn. Similarity of iris texture between identical twins. In *Computer Vision and Pattern Recognition Workshops (CVPRW), 2010 IEEE Computer Society Conference on*, pp. 22–29, June 2010.

10. A. Ariyaeeinia, C. Morrison, A. Malegaonkar, and S. Black. A test of the effectiveness of speaker verification for differentiating between identical twins. *Science & Justice*, 48(4):182–186, 2008.

11. P. J. Phillips, P. J. Flynn, K. W. Bowyer, R. W. V. Bruegge, P. J. Grother, G. W. Quinn, and M. Pruitt. Distinguishing identical twins by face recognition. In *Automatic Face Gesture Recognition and Workshops, 2011 IEEE International Conference on*, pp. 185–192, March 2011.

12. M. T. Pruitt, J. M. Grant, J. R. Paone, P. J. Flynn, and R. W. V. Bruegge. Facial recognition of identical twins. In *Biometrics, International Joint Conference on*, pp. 1–8, Washington, DC: IEEE, 2011.

13. T. H. N. Le, K. Luu, K. Seshadri, and M. Savvides. A facial aging approach to identification of identical twins. In *Biometrics: Theory, Applications and Systems (BTAS), 2012 IEEE Fifth International Conference on*, pp. 91–98, September 2012.

14. S. Chopra, R. Hadsell, and Y. LeCun. Learning a similarity metric discriminatively, with application to face verification. In *Computer Vision and Pattern Recognition, 2005. CVPR 2005. IEEE Computer Society Conference on*, Vol. 1, pp. 539–546, June 2005.

15. R. Hadsell, S. Chopra, and Y. LeCun. Dimensionality reduction by learning an invariant mapping. In *Computer Vision and Pattern Recognition, 2006 IEEE Computer Society Conference on*, Vol. 2, pp. 1735–1742, 2006.

16. M. Khalil-Hani and L. S. Sung. A convolutional neural network approach for face verification. In *High Performance Computing Simulation (HPCS), 2014 International Conference on*, pp. 707–714, Bologna, Italy: IEEE, July 2014.

17. K. B. Low and U. U. Sheikh. Learning hierarchical representation using siamese convolution neural network for human re-identification. In *Digital Information Management (ICDIM), 2015 Tenth International Conference on*, pp. 217–222, Jeju, South Korea: IEEE, October 2015.

18. Y. Lecun and F. J. Huang. Loss functions for discriminative training of energy-based models. In *Proceedings of the 10th International Workshop on Artificial Intelligence and Statistics*, Bridgetown, Barbados, 2005.

19. S. L. D.Yi, Z. Lei, and S. Z. Li. Learning face representation from scratch. *arXiv*, 1411.7923, 2014.

20. G. B. Huang and E. Learned-Miller. Labeled faces in the wild: Updates and new reporting procedures. Technical Report UM-CS-2014-003, University of Massachusetts, Amherst, MA, May 2014.

21. H. Lamba, A. Sarkar, M. Vatsa, R. Singh, and A. Noore. Face recognition for look-alikes: A preliminary study. In *2011 International Joint Conference on Biometrics (IJCB)*, pp. 1–6, Octber 2011.

22. D. E. King. Dlib-ml: A machine learning toolkit. *Journal of Machine Learning Research*, 10:1755–1758, 2009.

23. T. Hassner, S. Harel, E. Paz, and R. Enbar. Effective face frontalization in unconstrained images. In *Computer Vision and Pattern Recognition (CVPR), 2015 IEEE Conference on*, pp. 4295–4304, June 2015.

24. K. Simonyan and A. Zisserman. Very deep convolutional networks for large-scale image recognition. *CoRR*, abs/1409.1556, 2014.

25. Y. Jia, E. Shelhamer, J. Donahue, S. Karayev, J. Long, R. Girshick, S. Guadarrama, and T. Darrell. Caffe: Convolutional architecture for fast feature embedding. In *Proceedings of the 22Nd ACM International Conference on Multimedia*, MM '14, pp. 675–678, New York, ACM, 2014.

26. O. M. Parkhi, A. Vedaldi, and A. Zisserman. Deep face recognition. In *British Machine Vision Conference*, Swansea, UK, 2015.

27. K. Zhang, Z. Zhang, Z. Li, and Y. Qiao. Joint face detection and alignment using multitask cascaded convolutional networks. *IEEE Signal Processing Letters*, 23(10):1499–1503, 2016.

4

Tackling the Optimization and Precision Weakness of Deep Cascaded Regression for Facial Key-Point Localization

Yuhang Wu, Shishir K. Shah, and Ioannis A. Kakadiaris

CONTENTS

4.1	Introduction	85
4.2	Related Work	88
4.3	Method	89
	4.3.1 Optimized progressive refinement	89
	4.3.2 3D pose-aware score map	90
	4.3.3 DSL	91
	4.3.4 Proposal and refinement in dual pathway	93
	4.3.4.1 Features in the information pathway	93
	4.3.4.2 Score maps in the decision pathway	93
	4.3.4.3 Network structure	95
4.4	Experiments	96
	4.4.1 Databases and baselines	97
	4.4.2 Architecture analysis and ablation study	99
	4.4.3 Performance on challenging databases	101
	4.4.3.1 Robustness evaluation	101
4.5	Conclusion	104
Acknowledgment		105
References		105

4.1   Introduction

Facial key-point localization refers to detecting salient facial landmarks (e.g., eye corners, nose tip, etc.) on the human face. Multiple recently proposed pose-robust face-recognition systems [1–3] rely on accurate landmark annotation, which makes this problem important. Even though the boundaries of this domain have been consistently pushing forward in the recent years, localizing landmarks under large head-pose variations and strong occlusions remains challenging.

Current state-of-the-art approaches for facial key-point localization are based on cascaded regression [4–7]. The intent of the algorithm is to progressively minimize a difference ΔS between a predicted shape $\hat{S}$ and a ground-truth shape S in an incremental manner. This approach contains T stages, starting with an initial shape $\hat{S}^0$; the estimated shape $\hat{S}^t$ is gradually refined as:

$$\arg\min_{\mathbb{R}^t, \mathbb{F}^t} \sum_i ||\Delta S_i^t - \mathbb{R}^t(\mathbb{F}^t(\hat{S}_i^{t-1}, \mathbf{I}_i))||_2^2, \tag{4.1}$$

$$\hat{S}_i^t = \hat{S}_i^{t-1} + \Delta\hat{S}_i^{t-1}, \tag{4.2}$$

where i iterates overall training images. $\hat{S}_i^t$ is the estimated facial shape for image $\mathbf{I}_i$ in stage t; usually $\hat{S}_i^t$ can be represented as a $2L \times 1$ vector. L is the number of facial key points. $\mathbb{F}^t(\hat{S}_i^{t-1}, \mathbf{I}_i)$ is a mapping from image space to feature space. Because the obtained features are partially determined by $\hat{S}_i^{t-1}$, these features are called "shape-indexed features." $\mathbb{R}^t(\cdot)$ is a learned mapping from feature space to target parameter space. In deep cascaded regression [4,7–9], $\mathbb{F}^t(\cdot)$ can be used to denote all operations before the last fully connected layer. $\mathbb{R}^t(\cdot)$ represents the operations in the last fully connected layer whose input is an arbitrary dimensions feature vector ϕ_i^t, and output is the target parameter space.

Although cascaded regression is a useful framework for face alignment, several challenges need to be addressed when deriving a deep architecture. First, current deep cascaded regression is greedily optimized per each stage. The learned mapping $\mathbb{R}^t$ is not end-to-end optimal with respect to global-shape increment. When training a new mapping $\mathbb{R}^t$ for stage t, fixing the network parameters of previous stages leads to a stage-wise suboptimal solution. Different from cascaded face detection that can be easily formulated into a globally optimal structure [11], gradients that back-propagate from the later stages of the network are blocked because of reinitialization of shape parameters between stages. The second challenge arises from shape-indexed features: $\mathbb{F}^t(\hat{S}_i^{t-1}, \mathbf{I}_i)$. Shape-indexed features are extracted based on landmark locations [5,6]. However, how to effectively merge the information encoded in one-dimensional coordinate vectors into a two-dimensional (2D) image in an optimal way still remains an open problem. Even though some heuristic solutions (e.g., concatenating a three-dimensional (3D) geometric map with RGB channels [7] or highlighting pixel blocks [12]) alleviate the problem partially, the solution is not optimal from a gradient back-propagation perspective because the pixel values in the newly generated maps are assigned based on external parameters. The third challenge comes from information flow in deep architectures. The deep representation generated at the bottom layer ϕ_i^t, although highly discriminative and robust for object/face representation, loses too much spatial resolution after many pooling and convolutional layers. As a result, it cannot tackle pixel-level localization/classification tasks very well.

This phenomenon was recently named in the image segmentation field as spatial-semantic uncertainty [13]. Because in most deep regressions [4,7–9,14], where $\mathbb{R}^t$ solely relies on ϕ_i^t, precision of $\hat{S}_i^t$ may suffer from this structural limitation of deep networks.

To tackle the aforementioned challenges in deep cascaded regression models, we propose a globally optimized dual-pathway (GoDP) architecture where all inferences are conducted on 2D score maps to facilitate gradient back-propagation (Figure 4.1). Because there are very few landmark locations

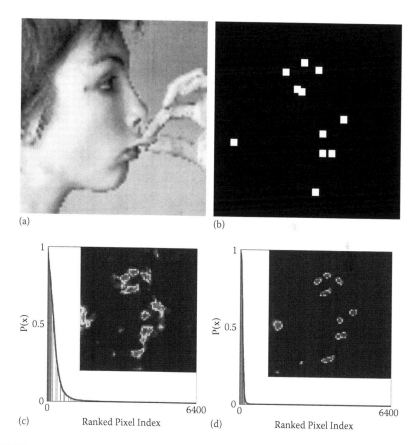

(a)   (b)

(c)   Ranked Pixel Index   (d)   Ranked Pixel Index

FIGURE 4.1

Score maps generated from different variations of DeconvNet. Pixels in the score maps indicate probabilities of visible facial key points. (a) Original image, (b) ground-truth mask, (c) DeconvNet [10], and (d) GoDP. We rank the pixel values in each score map and plot as the curve line in dark gray color underneath. The vertical lines in light gray color indicate the pixel values in the key-point candidate positions (3×3 white patches plotted in b). This comparison shows that the score map generated from GoDP is clear and discriminative.

activated on the 2D score maps, a distance-aware softmax function (DSL) that reduces the false alarms in the 2D score maps is proposed. To solve the spatial-semantic uncertainty problem of deep architecture, a dual-pathway model where shallow and deep layers of the network are jointly forced to maximize the possibility of highly specific candidate regions is proposed. As a result, our facial key-points localization model achieved state-of-the-art performance on multiple challenging databases. The key contributions of our work include:

- An off-the-shelf deep network that is able to generate high-quality 2D score maps for key-points localization.

- A new loss function designed for reducing false alarms in the 2D score maps.

- A heavily supervised proposal-refinement (PR) architecture to discriminatively extract spatial-semantic information from the deep network.

The rest of this chapter is organized as follows: In Section 4.2, we present the related work in deep cascaded regression and discuss the limitations of this method, and in Section 4.3, we introduce our proposed deep architecture and three critical components of this architecture, and in Section 4.4, we evaluate the proposed method in multiple challenging databases, and in Section 4.5, we present our conclusions.

4.2 Related Work

Most of the deep architectures used for face alignment are extended from the framework of cascaded regression. Sun et al. [8] first employed an AlexNet-like architecture to localize five fiducial points on faces. Later, Zhang et al. [9] invented a multitask framework demonstrating that a more robust landmark detector can be built through joint learning with correlated auxiliary tasks, such as head pose and facial expression, which outperformed several shallow architectures [15–18]. To conquer facial alignment problems under arbitrary head poses, Zhu et al. [6] and Jourabloo and Liu [4] employed a deformable 3D model to jointly estimate facial poses and shape coefficients online. These deep cascaded regression methods suppress multiple state-of-the-art shallow structures [19–21] and achieved remarkable performance on less-controlled databases such as AFLW [22] and AFW [23]. A common point of the previous architectures is that they require model reinitalization when switching stages. As a result, the parameters of each stage are optimized from a greedy stage-wise perspective, which is inefficient and suboptimal. Inspired by recent works [12,24,25] in human-pose estimation and face alignment, we employ 2D score maps as the targets for inference. This modification enables gradients back-propagation between stages, allows 2D feedback loops, and hence delivers an end-to-end model.

One fundamental challenge when employing 2D score maps for key-point localization is spatial-semantic uncertainty, which is critical, but has not been the focus of previous works on face alignment. Ghiasi and Fowlkes [13] pointed out that features generated from the bottom layers of deep networks, although encoding semantic information that is robust to image and human-identity variations, lack spatial resolution for tasks requiring pixel-level precision (e.g., image segmentation, key-points localization). As a result, Ghiasi and Fowlkes [13] proposed a Laplacian pyramid-like architecture that gradually refines the 2D score maps generated by the bottom layers through adding back features generated from top layers, which contain more spatial information. Motivated by similar observations, Newell et al. [26] proposed a heavily stacked structure by intensively aggregating shallow and deep convolutional layers to obtain better score map predictions. In our work, we build a more intuitive and tractable model without resorting to a stacked architecture. Because our network leads to a better spatial-semantic trade-off, we achieve state-of-the-art performance on facial key-point localization. Note that our architecture can be employed in any applications that require high precision in localization (e.g., human-pose estimation, object localization). Recently proposed hybrid inference methods (e.g., convolutional neural network (CNN)-based recurrent network [12], CNN-based conditional random field [27], and CNN-based deformable parts model [28]) can also be built on the high-quality score map generated by GoDP architecture.

4.3 Method

In this section, we first introduce three components of the proposed architecture. They are the basic elements that help us address multiple challenges in 2D score map-based inference. Then, we introduce our GoDP architecture.

4.3.1 Optimized progressive refinement

Because of optimization problems in traditional deep cascaded architecture, a global-optimization model is highly needed. However, the main difficulty in converting a cascaded regression approach into a globally optimized architecture is to back-propagate gradients between stages, where shape was usually used before to initialize new cascaded stages. In our work, we bypass the problem by representing landmark locations $\hat{S}_i^t$ through 2D score maps Ψ (we omit the index i for clarity), where information of landmark positions is summarized into probability values that indicate the likelihood of the existence of landmarks. In our work, Ψ denotes $(KL+1) \times W \times H$ score maps in stage t, where L is the number of landmarks, K is the number of subspaces (which will be introduced later), W and H are the width and height of the score maps. The extra $(KL+1)^{\text{th}}$ channel indicates the likelihood that a pixel belongs to background. Through this representation, gradients can pass through the score maps and be back-propagated from the latest stages of the cascaded model. Another insight of employing 2D probabilistic score maps is these outputs can be aggregated and summarized with convolutional features and create feedback loops, which can be represented as follows:

$$\boldsymbol{\Psi}^0 = \mathbb{F}_o^0(\mathbf{I}), \tag{4.3}$$

$$\mathbb{F}_b^{t-1}(\mathbf{I}, \boldsymbol{\Psi}^{t-1}) = \mathbb{F}_a^{t-1}(\mathbf{I}) \uplus \boldsymbol{\Psi}^{t-1}, \tag{4.4}$$

$$\Delta\boldsymbol{\Psi}^{t-1} = \mathbb{F}_c^{t-1}(\mathbb{F}_b^{t-1}(\mathbf{I}, \boldsymbol{\Psi}^{t-1})), \tag{4.5}$$

$$\boldsymbol{\Psi} = \boldsymbol{\Psi}^{t-1} + \Delta\boldsymbol{\Psi}^{t-1}, \tag{4.6}$$

where $\uplus$ denotes a layerwise concatenation in CNNs, $\mathbb{F}_o^0(\mathbf{I})$ represents the first $\boldsymbol{\Psi}^0$ generated from $\mathbf{I}$ after passing through several layers in CNN, and $\mathbb{F}_a^{t-1}(\cdot)$, $\mathbb{F}_b^{t-1}(\cdot)$, and $\mathbb{F}_c^{t-1}(\cdot)$ indicate different network operations with different parameter settings. Through the feedback loops, score maps generated by each stage can be directly concatenated with other convolutional layers through Equation 4.4, which behaves as a shape-indexed feature. In contrast to Zhu et al. [7] and Peng et al. [12], where score maps employed in the feedback loops are determined and synthesized through external parameters, in our architecture, $\boldsymbol{\Psi}^{t-1}$ in Equation 4.4 is fully determined by the parameters inside the network based on Equations 4.5 and 4.6. Therefore, our face alignment model can be optimized globally.

4.3.2   3D pose-aware score map

Unlike recent works that employ a deformable shape model [4,6], our model implicitly encodes 3D constraints to model complex appearance-shape dependencies on the human face across pose variations. We found that pose is a relative concept whose actual value is susceptible to sampling region, facial expressions, and other factors. As a result, it is difficult to learn an accurate and reliable mapping from image to the pose parameters without considering fiducial-points correspondence. Instead of estimating pose parameters [4,6,19,29] explicitly, we regard pose as a general domain index that encodes multimodality variations of facial appearance. Specifically, we used K score maps to indicate each landmark location, where K corresponds to the number of partitions of head pose. For each image, one out of K score maps is activated for each landmark. In this way, the network automatically encodes contextual information between the appearance of landmarks under different poses. At the final stage, the K score maps are merged into one score map through element-wise summation. In our implementation, K is equal to 3, and the subspace partition is determined by yaw variations.

4.3.3 DSL

Softmax loss function has been widely used in solving pixel-labeling problems in human-joint localization [25,30], image segmentation [10,13], and recently, facial key-point annotation [12]. One limitation of using softmax for key-point localization is that the function treats pixel labeling as an independent classification problem, which does not take into account the distance between the labeling pixel and the ground-truth key points. As a result, the loss function will assign equal penalty to the regions that lie very close to a key point and also the regions on the border of an image, which should not be classified as landmark candidates. Another drawback of this loss function is that it assigns equal weights to negative and positive samples, which may lead the network to converge into a local minimum, where every pixel is marked as background. This is a feasible solution from an energy perspective because the active pixels in the score maps are so sparse (only one pixel is marked as key-point per score map in maximum) that their weights play a small role in the loss function compared to the background pixels. To solve these problems, we modified the original loss function as follows. First, we assign larger cost when the network classifies a key-point pixel into background class; this helps the model stay away from local minima. Second, we assign different cost to the labeled pixels according to the distance between the labeled pixels and other key points, which makes the model aware of distances. This loss function can be formulated as follows:

$$\sum_x \sum_y m(x,y) w \sum_k t_k(x,y) log\left(\frac{e^{\psi_k(x,y)}}{\sum_{k'} e^{\psi_{k'}(x,y)}}\right), \qquad (4.7)$$

$$w = \begin{cases} \alpha, & k \in \{1 : KL\} & (4.8) \\ \beta log(d((x,y),(x',y')) + 1), & k = KL + 1, & (4.9) \end{cases}$$

where (x,y) are locations, $k \in \{1 : KL+1\}$ is the index of classes, $\psi_k(x,y)$ is the pixel value at (x,y) in the kth score map of Ψ, $t_k(x,y) = 1$ if (x,y) belongs to class k, and 0 otherwise. The binary mask $m(x,y)$ is used to balance the amount of key-point and background pixels employed in training. The weight w controls the penalty of foreground and background pixels. For a foreground pixel, we assign a constant weight α to w, whose penalty is substantially larger than nearby background pixels. For a background pixel, the distance $d((x,y),(x',y'))$ between the current pixel (x,y) and a key point (x',y') whose probability ranked the highest among the KL classes is taken into account. The result is that the loss function assigns the weights based on the distance between the current pixel and the most misleading foreground pixel among the score maps, which punishes false alarms adaptively. In Equation 4.9, we used a log function (base 10) to transform the distance into a weight and employed a constant β to control the magnitude of the cost. The shape of w is depicted in Figure 4.2. As a result, discrimination between the background and foreground pixels is encouraged according to the distance between a labeled pixel and a

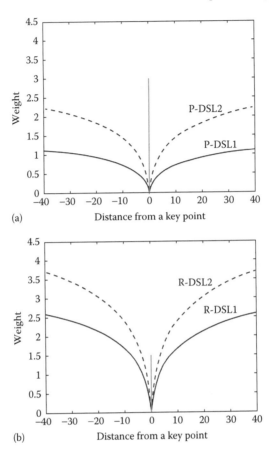

FIGURE 4.2
The shapes of the distance-aware softmax loss (DSL) employed in the decision pathway. The transformations of the functions after increasing the values of β are visualized through the dashed lines. The straight light gray-colored lines indicate the cost of missclassifying a key-point pixel to a background pixel, whereas the dark gray lines indicate the cost of missclassifying a background pixel to a key-point pixel. (L) DSL for proposal, (R) DSL for refinement.

specific key point. From a point of view of optimization, in back-propagation, because $d((x,y),(x',y'))$ is independent from $\psi_k(x,y)$, w will be a constant that can be directly computed through Equation 4.8 or 4.9.

When training the network, we first replace Equation 4.9 with a constant term (represented as β, which is less than α) and train the network with this degraded DSL (represented as SL in Figure 4.3) for the first six epochs. Then, Equation 4.9 is employed for further refinement. During training, inspired by curriculum learning [31], we gradually increase the value of β and encourage the network to discriminate pixels closer to the key-point locations.

4.3.4 Proposal and refinement in dual pathway

To better exploit the spatial and semantic information encoded in a deep network, we propose a dual-pathway architecture as shown in Figure 4.3. Derived from DeconvNet [10], the unique design of the proposed architecture includes separate pathways used for generating discriminative features and making decisions. We designate them as "information pathway" and "decision pathway." In the decision pathway, the depth of each layer is strictly kept as $KL + 1$ where each channel corresponds to a score map ψ_k. In the information pathway, depths of layers are unconstrained to enrich task-relevant information.

4.3.4.1 Features in the information pathway

The design of the information pathway is built on the findings that feature maps generated from the deep layers of the network contain robust information that is invariant to the changing of image conditions, but lack enough resolution to encode exact key-point locations. Although the feature maps of shallow layers contain enough spatial information to localize the exact position of each key point, they also contain a large amount of irrelevant noise. To handle this dilemma, we build a structure such that the features extracted from shallow layers are used to propose candidate regions, while the features extracted from deep layers help to filter out false alarms and provide structural constraints. This is accomplished by imposing different losses to supervise shallow- and deep-level features generated from shallow and deep layers. We adjust the parameters of DSL in the decision pathway and enforce a large penalty when the shallow-level features fail to assign large positive probabilities to key-point locations, but give a smaller cost when they misidentify a background into a key-point candidate. This is a high detection-rate policy to supervise shallow-level features. In contrast, we adopt a low false-alarm policy to supervise deep-level features: we enforce high penalty when deep-level features misidentity a background pixel as key point, but slightly tolerate the error in the other way around. The results are shown in Figure 4.3. After each shallow-level proposal, the contrast between background and foreground is increased, whereas, after each deep-level refinement, the background noise is suppressed. As a result, the key-point regions are gradually shrunk and highlighted.

4.3.4.2 Score maps in the decision pathway

In the decision pathway, $\mathbf{\Psi}^0$ is first initialized with the output of the second deconvolution layers, where high-level information is well-preserved. Then, the probabilistic corrections $\Delta\mathbf{\Psi}^{t-1}$ generated from the shallow-level and deep-level layers of the network are computed and added to the decision pathway with the supervision of multiple DSLs.

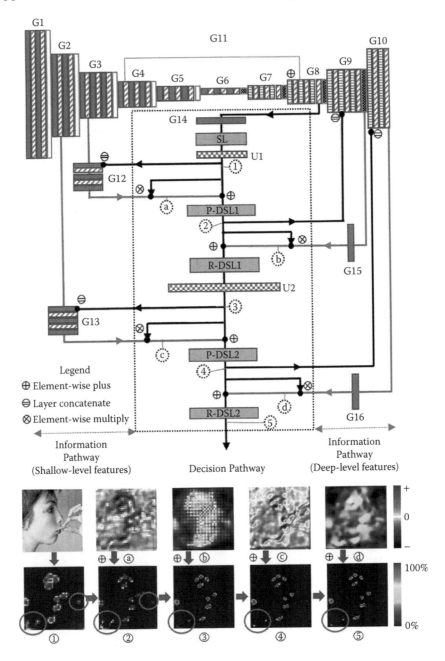

FIGURE 4.3

The architecture of the proposed globally optimized dual-pathway (GoDP) model. Based on a naive DeconvNet [10], we derive a precise key-point detector by discriminatively extracting spatial and semantic information from shallow and deep layers of the network.   (*Continued*)

As shown in Figure 4.3, during inference, score maps are first initialized on the decision pathway through Equation 4.3, and then concatenated with the layers in the information pathway through Equation 4.4. These newly formed features are processed and aggregated into the decision pathway using Equation 4.5, and the score maps in the decision pathway are updated by Equation 4.6. The same process repeats several times to generate the final score maps. The intention of this architecture is identical to cascaded regression, where in each stage, features are generated and contribute to reduce residual errors between predicted key-point locations and ground-truth locations. The predicted locations then get updated and are used to reinitialize a new stage. The difference is our 2D inference model fully exploits the information encoded in a single network instead of resorting to a stacked architecture.

4.3.4.3 Network structure

In Figure 4.3, we employed a standard DeconvNet architecture containing 10 groups of layers (G1, G2, G3, G4, G5, G6, G7, G8, G9, G10) as feature source. Each group contains two or three convolutional/deconvolutional layers, batch normalization layers, and one pooling/unpooling layer. We added a hyperlink to connect G4 and G8 to avoid information bottleneck. The decision pathway is derived from the layer of G8, before unpooling. Bilinear upsampling layers are denoted as U1 and U2. Loss layer SL represents a degraded DSL (introduced in Section 4.3.3, represented as SL). We use P-DSL to represent DSL used for supervising key-point candidate proposal. We use R-DSL to represent DSL used for supervising candidate refinement. Shapes of these DSLs are plotted in Figure 4.2. The layers G12, G13, G14, G15, and G16 are additional groups of layers used to convert feature maps from information pathway to score maps in the decision pathway. The layers G12 and G13 contain three convolutional and two batch normalization layers. The layers G14, G15, and G16 include one convolutional layer. The settings of convolutional layers in G12 and G13 are the same: width 3, height 3, stride 1, pad 1 except the converters (last layer of G12 and G13), which connect the information pathway and the decision pathway, whose kernel size is 1×1. The other converters G14, G15, and G16 have the same kernel size: 1×1.

FIGURE 4.3 (Continued)
The framework is motivated by cascaded regression, which contains residual error corrections and error feedback loops. Moreover, GoDP is end-to-end trainable, fully convolutional, and optimized from a pixel-labeling perspective instead of traditional regression. Under the architecture, we visualize the $(KL+1)^{th}$ score map sampled through the network, which indicates the probability of key-point locations. The letter and number below each score map indicate the corresponding position in the network architecture. We highlight background regions with light-gray colored circles to indicate how the proposal and refinement technique deal with background noises.

4.4 Experiments

In our experiments, we train the network from scratch. For each score map, there is only one pixel at most that is marked as key-point pixel (depending on the visibility of the key point). We employ different sampling ratios for the background pixels that are nearby or further from the key point. The threshold for differentiating nearby and far-away is measured by pixel distance on the score maps. In this chapter, we employ three pixels as the threshold. At the beginning, the network is trained with features generated from shallow-level layers only, which means the network has three loss functions instead of five in the first three epochs. After training the network for three epochs, we fine-tune the network with all five loss functions for another three epochs. In these six epochs, we employ a degraded DSL (SL) as explained in Section 4.3.2, then DSL is used and the whole architecture is as shown in Figure 4.3. The learning rate is gradually reduced from 10^{-3} to 10^{-7} during the whole training process. We employ the stochastic gradient descent (SGD) method to train the network. The input size of the network is 160×160 (gray scale) and the output size of the score map is 80×80. It takes 3 days to train on one NVIDIA Titan X. The detailed parameter settings in training are shown in Tables 4.1 through 4.5.

TABLE 4.1
Parameters of SL

Stage	Sampling ratio: Far-away pixels	Sampling ratio: Nearby pixels	Value of α	Value β	Type of loss	Epoch
1	0.005	0.1	1	0.2	SL	3
2	0.005	0.1	1	0.2	SL	3
3	0.005	0.1	1	0.2	SL	3

TABLE 4.2
Parameters of P-DSL1

Stage	Sampling ratio: Far-away pixels	Sampling ratio: Nearby pixels	Value of α	Value β	Type of loss	Epoch
1	0.005	0.1	1	0.2	SL	3
2	0.001	0.2	3	0.1	SL	3
3	0.001	0.15	3	0.6	DSL	3

TABLE 4.3
Parameters of R-DSL1

Stage	Sampling ratio: Far-away pixels	Sampling ratio: Nearby pixels	Value of α	Value β	Type of loss	Epoch
1	—	—	—	—	—	—
2	0.01	0.05	1	0.3	SL	3
3	0.01	0.05	1.5	1	DSL	3

TABLE 4.4
Parameters of P-DSL2

Stage	Sampling ratio: Far-away pixels	Sampling ratio: Nearby pixels	Value of α	Value β	Type of loss	Epoch
1	0.005	0.1	1	0.2	SL	3
2	0.001	0.2	3	0.1	SL	3
3	0.001	0.15	3	0.6	DSL	3

TABLE 4.5
Parameters of R-DSL2

Stage	Sampling ratio: Far-away pixels	Sampling ratio: Nearby pixels	Value of α	Value β	Type of loss	Epoch
1	—	—	—	—	—	—
2	0.01	0.05	1	0.3	SL	3
3	0.01	0.05	1.5	1	DSL	3

4.4.1 Databases and baselines

Three highly challenging databases are employed for evaluation: AFLW [22], AFW [23], and UHDB31 [32]. The detailed experimental settings are summarized in Table 4.6. We strictly follow the training and testing protocol as in Zhu et al. [6] and conduct our experiment on AFLW-PIFA (3901 images for training, 1299 images for testing, 21 landmarks annotated in each image) and ALFW-Full (20,000 training, 4,386 testing, 19 landmarks annotated in each image). We note the models trained on AFLW-PIFA as $M1$ and the models trained on AFLW-Full as $M2$. For evaluating on the AFW database (468 images for testing six landmarks annotated in each image), $M2$ is used. We picked six estimated landmarks out of 19 to report the performance. To evaluate the accuracy of algorithms under frontal faces, $M2$ is

TABLE 4.6

Detailed experiment settings of our algorithm

Evaluation name	Training set	# of Training samples	Trained model	Testing set	# of Testing samples	Point	Normalizating factor	Settings
AFLW-PIFA	AFLW	3,901	*M1*	AFLW	1,299	21	Face Size	Following [6]
AFLW-Full	AFLW	20,000	*M2*	AFLW	4,386	19	Face Size	Following [6]
AFLW-F	AFLW	—	*M2*	AFLW	1,314	19	Face Size	Ours
AFW	AFLW	—	*M2*	AFW	468	6 out of 19	Face Size	Ours
UHDB31	AFLW	—	*M2*	UHDB31	1,617	9 out of 19	Face Size	Ours

also employed. Different from Zhu et al. [6], all 1314 images out of 4386 in the AFLW-Full database with 19 visible landmarks are considered as frontal faces and used for testing. Results are shown in Table 4.10 under the label AFLW-F. The database UHDB31 is a lab-environment database, which contains 1617 images, 77 subjects, and 12 annotated landmarks for each image. This is a challenging database, including 21 head poses, combining seven yaw variations: $[-90°:+30°:90°]$ and three pitch variations: $[-30°:+30°:30°]$. We employed 9 landmarks (ID: 7,9,10,12,14, 15,16, 18,20 in Kostinger et al. [22]) to compute landmark errors. Model $M2$ is employed for evaluating.

Multiple state-of-the-art methods (CDM [18], RCPR [15], CFSS [21], ERT [33], SDM [17], LBF [20], PO-CR [5], CCL [6], HF [14], PAWF [4], and 3DDFA [7]) are selected as baselines. In our implementation, Hyperface (HF) is trained without the loss of gender. The network architecture remains the same. The performance of 3DDFA and PAWF are reported based on the code provided by their authors. We employed normalized mean error (NME) to measure the performance of algorithms as in Zhu et al. [6]. Same as Zhu et al. [6], the bounding box defined by AFLW is used to normalize the mean error of landmarks and initialize the algorithm. When the AFLW bounding box is not available (e.g., on UHDB31 and AFW database) or not rectangle (AFLW-PIFA), we use the bounding-box generator provided by the authors of AFLW to generate a new bounding box based on the visible landmarks. For the AFLW-PIFA database, after we generate new bounding boxes, we enlarge them by 15% to guarantee the ears are included, while the NME is computed using the original size of the bounding boxes.

4.4.2 Architecture analysis and ablation study

Along with the development of the deep network, the network structures become complex, which might make the functionality of individual modules unclear to the reader. To evaluate the capabilities of various networks for generating discriminative score maps, we analyzed new connections/structures of recent architectures on the DeconvNet platform [10] to control uncertainty. The hourglass network (HGN) [26] is a recent extension of DeconvNet. The core contribution of hourglass net is that it aggregates features from shallow to deep layers through hyperconnections, which blends the spatial and semantic information for discriminative localization. Different from our supervised proposal and refinement architecture, the information fusion of HGN is conducted in an unsupervised manner. Our implementation of hourglass net is based on DeconvNet, we add three hyperlinks to connect shallow and deep layers but remove residual connections [34]. This model is selected to be our baseline. The detailed network settings for our implementation can be viewed in Figure 4.4.

In this experiment, we first employ a landmark mask to separate foreground and background pixels as shown in Figure 4.1. Then we compute the mean probability of foreground and background pixels based on the mask. We average the mean probability over all the testing images on the PIFA-AFLW

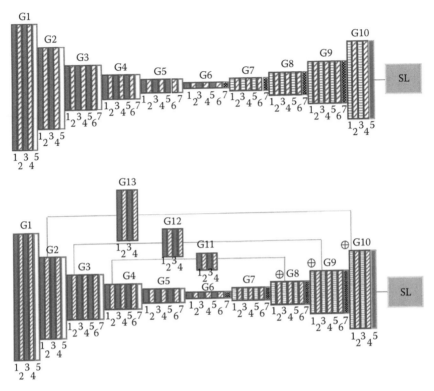

FIGURE 4.4
Network settings: (T) DeconvNet, (B) DeconvNet with hourglass [26]
connections.

database and obtain the numbers in Table 4.7. We observed that GoDP
performs significantly better in discriminating foreground and background
pixels than other structures and has a smaller landmark detection error. We
also evaluated our architecture without DSL (DSL in Table 4.7) and another
architecture without both DSL and PR architecture (degraded DSL every-
where with the same parameters). The result is as shown in Table 4.7 with
the name GoDP-DSL-PR. Table 4.7 shows that DSL is critical for training
a highly discriminative key-point detector and also contributes to regularize
our PR architecture. Additionally, we observe that the hyperlinks introduced
in hourglass net suppress background noise in DeconvNet.

In the next experiment, we trained GoDP to detect occluded landmarks.
In stage 3, we used the coordinates of all landmarks as the ground truth of
the last two DSLs (previous DSL/SL are trained with visible landmarks),
fine-turned from stage 2, and trained the whole network for three epochs. The
results are shown in Table 4.7 with the name **GoDP(A)**. To compare with
HGN, we trained the HGN to detect occluded landmarks, the result is as
shown as HGN(A). We observe that GoDP(A) can better detect both visible

TABLE 4.7

Performance on PIFA-AFLW database

Method	MPK	MPB	NME-Vis	NME-All
DeconvNet [10]	51.83	4.50	4.13	8.36
HGN [26]	28.38	0.96	3.04	11.05
GoDP−DSL−PR	31.79	1.01	3.35	13.30
GoDP−DSL	26.15	0.86	3.87	13.20
GoDP	39.78	1.30	2.94	11.17
GoDP(A)−DSL	99.37	99.92	3.37	5.75
HGN(A)	48.42	14.19	3.08	5.04
GoDP(A)	47.59	6.74	**2.86**	**4.61**

Note: MPB (%), mean probability of background pixels (small is better); MPK (%), mean probability of key-point candidate pixels (large is better); NME-Vis, NME (%) of visible landmarks; NME, represents NME (%) of all 21 landmarks.

and invisible (all) landmarks if we train the network in this manner. Since then, we use GoDP(A) in the following experiments for comparison.

4.4.3 Performance on challenging databases

We compared GoDP(A) with state-of-the-art cascaded-regression based key-point detectors. Because we strictly follow the experimental protocol as in Zhu et al. [6], we directly cite their numbers in Table 4.8. GoDP(A) the lowest NME in all comparisons: Tables 4.8 through 4.10, and Figure 4.5. Table 4.9 indicates we obtain impressive results on both frontal and profile faces, which demonstrates GoDP is a highly accurate deep network–based detector. A qualitative comparison can be found in Figure 4.6. However, as shown in Figure 4.5(R), GoDP is a discriminative model and its performance is not as consistent as 3DDFA in terms of detecting both visible and invisible landmarks. This trade-off between detection accuracy and shape rationality requires more exploration in future work.

4.4.3.1 Robustness evaluation

To further review the properties of GoDP, we compared robustness of GoDP(A) and HF (regression-based method) under different bounding-box initializations. This is important because bounding boxes generated by real-face detectors always vary in size and position. We artificially add Gaussian noise to the provided bounding boxes of AFLW-Full. The noise is generated based on the size of bounding boxes, where σ controls the intensity of the Gaussian noise. The noise is added based on the size and location of bounding boxes, and the results are as shown in Figure 4.7, which discloses GoDP is more robust to variations of bounding-box sizes, but sensitive to translation errors.

One explanation of why GoDP is more robust to variations of bounding-box sizes, but sensitive to translation errors is that because GoDP is a

TABLE 4.8

NME (%) of all annotated landmarks

Baseline	Nondeep Learning methods								Deep cascaded R.		Deep end-to-end	
	CDM	RCPR	CFSS	ERT	SDM	LBF	PO-CR	CCL	PAWF	3DDFA	HF	GoDP(A)
AFLW-PIFA	8.59	7.15	6.75	7.03	6.96	7.06	—	5.81	6.00	6.38	—	4.61
AFLW-Full	5.43	3.73	3.92	4.35	4.05	4.25	5.32	2.72	—	4.82	4.26	1.84

TABLE 4.9

Performance of GoDP(A)/HF on 21 views of UHDB31

30°	**2.0**/5.2	**2.1**/4.9	**1.8**/5.8	**1.6**/4.2	**1.8**/5.5	**2.1**/5.2	**1.8**/5.3
0°	**1.8**/3.6	**1.8**/3.0	**1.8**/3.2	**1.2**/2.2	**1.5**/3.0	**1.7**/3.3	**1.7**/3.7
−30°	**2.7**/5.3	**2.1**/5.3	**1.8**/4.3	**1.6**/3.3	**1.6**/3.8	**2.1**/5.0	**2.4**/6.0
	−90°	−60°	−30°	0°	30°	60°	90°

Note: NME (%) of visible landmarks is reported. Columns correspond to pitch variations, rows correspond to yaw variations.

TABLE 4.10

NME (%) of visible landmarks

	Deep cascaded R.		Deep end-to-end	
Evaluation	PAWF	3DDFA	HF	GoDP(A)
AFLW-PIFA	4.04	5.42	—	**2.86**
AFLW-Full	—	4.52	3.60	**1.64**
AFLW-F	—	4.13	2.98	**1.48**
AFW	4.13	3.41	3.74	**2.12**

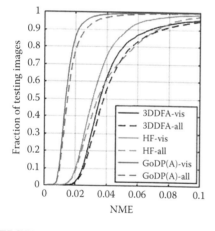 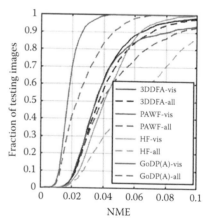

FIGURE 4.5

NME (%) increasing on all 19 landmarks of AFLW-Full database 4386 images under noised bounding-box initializations. The σ is measured in percentage. (L) GoDP(A), (R) Hyperface.

detection-based method, it is unable to predict any key points outside the response region, but regression-based methods can. One solution to compensate for this limitation in the future is through randomly initializing multiple bounding boxes as in Kazemi and Sullivan [33] and predicting landmark locations using median values.

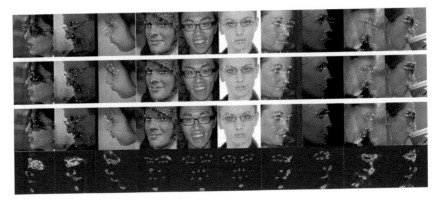

FIGURE 4.6
The CED of deep learning-based methods. vis/all, the error of visible/all landmarks. (L) AFLW-Full: 4386 images, (R) UHDB31: 1617 images.

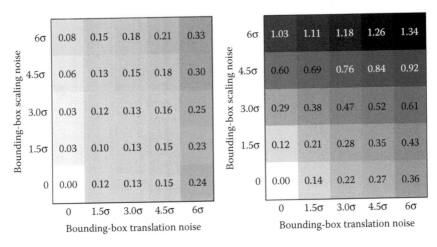

FIGURE 4.7
Qualitative results on AFLW-Full database. (T) Hyperface [14], (M) 3DDFA [7], (B) GoDP(A) with score maps.

4.5   Conclusion

We propose an efficient deep architecture that is able to localize facial key points precisely. The architecture transforms the traditional regression problem into a 2D-detection problem. We designed a new loss function and a unique proposal-refinement technique to deal with the challenges that come up with this new problem and thereby successfully tackle the optimization and precision weakness of deep-cascaded regression. In the future, we will explore how to embed the global-shape constraints into the neural network to regularize the estimation.

Acknowledgment

This material is based on work supported by the U.S. Department of Homeland Security under Grant Award Number 2015-ST-061-BSH001. This grant is awarded to the Borders, Trade, and Immigration Institute: A DHS Center of Excellence led by the University of Houston and includes support for the project "Image and Video Person Identification in an Operational Environment: Phase I" awarded to the University of Houston. The views and conclusions contained in this document are those of the authors and should not be interpreted as necessarily representing the official policies, either expressed or implied, of the U.S. Department of Homeland Security.

Reprinted from Image and Vision Computing, Vol 71, Y. Wu, S. K. Shah, I. A. Kakadiaris, GoDP: Globally Optimized Dual Pathway deep network architecture for facial landmark localization in-the-wild, Copyright 2018, with permission from Elsevier. Note that this book chapter is partially reprinted from the original paper.

References

1. T. Hassner, I. Masi, J. Kimand J. Choi, S. Harel, P. Natarajan, and G. Medioni. Pooling faces: Template based face recognition with pooled face image. In *Proceedings of the IEEE Conference on Computer Vision and Pattern Recognition Workshops*, Las Vegas, NV, June 26–July 1, 2016.

2. I. A. Kakadiaris, G. Toderici, G. Evangelopoulos, G. Passalis, X. Zhao, S. K. Shah, and T. Theoharis. 3D-2D face recognition with pose and illumination normalization. *Computer Vision and Image Understanding*, 154:137–151, 2017.

3. Y. Taigman, M. Yang, M. Ranzato, and L. Wolf. Deepface: Closing the gap to human-level performance in face verification. In *Proceedings of the IEEE Conference on Computer Vision and Pattern Recognition*, Columbus, OH, June 24–27, 2014.

4. A. Jourabloo and X. Liu. Large-pose face alignment via CNN-based dense 3D model fitting. In *Proceedings of the IEEE Conference on Computer Vision and Pattern Recognition*, Las Vegas, NV, June 26–July 1, 2016.

5. G. Tzimiropoulos. Project-out cascaded regression with an application to face alignment. In *Proceedings of the IEEE Conference on Computer Vision and Pattern Recognition*, pp. 3659–3667, Boston, MA, June 7–12, 2015.

6. S. Zhu, C. Li, C. C. Loy, and X. Tang. Unconstrained face alignment via cascaded compositional learning. In *Proceedings of the IEEE Conference on Computer Vision and Pattern Recognition*, Las Vegas, NV, June 26–July 1, 2016.

7. X. Zhu, Z. Lei, X. Liu, H. Shi, and S. Z. Li. Face alignment across large poses: A 3D solution. In *Proceedings of the IEEE Conference on Computer Vision and Pattern Recognition*, Las Vegas, NV, June 26–July 1, 2016.

8. Y. Sun, X. Wang, and X. Tang. Deep convolutional network cascade for facial point detection. In *Proceedings of the Computer Vision and Pattern Recognition*, Portland, OR, June 25–27, 2013.

9. Z. Zhang, P. Luo, C. C. Loy, and X. Tang. Facial landmark detection by deep multi-task learning. In *Proceedings of the European Conference on Computer Vision*, Zurich, Switzerland, September 6–12, 2014.

10. H. Noh, S. Hong, and B. Han. Learning deconvolution network for semantic segmentation. In *Proceedings of the IEEE International Conference on Computer Vision*, Santiago, Chile, December 13–16, 2015.

11. H. Qin, J. Yan, X. Li, and X. Hu. Joint training of cascaded CNN for face detection. In *Proceedings of the Computer Vision and Pattern Recognition*, Las Vegas, NV, June 26–July 1, 2016.

12. X. Peng, R. S. Feris, X. Wang, and D. N. Metaxs. A recurrent encoder-decoder for sequential face alignment. In *Proceedings of the European Conference on Computer Vision*, Amsterdam, the Netherlands, October 11–14, 2016.

13. G. Ghiasi and C. C. Fowlkes. Laplacian pyramid reconstruction and refinement for semantic segmentation. In *Proceedings of the European Conference on Computer Vision*, Amsterdam, the Netherlands, October 11–14, 2016.

14. R. Ranjan, V. M. Patel, and R. Chellappa. Hyperface: A deep multi-task learning framework for face detection, landmark localization, pose estimation, and gender recognition. *arXiv:1603.01249*, 2016.

15. X. P. Burgos-Artizzu, P. Perona, and P. Dollár. Robust face landmark estimation under occlusion. In *Proceedings of the IEEE International Conference on Computer Vision*, Sydney, Australia, December 3–6, 2013.

16. X. Cao, Y. Wei, F. Wen, and J. Sun. Face alignment by explicit shape regression. In *Proceedings of the IEEE Conference on Computer Vision and Pattern Recognition*, Providence, RI, June 16–21, 2012.

17. X. Xiong and F. De la Torre. Supervised descent method and its applications to face alignment. In *Proceedings of the IEEE Conference on Computer Vision and Pattern Recognition*, pp. 532–539, Portland, OR, June 25–27, 2013.

18. X. Yu, J. Huang, S. Zhang, W. Yan, and D. N. Metaxas. Pose-free facial landmark fitting via optimized part mixtures and cascaded deformable shape model. In *Proceedings of the IEEE International Conference on Computer Vision*, Sydney, Australia, December 3–6, 2013.

19. A. Jourabloo and X. Liu. Pose-invariant 3D face alignment. In *Proceedings of the International Conference on Computer Vision*, Santiago, Chile, December 13–16, 2015.

20. S. Ren, X. Cao, Y. Wei, and J. Sun. Face alignment at 3000 FPS via regressing local binary features. In *Proceedings of the IEEE Conference on Computer Vision and Pattern Recognition*, Columbus, OH, June 24–27, 2014.

21. S. Zhu, C. Li, C. C. Loy, and X. Tang. Face alignment by coarse to fine shape searching. In *Proceedings of the IEEE Conference on Computer Vision and Pattern Recognition*, Boston, MA, June 7–12, 2015.

22. M. Kostinger, P. Wohlhart, P. M. Roth, and H. Bischof. Annotated facial landmarks in the wild: A large-scale, real-world database for facial landmark localization. In *Proceedings of the IEEE International Workshop on Benchmarking Facial Image Analysis Technologies*, Barcelona, Spain, November 13, 2011.

23. X. Zhu and D. Ramanan. Face detection, pose estimation, and landmark localization in the wild. In *Proceedings of the IEEE Conference on Computer Vision and Pattern Recognition*, Providence, RI, June 16–21, 2012.

24. A. Bulat and G. Tzimiropoulos. Convolutional aggregation of local evidence for large pose face alignment. In *Proceedings of the British Machine Vision Conference*, York, UK, September 19–22, 2016.

25. S. Wei, V. Ramakriashna, T. Kanade, and Y. Sheikh. Convolutional pose machines. In *Proceedings of the IEEE Conference on Computer Vision and Pattern Recognition*, Las Vegas, NV, June 26–July 1, 2016.

26. A. Newell, K. Yang, and J. Deng. Stacked hourglass networks for human pose estimation. In *Proceedings of the European Conference on Computer Vision*, Amsterdam, the Netherlands, October 11–14, 2016.

27. X. Chu, W. Ouyang, H. Li, and X. Wang. CRF-CNN: Modelling structured information in human pose estimation. In *Proceedings of the Neural Information Processing Systems*, Barcelona, Spain, December 5–10, 2016.

28. W. Yang, W. Ouyang, H. Li, and X. Wang. End-to-end learning of deformable mixture of parts and deep convolutional neural networks for human pose estimation. In *Proceedings of the European Conference on Computer Vision*, Amsterdam, the Netherlands, October 11–14, 2016.

29. Y. Wu, X. Xu, S. K. Shah, and I. A. Kakadiaris. Towards fitting a 3D dense facial model to a 2D image: A landmark-free approach. In *Proceedings of the International Conference on Biometrics: Theory, Applications and Systems*, Arlington, VA, September 8–11, 2015.

30. X. Chu, W. Ouyang, H. Li, and X. Wang. Structured feature learning for pose estimation. In *Proceedings of the IEEE Conference on Computer Vision and Pattern Recognition*, Las Vegas, NV, June 26–July 1, 2016.

31. Y. Bengio, J. Louradour, R. Collobert, and J. Weston. Curriculum learning. In *Proceedings of the International Conference on Machine Learning*, Montreal, Canada, June 14–18, 2009.

32. Y. Wu, S. K. Shah, and I. A. Kakadiaris. Rendering or normalization? An analysis of the 3D-aided pose-invariant face recognition. In *Proceedings of the IEEE International Conference on Identity, Security and Behavior Analysis*, Sendai, Japan, February 29–March 2, 2016.

33. V. Kazemi and J. Sullivan. One millisecond face alignment with an ensemble of regression trees. In *Proceedings of the IEEE Conference on Computer Vision and Pattern Recognition*, Columbus, OH, June 24–27, 2014.

34. K. He, X. Zhang, S. Ren, and J. Sun. Deep residual learning for image recognition. In *Proceedings of the IEEE Conference on Computer Vision and Pattern Recognition*, Las Vegas, NV, June 26–July 1, 2016.

5

Learning Deep Metrics for Person Reidentification

Hailin Shi, Shengcai Liao, Dong Yi, and Stan Z. Li

CONTENTS

5.1   Introduction 109
5.2   Deep Metric Learning 111
    5.2.1   Neural network architecture 111
    5.2.2   Metric and cost function 112
    5.2.3   Performance 114
5.3   Training Against Large Intraclass Variations 116
    5.3.1   Moderate positive mining 116
    5.3.2   Constrained metric embedding 118
    5.3.3   Performance 119
5.4   Summary and Future Directions 122
References 122

5.1   Introduction

Person reidentification aims to match the pedestrian images of the same identity from disjoint camera views. The major challenges come from the large intraclass variations in poses, lightening, occlusion, and camera views in the pedestrian data. The conventional framework of person reidentification generally includes two parts, feature extraction and similarity learning. Feature extraction is to extract hand-crafted features from pedestrian images. Similarity learning is to learn appropriate similarity or distance metrics from pairs of pedestrian samples. These sorts of approaches are usually referred to as *traditional methods*. Some of them focus on the first part, that is, feature computation for pedestrian image [1,2], whereas some others attempt to improve the second part, metric learning for reidentification [3–10] or to improve both parts together [11–14].

In recent years, the training data of person reidentification has increased steadily [15,16], and the computational resources have been greatly improved by GPU implementations [17,18] and distributed computing clusters [19]. As a result of these promotions, an increasing number of deep learning–based approaches are proposed for person reidentification. Some of the deep learning methods [15,20] employ the patch-based matching algorithm that compares the pedestrian pair early in the convolutional neural network (CNN), then summarizes the comparison by using a softmax classifier to output the decision whether the input pair belongs to the same subject. The early comparison in the CNN exploits the spatial correspondence in feature maps. Gate Siamese CNNs [21] perform the middle-level comparison in a similar way. Su et al. [22] trained the network with the assistance of pedestrian attributes. Varior et al. [23] proposed a Siamese Long Short-Term Memory architecture to accomplish the task in a sequential patch manner.

Compared with training by softmax, some other deep learning methods propose to let the network learn discriminative representations of pedestrian via appropriate metrics and cost functions. These methods, generally referred to as *deep metric learning*, unify the two parts (features computing and metric learning) of person reidentification into an integrated framework. This framework is expected to give better generalization ability than the softmax classifier, which is critical to perform the person-reidentification task in which the training and test sets do not share galleries.

The CNN extracts discriminative features from the input images, and the metric component compares the features with the learned metric. Previously, we proposed a Deep Metric Learning (DML) method [24], which first adopted a Siamese network to extract the features of the input pair; then, the features were compared by using the metrics such as the Euclidean distance, the cosine similarity, or the absolute difference, and finally, the cost function (the binomial deviance or Fisher criterion) was computed to train the network by stochastic gradient descent. Recently, we proposed another approach called Embedding Deep Metric (EDM) [25], which implemented the Mahalanobis distance in a single fully connected (FC) layer after CNNs and improved the generalization ability of the metric through the regularization that makes a balance between the Mahalanobis and Euclidean distance. Besides, EDM suggests that selecting suitable positive training samples is important for EDM as well. Considering pedestrian data has large intraclass variations (illumination, pose, occlusion, etc.), the feature is distributed as highly curved manifolds, thus training with the Euclidean distance in the global range is suboptimal. Based on the analysis, Moderate Positive Mining is introduced to choose suitable positive samples for better training against large variations.

In this chapter, the focus is mainly on the two representative methods, DML (Section 5.2) and EDM (Section 5.3), of which the major contributions are EDM and Moderate Positive Mining for learning deep neural networks for person reidentification.

5.2   Deep Metric Learning

In this section, we introduce the method of DML (i.e., the deep metric learning for practical person reidentification [24]). This work employs a Siamese network to extract the features from a pair of input pedestrian images and train the network with various choices of metrics and cost functions on the top of the network. The network jointly learns the image feature and discriminative metric in an unified framework. Specifically, the combination of cosine similarity and binomial deviance performs the best and shows robustness to large variations. For the purpose of practical person reidentification, the network is trained in one data set, but tested across different other data sets.

5.2.1   Neural network architecture

The network has a symmetric architecture, referred to as Siamese network, which is shown in Figure 5.1. The two subnetworks consist of convolutional layers with shared weights. Therefore, the whole network is in a light-weight fashion and able to handle well the general task of cross-dataset person reidentification. The detail structure of the subnetwork is shown in Figure 5.2. It contains three branches of a convolutional network. Each of the branches, which are in charge of a fixed patch of image from top to bottom, respectively, is composed by two convolutional layers and two max-pooling layers. The number of filters is 64 for both convolutional layers. The filter sizes are 7×7 and 5×5, respectively. The nonlinearity is introduced by the rectified linear unit [17] function. Every max-pooling layer also processes a cross-channel

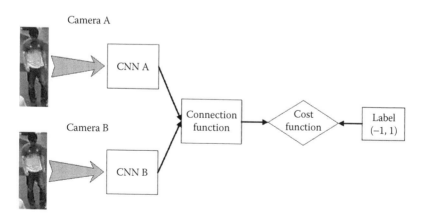

FIGURE 5.1
The framework of DML Siamese network, including CNNs, connection function, and cost function.

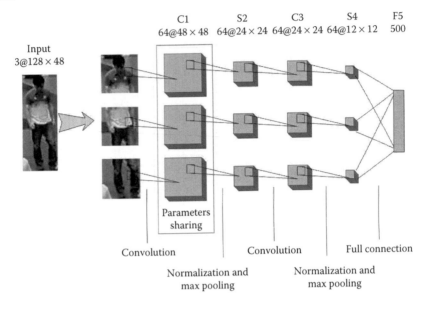

FIGURE 5.2
The details of the subnetwork in DML network.

normalization operation so the model can be more steady numerically. The weights of the first convolutional layer are shared between the branches, whereas the second convolutional layer is build in an unshared way. This is because the first convolutional layer extracts the common low-level image features directly from the inputs, and the following layers are expected to learn high-level patch-specific features. At the end, an FC layer is disposed to receives the three branches and outputs a 500-dimension feature vector.

The two CNN subnetworks are connected by a connection function (i.e., metric). Given a pair of pedestrian images, the Siamese network extracts the features from the images, and the similarity is computed by the connection metric. In the training stage, the network predicts a label $l = \pm 1$ denoting whether the image pair belongs to the same subject, compares the prediction with the ground-truth labeling, and computes and back-propagates the gradient to update the network weights; in the test stage, however, the network directly outputs the similarity score for the ranking evaluation.

5.2.2   Metric and cost function

There are various choices to be used as the metric and the cost function. DML selects three of them as the candidates for metric: the Euclidean distance, cosine similarity, and absolute difference. The definition of the metrics are given as:

$$S_{euc}(x,y) = -\|x - y\|_2,$$ (5.1)

$$S_{cos}(x,y) = \frac{x^T y}{\|x\|\|y\|},$$ (5.2)

$$S_{abs}(x,y) = -\|x - y\|_1,$$ (5.3)

where x and y denotes the input pair. The distance metrics are converted to similarity by the negation. Among these metrics, the Euclidean distance has a convenience of simple derivation, but the unbounded gradient would lead to blowing up of training; the cosine similarity has a bounded support and the invariance to the scaling of features; and the absolute distance is not differentiable in some situations. According to the good property of cosine similarity, DML adopts the cosine similarity to perform the connection metric.

As for the loss function, DML adopts two candidates in the experiments: the binomial deviance and the Fisher criterion. The formulations are given as:

$$L_{dev} = \sum_{i,j} W \odot ln(e^{-\alpha(S-\beta)\odot M} + 1),$$ (5.4)

$$L_{fisher} = -\frac{(\sum_{i,j} P \odot S)^2}{\sum_{i,j}(S - \bar{S})^2},$$ (5.5)

where $\odot$ denotes the element-wise matrix product, and

$$S = [S_{ij}]_{n\times n}, \qquad S_{ij} = S(x_i, x_j),$$ (5.6)

$$M = [M_{ij}]_{n\times n}, \quad M_{ij} = \begin{cases} 1, & \text{positive pair} \\ -1, & \text{negative pair} \\ 0, & \text{neglected pair}, \end{cases}$$ (5.7)

$$W = [W_{ij}]_{n\times n}, \quad W_{ij} = \begin{cases} 1/n_p, & \text{positive pair} \\ 1/n_n, & \text{negative pair} \\ 0, & \text{neglected pair}, \end{cases}$$ (5.8)

$$P = [P_{ij}]_{n\times n}, \quad P_{ij} = \begin{cases} 1/n_p, & \text{positive pair} \\ -1/n_n, & \text{negative pair} \\ 0, & \text{neglected pair}, \end{cases}$$ (5.9)

where:

S_{ij} is the similarity of sample x_i and x_j
$\bar{S}$ is the mean of S
M_{ij} indicates whether x_i and x_j belong to the same identity
n_p and n_n is the number of positive and negative pairs, respectively
α and β are the hyperparameters of binomial deviance

Through minimizing the binomial deviance (Equation 5.4) and Fisher criterion (Equation 5.5), the network learns to reduce the distances of positive pairs while englarging the distances of negative pairs.

It is worth noting that the binomial deviance cost gives more attention to the samples near the decision boundary or misclassified, whereas the Fisher criterion is focused on all the entries of the similarity matrix S equally.

5.2.3   Performance

By calculating the gradient of cost function with respect to the sample and weight, the network is updated via stochastic gradient descent until the training converges. The preliminary experiment is conducted on VIPeR [26] to compare the binomial deviance and the Fisher criterion. Then, the main evaluation is accomplished in both intra- and cross-dataset manners with the data sets VIPeR, PRID [27], i-LIDS [28], and CUHK Campus [5].

VIPeR is a challenging and widely used data set for person reidentification. It includes 632 subjects, with two pedestrian images per subject captured by two different cameras in an outdoor environment. There are large variations in poses, illuminations, and camera angles among these images. The PRID data set is similar to VIPeR and contains the pedestrian images from two disjoint cameras, which include 385 and 749 subjects, respectively. There are 200 overlapping subjects between them. The experiment randomly selects 100 of them for training and the remaining for test. For the VIPeR data set, the 632 subjects are divided into two subsets with equal number for either training or test. The i-LIDS data set contains 119 subjects with a total 476 of images from multiple disjoint cameras. There are four images for each subject on average. Most of the images show large variations of illumination and resolution. The CUHK Campus data set is relatively large in scale, including 1816 subjects and 7264 pedestrian images in total. Each subject has four images, captured with two camera views in a campus environment. Camera A captures the frontal view or back view of pedestrians, whereas camera B captures the side views. The image resolution varies among these data sets. For example, CUHK Campus consists of high-resolution pedestrian images, whereas the others are in relatively low resolution. All the images are normalized to 128×48 RGB for the training and test.

The preliminary experiment results are listed in Table 5.1. At all ranks, the binomial deviance gives better accuracy than the Fisher criterion.

TABLE 5.1

Comparison of recognition rate of binomial deviance and Fisher criterion on VIPeR under Dev. view

Rank	1 (%)	5 (%)	10 (%)	15 (%)	20 (%)	25 (%)	30 (%)	50 (%)
Binomial deviance	34.49	60.13	74.37	80.70	84.18	88.61	91.14	96.84
Fisher Criterion	14.24	35.13	47.15	56.96	62.66	67.41	71.84	80.06

TABLE 5.2

Performance comparison on VIPeR

Rank	1 (%)	5 (%)	10 (%)	15 (%)	20 (%)	25 (%)	30 (%)	50 (%)
LAFT [5]	29.6	–	69.3	–	–	88.7	–	96.8
Salience [8]	30.2	52.3	–	–	–	–	–	–
DML	34.4	62.2	75.9	82.6	87.2	89.7	92.3	96.5

TABLE 5.3

Performance comparison on PRID

Rank	1 (%)	5 (%)	10 (%)	15 (%)	20 (%)	25 (%)	30 (%)	50 (%)
Descr. Model [27]	4	–	24	–	37	–	–	56
RPML [29]	15	–	42	–	54	–	–	70
DML	17.9	37.5	45.9	50.7	55.4	59.3	63.1	71.4

As mentioned in the previous section, the superiority of binomial deviance can be attributed to the focus on the hard samples near the decision boundary.

According to the good performance of binomial deviance, the final DML model adopts the combination of cosine similarity and binomial deviance for training. The intradata set evaluation is conducted on VIPeR and PRID. The performance of DML is listed in Tables 5.2 and 5.3. DML improves the previous rank-1 identification rate by 4% and 3% on VIPeR and PRID, respectively. One can refer to [24] for the details of performance comparison with the state-of-the-art methods.

In the cross-dataset evaluation, DML is trained on i-LIDS, CUHK Campus, or the fusion of both, and tested on VIPeR and PRID. For both tests on VIPeR and PRID, the best results of DML are obtained through the training on the fusion of the two data sets, i-LIDS and CUHK Campus (Table 5.4), outperforming the previous methods. One can refer to the experiment part of DML [24] for the comparison details.

TABLE 5.4

Cross-dataset evaluation of DML on VIPeR and PRID

Training set	Test set	1 (%)	10 (%)	20 (%)	30 (%)
i-LIDS	VIPeR	11.61	34.43	44.08	52.69
CUHK	VIPeR	16.27	46.27	59.94	70.13
i-LIDS + CUHK	VIPeR	17.72	48.80	63.35	72.85
i-LIDS	PRID	8.0	25.5	38.9	45.6
CUHK	PRID	7.6	23.4	30.9	36.1
i-LIDS + CUHK	PRID	13.8	35.4	45.0	51.3

5.3 Training Against Large Intraclass Variations

In this section, we present the work of EDM for person reidentification [25]. Similar to DML, EDM also employs the framework of a Siamese network and further implements the Mahalanobis distance into a single FC layer with a special regularization. Besides, EDM proposes the Moderate Positive Mining strategy to train better networks against large intraclass variations in the pedestrian data.

5.3.1 Moderate positive mining

Compared with face recognition, person reidentification is a much more difficult task because of the large intraclass variations that are specific in pedestrian data. These variations mainly come from the factors of illumination, occlusion, pose, background, misalignment co-occurrence of people, appearance changing, and so on. Figure 5.3a shows some examples of hard

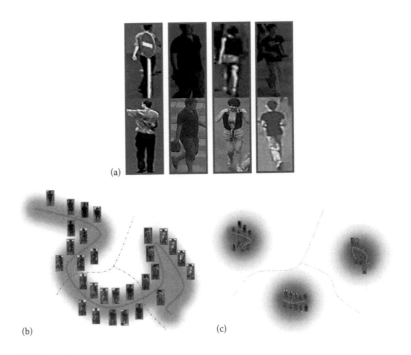

FIGURE 5.3
(a) Some hard positive cases of reidentification. They are influenced by the factors of occlusion, illumination, pose, etc. (b) Illustration of the highly curved manifold of three identities. (c) The purpose of Moderate Positive Mining is to reduce the intraclass variance while preserving the intrinsic graphical structure.

positives, which are even difficult for humans to distinguish. As a result, the pedestrian data are distributed as highly curved manifolds when they are mapped into feature space (Figure 5.3b). Therefore, using the Euclidean distance in the global range is inappropriate for both training and testing CNNs. However, the geodesic distance is not available because the distribution is unknown. In light of manifold learning methods [30–32], EDM proposes to use the Euclidean distance in the local range and the graphical relationship to approximate the geodesic distance for training CNNs. The goal is to reduce the intraclass variance along the manifold for the supervised learning, while preserving the intrinsic graphical structure (Figure 5.3c).

To accomplish the geodesic distance approximation by local Euclidean, the definition of *local* becomes critical. Based on the batch-training framework of deep CNNs, EDM proposes to select moderate positive samples in an adaptive way throughout the training process. This training sample-selection strategy, namely Moderate Positive Mining, is performed to provide suitable positive training pairs to the CNN learning.

Specifically, given two image sets, $\mathcal{I}_1$ and $\mathcal{I}_2$, captured by two disjoint cameras, we denote the positive pair (same identity) as $\{\mathbf{I}_1, \mathbf{I}_2^p | \mathbf{I}_1 \in \mathcal{I}_1, \mathbf{I}_2^p \in \mathcal{I}_2\}$, and the negative pair (different identities) as $\{\mathbf{I}_1, \mathbf{I}_2^n | \mathbf{I}_1 \in \mathcal{I}_1, \mathbf{I}_2^n \in \mathcal{I}_2\}$, and the CNN as $\mathbf{\Psi}(\cdot)$. $d(\cdot, \cdot)$ is the Mahalanobis or Euclidean distance. The Moderate Positive Mining is performed as follows.

Algorithm 5.1: Moderate positive mining

Input: Randomly select an anchor sample $\mathbf{I}_1$, its positive samples $\{\mathbf{I}_2^{p_1}, \ldots, \mathbf{I}_2^{p_k}\}$, and negative samples $\{\mathbf{I}_2^{n_1}, \ldots, \mathbf{I}_2^{n_k}\}$ to form a mini-batch.

Step 1 Input the images into the network for obtaining the features, and compute their distances $\{d(\mathbf{\Psi}(\mathbf{I}_1), \mathbf{\Psi}(\mathbf{I}_2^{p_1})), \ldots, d(\mathbf{\Psi}(\mathbf{I}_1), \mathbf{\Psi}(\mathbf{I}_2^{p_k}))\}$ and $\{d(\mathbf{\Psi}(\mathbf{I}_1), \mathbf{\Psi}(\mathbf{I}_2^{n_1})), \ldots, d(\mathbf{\Psi}(\mathbf{I}_1), \mathbf{\Psi}(\mathbf{I}_2^{n_k}))\}$;

Step 2 Mine the hardest negative sample

$$\hat{\mathbf{I}}_2^n = argmin_{j=1\ldots k}\{d(\mathbf{\Psi}(\mathbf{I}_1), \mathbf{\Psi}(\mathbf{I}_2^{n_j}))\};$$

Step 3 From the positive samples, choose those $\tilde{\mathbf{I}}_2^{p_m}$ satisfying

$$d(\mathbf{\Psi}(\mathbf{I}_1), \mathbf{\Psi}(\tilde{\mathbf{I}}_2^{p_m})) \leq d(\mathbf{\Psi}(\mathbf{I}_1), \mathbf{\Psi}(\hat{\mathbf{I}}_2^n));$$

Step 4 Mine the hardest one among these chosen positives as our moderate positive sample

$$\hat{\mathbf{I}}_2^p = argmax_{\tilde{\mathbf{I}}_2^{p_m}}\{d(\mathbf{\Psi}(\mathbf{I}_1), \mathbf{\Psi}(\tilde{\mathbf{I}}_2^{p_m}))\}.$$

If none of the positives satisfies the condition in **Step 3**, choose the positive with the smallest distance as the moderate positive sample.

Output: The moderate positive sample $\hat{\mathbf{I}}_2^p$.

First, an anchor sample and its positive samples and negative samples (equally sized) are randomly selected from the data set to form a mini-batch;

then, the algorithm mines the hardest negative sample, and chooses the positive samples that have smaller distances than the hardest negative; finally, it mines the hardest one among these chosen positives as the moderate positive sample. Obviously, the *moderate positive* is defined adaptively within each subject, whereas their hard negatives are also involved in case the positives are too easy or too hard to be mined. Once the moderate positive samples are found, they are provided along with the negatives to the CNN for learning.

5.3.2 Constrained metric embedding

EDM adopts the Mahalanobis distance as the connection metric to evaluate the similarity of input pairs. Instead of computing the distance explicitly, EDM implements the Mahalanobis distance into a single FC layer. Therefore, the metric matrix can be learned jointly with the CNN weights. Besides, the weight learning of the FC layer is constrained by a special regularization for better generalization ability. This unit, defined as the metric-learning layers of EDM, is shown in Figure 5.4.

By decomposing the metric matrix with its positive semi-definite property, the Mahalanobis distance is developed as:

$$d(\mathbf{x}_1, \mathbf{x}_2) = \sqrt{(\mathbf{x}_1 - \mathbf{x}_2)^T \mathbf{M} (\mathbf{x}_1 - \mathbf{x}_2)} \tag{5.10}$$

$$= \sqrt{(\mathbf{x}_1 - \mathbf{x}_2)^T \mathbf{W} \mathbf{W}^T (\mathbf{x}_1 - \mathbf{x}_2)},$$

$$= \|\mathbf{W}^T (\mathbf{x}_1 - \mathbf{x}_2)\|_2. \tag{5.11}$$

The matrix $\mathbf{W}$ corresponds to the weight of the FC layer. Furthermore, the metric-learning layer is improved by a constraint that pushes the matrix

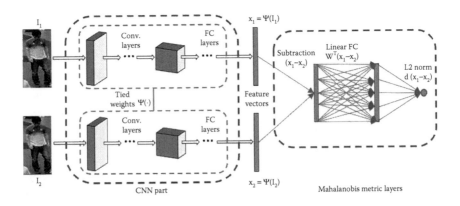

FIGURE 5.4
The overview of EDM framework. At the right part, the metric-learning layers compute the distance of two samples. $\mathbf{x}_1$ and $\mathbf{x}_2$ are the feature vectors extracted by the CNN from the images.

$\mathbf{W}\mathbf{W}^T$ close to the identity matrix $\mathbf{I}$. Then, the network is trained via a combination of contrastive cost (L) and the regularization term, formed as:

$$\hat{L} = L + \frac{\lambda}{2}\|\mathbf{W}\mathbf{W}^T - \mathbf{I}\|_F^2. \tag{5.12}$$

When the value of λ becomes very large, the metric falls to the naive Euclidean distance, which is less discriminative than the constraint-free Mahalanobis distance, but may be more robust to unseen test data. The purpose of the constraint is to find an optimal balance between the Euclidean and Mahalanobis distance.

5.3.3 Performance

The validation set of CUHK03 [15] is used to confirm the effectiveness of Moderate Positive Mining and the constrained deep metric embedding. Then, the performance of EDM is reported with the tests on CUHK03, CUHK01 [33], and VIPeR.

The CUHK03 data set contains 1369 subjects, each of which has around 10 images. The images are captured with six surveillance cameras over months, with each person observed by two disjoint camera views and having an average of 4.8 images in each view. The common protocol randomly selects 1169 subjects for training, 100 for validation, and 100 for test with single-shot setting. The CUHK01 data set contains 971 subjects, with four images per subject from two disjoint cameras. It is divided into a training set of 871 subjects and a test set of 100. These images are captured in the similar environment, but with nonoverlapping identities of those in CUHK03. As mentioned in the previous section, VIPeR is a challenging data set for deep learning methods because of its small size and large variations in background, illumination, and viewpoint. Nonetheless, VIPeR is used to evaluate the robustness of EDM. All these pedestrian images are normalized into 128×64 RGB for the training and test.

The effectiveness of Moderate Positive Mining is proved on the validation set. The CNN is deployed and trained in a similar way with the DML method. The network is first pretrained with a softmax classifier as the baseline model, of which the outputs correspond to the training subjects. Then, the softmax is discarded, and the network is further trained with the metric-learning layer. The performance is shown in Figure 5.5a, in which the comparison includes the different combinations of Moderate Positive Mining and Hard Negative Mining [34]. The collaboration of two mining strategies achieves the best result (light gray-colored line), whereas the absence of Moderate Positive Mining leads to a significant derogation of performance (dark-gray). This reflects that the manifold is badly learned if all the positives are used undiscriminatingly. If no mining strategy is used (black), the network gives a very low identification rate at low ranks, even worse than the softmax baseline (black). This indicates that Moderate Positive Mining and Hard Negative Mining are both crucial for training. From Figure 5.5a we can also see that the performances of the three

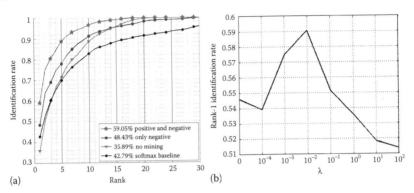

(a) (b)

FIGURE 5.5

(a) Accuracies under different mining settings on CUHK03 validation. (b) The rank-1 identification rates with different λ of the weight constraint.

metric-embedded networks are much better than the softmax-based baseline network. Especially, the identification rates of the former three are near 100% after rank 20, whereas the baseline network remains at a lower identification rate than the other three. This indicates that the training with the metric layers is the basic contributor of the improvement.

Also on the CUHK03 validation, the varying value of λ leads to different rank-1 accuracies (Figure 5.5b). If λ is too small, the metric is free of constraint, leading to low generalization ability, whereas when λ is too large, the metric is suffered by under-fitting. The best result is achieved at $\lambda = 10^{-2}$.

As reported in [25], because of the improvements from Moderate Positive Mining and the constrained metric embedding, the network achieves competitive results with the state-of-the-art methods (including IDLA [20], FPNN [15], LOMO-XQDA [12], KISSME [4], DeepFeature [35], Siamese LSTM [23], Gated S-CNN [21], and SSDAL+XQDA [22]) on the test sets of CUHK03 and VIPeR (Tables 5.5 and 5.7). With much less weights, the EDM network improves the previous best rank-1 identification accuracy by 4% on CUHK01 (Table 5.6). Figure 5.6 shows the CMC curves and rank-1 accuracies of EDM and partial competitors on the three data sets.

TABLE 5.5

Performance comparison on CUHK03

Rank	1 (%)	5 (%)	10 (%)	15 (%)	20 (%)
KISSME [4]	14.17	37.47	52.20	62.07	69.38
FPNN [15]	20.65	50.94	67.01	76.19	83.00
LOMO-XQDA [12]	52.20	82.23	92.14	94.74	96.25
IDLA [20]	54.74	86.50	93.88	97.68	98.10
Siamese LSTM [23]	57.3	80.1	88.3	–	–
Gated S-CNN [21]	68.1	88.1	94.6	–	–
EDM	61.32	88.90	96.44	99.04	99.94

TABLE 5.6

Performance comparison on CUHK01

Rank	1 (%)	5 (%)	10 (%)
FPNN [15]	27.87	59.64	73.53
KISSME [4]	29.40	60.18	74.44
IDLA [20]	65	89	94
EDM	69.38	91.03	96.84

TABLE 5.7

Performance comparison on VIPeR

Rank	1 (%)	5 (%)	10 (%)	15 (%)
IDLA [20]	34.81	63.61	75.63	80.38
DeepFeature [35]	40.47	60.82	70.38	78.32
mFilter+LADF [6]	43.39	73.04	84.87	90.85
SSDAL+XQDA [22]	43.5	71.8	81.5	—
Siamese LSTM [23]	42.4	68.7	79.4	—
Gated S-CNN [21]	37.8	66.9	77.4	—
EDM	40.91	67.41	79.11	86.08

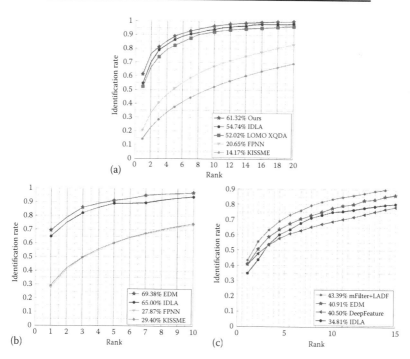

FIGURE 5.6

The CMC curves and rank-1 accuracies on (a) CUHK03, (b) CUHK01, and (c) VIPeR.

5.4   Summary and Future Directions

Given the CNN framework for visual tasks, the deep metric-embedding methods encounter three major issues: training sample selection, metric design, and cost function. The selection of appropriate metric and cost function is critical for training the network effectively. Besides, because the large variations exist in the intrapersonal data, selecting a suitable positive training sample is also critical. How to deal with these issues is the central topic of DML for person reidentification because the pedestrian data are suffered by large intraclass variations, which lead to greater difficulty in comparison with face recognition.

In this chapter, we introduced two recent methods based on DML for person reidentification. DML and EDM are dedicated to handle the issues via the proposed network and training strategy. However, there is still room for improvement on the benchmark performance of both intra- and cross-dataset evaluation. In light of the suggestions from DML and EDM, there are certain research challenges to be addressed based on them. For example, the robustness to pose could be improved by involving body geometry information and pose normalization; the Fisher criterion in DML needs to be modified to adapt the heteroscedastic distributions; and although the idea of Moderate Positive Mining is intuitive, it is important to formulate the correspondence between the sample selection and the target performance. Besides, more attempts are expected for the improvement of loss function. For example, DeepFeature [35] employed triplet loss for training a deep CNN for person reidentification in their work. Another issue for DML is the limited amount of pedestrian training data. DTML [36] accomplished the deep metric learning via transfer learning technique to cope with the small data set size problem. These open issues may lead to the major directions in developing deep metrics for person reidentification in the future.

References

1. Y. Yang, J. Yang, J. Yan, S. Liao, D. Yi, and S. Z. Li. Salient color names for person re-identification. In *Computer Vision–ECCV 2014*, pp. 536–551. Springer, Zurich, Switzerland, 2014.

2. R. Zhao, W. Ouyang, and X. Wang. Learning mid-level filters for person re-identification. In *Computer Vision and Pattern Recognition (CVPR), 2014 IEEE Conference on*, pp. 144–151. IEEE, 2014.

3. S. Paisitkriangkrai, C. Shen, and A. van den Hengel. Learning to rank in person re-identification with metric ensembles. In *Proceedings of the IEEE Conference on Computer Vision and Pattern Recognition*, pp. 1846–1855, 2015.

4. M. Koestinger, M. Hirzer, P. Wohlhart, P. M. Roth, and H. Bischof. Large scale metric learning from equivalence constraints. In *Computer Vision and Pattern Recognition (CVPR), 2012 IEEE Conference on*, pp. 2288–2295. IEEE, 2012.

5. W. Li and X. Wang. Locally aligned feature transforms across views. In *Proceedings of the IEEE Conference on Computer Vision and Pattern Recognition*, pp. 3594–3601, 2013.

6. Z. Li, S. Chang, F. Liang, T. S. Huang, L. Cao, and J. R. Smith. Learning locally-adaptive decision functions for person verification. In *Computer Vision and Pattern Recognition (CVPR), 2013 IEEE Conference on*, pp. 3610–3617. IEEE, 2013.

7. N. Martinel, C. Micheloni, and G. L. Foresti. Saliency weighted features for person re-identification. In *Computer Vision-ECCV 2014 Workshops*, pp. 191–208. Springer, Zurich, Switzerland, 2014.

8. R. Zhao, W. Ouyang, and X. Wang. Person re-identification by salience matching. In *Computer Vision (ICCV), 2013 IEEE International Conference on*, pp. 2528–2535. IEEE, 2013.

9. Z. Zhang and V. Saligrama. Prism: Person re-identification via structured matching. In *IEEE Transactions on Circuits and Systems for Video Technology*, 2016.

10. Z. Zhang, Y. Chen, and V. Saligrama. Group membership prediction. In *Proceedings of the IEEE International Conference on Computer Vision*, pp. 3916–3924, 2015.

11. S. Khamis, C.-H. Kuo, V. K. Singh, V. D. Shet, and L. S. Davis. Joint learning for attribute-consistent person re-identification. In *Computer Vision-ECCV 2014 Workshops*, pp. 134–146. Springer, Zurich, Switzerland, 2014.

12. S. Liao, Y. Hu, X. Zhu, and S. Z. Li. Person re-identification by local maximal occurrence representation and metric learning. In *Proceedings of the IEEE Conference on Computer Vision and Pattern Recognition*, pp. 2197–2206, 2015.

13. F. Xiong, M. Gou, O. Camps, and M. Sznaier. Person re-identification using kernel-based metric learning methods. In *Computer Vision–ECCV 2014*, pp. 1–16. Springer, Zurich, Switzerland, 2014.

14. Z. Zhang, Y. Chen, and V. Saligrama. A novel visual word co-occurrence model for person re-identification. In *Computer Vision-ECCV 2014 Workshops*, pp. 122–133. Springer, Zurich, Switzerland, 2014.

15. W. Li, R. Zhao, T. Xiao, and X. Wang. Deepreid: Deep filter pairing neural network for person re-identification. In *Proceedings of the IEEE Conference on Computer Vision and Pattern Recognition*, pp. 152–159, 2014.

16. L. Zheng, L. Shen, L. Tian, S. Wang, J. Wang, and Q. Tian. Scalable person re-identification: A benchmark. In *Computer Vision, IEEE International Conference on*, 2015.

17. A. Krizhevsky, I. Sutskever, and G. E. Hinton. Imagenet classification with deep convolutional neural networks. In *Advances in Neural Information Processing Systems*, pp. 1097–1105, South Lake Tahoe, Nevada, 2012.

18. Y. Jia, E. Shelhamer, J. Donahue, S. Karayev, J. Long, R. Girshick, S. Guadarrama, and T. Darrell. Caffe: Convolutional architecture for fast feature embedding. In *Proceedings of the 22nd ACM International Conference on Multimedia*, pp. 675–678. ACM, Orlando, Florida, 2014.

19. J. Dean, G. Corrado, R. Monga, K. Chen, M. Devin, M. Mao, A. Senior et al. Large scale distributed deep networks. In *Advances in Neural Information Processing Systems*, pp. 1223–1231, 2012.

20. E. Ahmed, M. Jones, and T. K. Marks. An improved deep learning architecture for person re-identification. In *Proceedings of the IEEE Conference on Computer Vision and Pattern Recognition*, pp. 3908–3916, 2015.

21. R. R. Varior, M. Haloi, and G. Wang. Gated siamese convolutional neural network architecture for human re-identification. In *European Conference on Computer Vision*, pp. 791–808. Springer, Amsterdam, Netherlands, 2016.

22. C. Su, S. Zhang, J. Xing, W. Gao, and Q. Tian. Deep attributes driven multi-camera person re-identification. In *European Conference on Computer Vision*, pp. 475–491. Springer, Amsterdam, Netherlands, 2016.

23. R. R. Varior, B. Shuai, J. Lu, D. Xu, and G. Wang. A siamese long short-term memory architecture for human re-identification. In *European Conference on Computer Vision*, pp. 135–153. Springer, Amsterdam, Netherlands, 2016.

24. D. Yi, Z. Lei, and S. Z. Li. Deep metric learning for practical person re-identification. CoRR, abs/1407.4979, 2014.

25. H. Shi, Y. Yang, X. Zhu, S. Liao, Z. Lei, W. Zheng, and S. Z. Li. Embedding deep metric for person re-identification: A study against large variations. In *European Conference on Computer Vision*, pp. 732–748. Springer, Amsterdam, Netherlands, 2016.

26. D. Gray, S. Brennan, and H. Tao. Evaluating appearance models for recognition, reacquisition, and tracking. In *Proceedings of the IEEE International Workshop on Performance Evaluation for Tracking and Surveillance (PETS)*, Vol. 3. Citeseer, 2007.

27. M. Hirzer, C. Beleznai, P. M. Roth, and H. Bischof. Person re-identification by descriptive and discriminative classification. In *Scandinavian conference on Image analysis*, pp. 91–102. Springer, Ystad Saltsjöbad, Sweden, 2011.

28. W.-S. Zheng, S. Gong, and T. Xiang. Person re-identification by probabilistic relative distance comparison. In *Computer vision and pattern recognition (CVPR), 2011 IEEE conference on*, pp. 649–656. IEEE, 2011.

29. M. Hirzer, P. M. Roth, M. Köstinger, and H. Bischof. Relaxed pairwise learned metric for person re-identification. In *European Conference on Computer Vision*, pp. 780–793. Springer, Firenze, Italy, 2012.

30. J. B. Tenenbaum, V. De Silva, and J. C. Langford. A global geometric framework for nonlinear dimensionality reduction. *Science*, 290(5500):2319–2323, 2000.

31. S. T. Roweis and L. K. Saul. Nonlinear dimensionality reduction by locally linear embedding. *Science*, 290(5500):2323–2326, 2000.

32. M. Belkin and P. Niyogi. Laplacian eigenmaps for dimensionality reduction and data representation. *Neural computation*, 15(6):1373–1396, 2003.

33. W. Li, R. Zhao, and X. Wang. Human reidentification with transferred metric learning. In *ACCV (1)*, pp. 31–44, Daejeon, Korea, 2012.

34. F. Schroff, D. Kalenichenko, and J. Philbin. Facenet: A unified embedding for face recognition and clustering. In *Proceedings of the IEEE Conference on Computer Vision and Pattern Recognition*, 2015.

35. S. Ding, L. Lin, G. Wang, and H. Chao. Deep feature learning with relative distance comparison for person re-identification.*Pattern Recognition*, 48(10):2993–3003, 2015.

36. J. Hu, J. Lu, and Y.-P. Tan. Deep transfer metric learning. In *Proceedings of the IEEE Conference on Computer Vision and Pattern Recognition*, pp. 325–333, 2015.

6

Deep Face-Representation Learning for Kinship Verification

Naman Kohli, Daksha Yadav, Mayank Vatsa, Richa Singh, and Afzel Noore

CONTENTS

6.1   Introduction 128
    6.1.1   Research problem 133
6.2   Deep Learning Algorithms for Kinship Verification Using
    Representation Learning Framework 133
    6.2.1   Stacked denoising autoencoder 134
    6.2.2   Deep belief networks 135
    6.2.3   Kinship verification via representation learning
        framework 135
        6.2.3.1   Kinship verification via supervised training of
                extracted features/kinship classification 137
6.3   Experimental Evaluation 138
    6.3.1   Kinship data sets 138
    6.3.2   Architectural details of the deep learning KVRL
        framework 139
    6.3.3   Experimental protocol 140
    6.3.4   Performance results on five existing kinship
        databases 140
        6.3.4.1   Cornell kinship data set 142
        6.3.4.2   KinFaceW-I and KinFaceW-II databases 142
        6.3.4.3   UB KinFace database 143
        6.3.4.4   WVU kinship database 143
6.4   Self-Kinship Problem (Age-Invariant Face Verification) 145
    6.4.1   Self-kinship experimental results 146
6.5   Summary and Future Research 147
References 147

6.1   Introduction

Kinship refers to the sharing of selected genetic characteristics and features between members of a family. Kinship and its characteristics have been widely studied in diverse scientific disciplines such as anthropology, psychology, neuroscience, and computer vision (as displayed in Figure 6.1). The study of kinship is valuable from various perspectives. Kinship was the central idea of anthropology research for more than a century [1]. The research on kinship systems and their anthropological aspect was pioneered by Morgan [2] and has gathered significant attention over the years. Moreover, researchers have evaluated the effect of nature versus nurture [3] and have observed social bonds arising in specific cultural context. The entire area of *new genetics* has brought kinship to the forefront because of the prevailing understanding of disease etiology [4].

Looking at kinship from a social perspective, organisms tend to form groups based on kinship relations. It was suggested that kin-group characteristics affect the cooperation among the kin through residence and mating patterns [5]. There have been several studies that aim to unravel the process of kinship recognition among organisms. This ability is observed across all primates in terms of living in groups, forming bonds with offspring, and recognition of kin [6]. Taylor and Sussman [7] studied kinship affinities among Lemur catta and established that several factors such as proximity, grooming, and mating patterns are dependent on kinship among the lemurs.

The cognitive process of humans to identify kinship relations and exploration of forming kin-based bonds has been a popular topic of interest in neuroscience. It has been established that humans recognize and rely on kinship affinities that trigger cognitive and social interactions, and family support networks, including an understanding of common genetic diseases for possible medical interventions. Likewise, the bonds between children and mother have been an area of huge interest in the field of psychology and neuroscience [8–10].

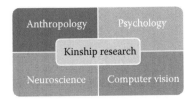

FIGURE 6.1
Kinship has been explored in various diverse research areas such as anthropology, psychology, neuroscience, and computer vision.

Kinship determination in humans is performed by two techniques: DNA testing and verification via facial cues. DNA testing is the traditional way to determine the similarity of genetic material between two individuals. The result of this technique is highly accurate and is age invariant. It is also used for locating genetic relatives and gathering family genealogical details. However, this reliable process is intrusive and may require several hours for processing. On the other hand, relatively new computational approaches have been proposed to verify kinship using facial images, leading to enhanced interest in this field. Faces are valuable visual stimuli, and it has been observed that humans have excellent face-processing and cognition skills [11]. Faces can convey ethnicity, gender, age, and emotions of an individual. Therefore, faces are also being used to determine kin relationship between individuals. In this chapter, kinship refers to at most one-generation direct blood descendants. It has been demonstrated that these descendants share similarities in face structure as well. However, kinship information is also present in second-level relationships such as grandmother–grandchild. These similarities have been called *familial traits* by Hogben [12]. Some examples of kin relationship are shown in Figure 6.2.

The hypothesis that similarity among human faces can be a cue for kinship was first formulated by Daly and Wilson [13]. Since then, facial similarity/resemblance has been used to determine kinship in various research experiments [14–18]. In these experiments, participants were presented with the face images and were asked to judge if a kin relationship existed. Maloney and Martello [19] examined the relation between *similarity* and *kinship detection* among siblings and concluded that observers look for similarity in judging kinship among children. Martello and Maloney [20] concluded that the upper portion of the face has more discriminating power as compared to the lower half in kinship recognition. In a different study, to determine the effect of lateralization on allocentric kin recognition, they concluded that the right half-portion

Brother–Brother   Brother–Sister   Sister–Sister   Mother–Daughter

Father–Daughter   Father–Son   Mother–Son

FIGURE 6.2
Common examples of lineal kin relationships.

of the face is equal to the left half-portion of the face [21]. Kaminski et al. [22] demonstrated that humans can identify kin generations apart with a varying success ratio. They also deduced that children resemble their mothers more as compared to their fathers, a result which is also presented in [23] and [24].

Various neurological studies have been conducted to examine the reasons behind the ability of humans to detect genetic relatedness. Lieberman et al. [25] proposed the existence of Kinship Model Index (KI_i), which allows humans to detect kin. They determined two cues: (1) duration of coresidence and (2) Maternal Perinatal Association (MPA) to be used by humans to compute KI_i. Kaminski et al. [26] established that later-born children are better at kin recognition as compared to first-born, which further substantiated the effect of MPA given by [25]. Platek and Kemp [27] performed a study using functional magnetic resonance imaging to investigate the differences among humans between viewing kin images as compared to other classes of face (self, unfamiliar). Their findings suggest that the preexisting facial network learned by humans from their birth is used to discriminate between kin and nonkin with the presence of an established neurocognitive system.

Kinship verification has also gathered enthusiasm from computer-vision and machine-learning communities. Automatic kinship verification using facial images has several applications such as:

1. Locating relatives in public databases,

2. Determining kin of a victim or suspect by law enforcement agencies and screening asylum applications where kinship relationships are to be determined,

3. Organizing and resolving identities in photo albums, and

4. Boosting automatic face recognition capabilities.

The problem of kinship verification is particularly challenging because of the large intra-class variations among different kin pairs and different kin relations. These diverse variations in the kin pairs can be attributed to various aspects such as age gap in the kin pairs, differences in gender of the kin pair (for instance, mother–son and father–daughter), and variations resulting from ethnicity (such as biracial kin pairs). At the same time, look-alikes decrease the interclass variation among the facial images of kin. For instance, a doppelganger of an individual may have high facial similarity with that person, but is unlikely to share the same genetic material.

The research in kinship verification by using facial images commenced in 2010 by Fang et al. [28]. They collected the first kin face pair data set, Cornell KinFace database, consisting of 286 subjects. They also proposed an algorithm for facial-feature extraction and forward-selection methodology for verifying kin pairs. A pictorial structure model with springlike connections was used along with color features, length between facial parts, and gradient histogram for verifying kin. They demonstrated kinship-verification performance of 70.67% on the Cornell KinFace database. Since then, several algorithms

TABLE 6.1

Overview of kinship verification algorithms published from 2010 to 2017

Year	Kinship verification algorithm	Database	Accuracy (%)
2010	Pictorial structure model [28]	Cornell KinFace	70.67
2011	Transfer learning [29]	UB KinFace	60.00
	Transfer subspace learning [30]	UB KinFace	69.67
	Spatial pyramid learning–based (SPLE) kinship [31]	Private Database	67.75
2012	Attributes LIFT learning [32]	UB KinFace	82.50
	Self-similarity representation of Weber faces [33]	UB KinFace	69.67
		IIITD Kinship	75.20
	Product of likelihood ratio on salient features [34]	Private Database	75.00
	Gabor based gradient oriented pyramid [35]	Private Database	69.75
2013	Spatio-temporal features [36]	UvA-NEMO Smile	67.11
2014	Multiview neighborhood repulsed metric learning [37]	KinFaceW-I	69.90
		KinFaceW-II	76.50
	Discriminative multimetric learning [38]	Cornell KinFace	73.50*
		UB KinFace	74.50
		KinFaceW-I	72.00*
		KinFaceW-II	78.00*
	Discrimination via gated autoencoders [39]	KinFaceW-I	74.50
		KinFaceW-II	82.20
	Prototype discriminative feature learning [40]	Cornell KinFace	71.90
		UB KinFace	67.30
		KinFaceW-I	70.10
		KinFaceW-II	77.00
2015	Inheritable Fisher vector [41]	KinFaceW-I	73.45
		KinFaceW-II	81.60
2016	VGG-PCA [42]	FIW Database	66.90
	Ensemble similarity learning [43]	KinFaceW-I	78.60
		KinFaceW-II	75.70
2017	KVRL [44]	Cornell KinFaceW	89.50
		UB KinFace	91.80
		KinFaceW-I	96.10
		KinFaceW-II	96.20
		WVU Kinship	90.80

*Represents that the value is taken from receiver operating characteristic (ROC) curve in the paper.

have been proposed for detecting kin using different machine learning techniques and Table 6.1 summarizes the papers published from 2010 to 2017.

Xia et al. [29] introduced a new publicly available UB KinFace database, which consists of 200 groups, each containing one image of child and parents when they were young and old. An experimental protocol where the negative kin pairs were randomly created was discussed. They proposed a transfer subspace method for kinship verification where the images of parents when they were young were used as an intermediate distribution toward which the images of old parents and images of young children can be bridged. They showed kinship-verification accuracy of 60% on the UB KinFace Database. In another paper, Shao et al. [30] used Gabor filters alongside metric learning and transfer subspace learning on the same database. They reported an accuracy of 69.67% for kinship verification. Xia et al. [32] also employed the algorithm alongside understanding semantic relevance in the associate metadata to identify kin pairs in images.

Metric-learning approaches involve learning a distance function for a task. Lu et al. [37] proposed a multiview neighborhood repulsed metric-learning approach, where a metric was learned such that kin pairs are closer to each other as compared to nonkin pairs. They also introduced two new databases, KinFaceW-I and KinFaceW-II, for promoting research in automatic-kinship verification and developed fixed protocols for these databases. Each pair of images in KinFaceW-I is acquired from different photo, whereas in KinFaceW-II, it is acquired from the same photo. An accuracy of 69.90% is shown on KinFaceW-I and 76.50% is shown on KinFaceW-II. Yan et al. [38] jointly learned multiple distance metrics on different features extracted from a pair of images. The correlation of different features belonging to the same sample is maximized alongside the probability of kin images belonging together. They reported an accuracy of 72.00% on KinFaceW-I and 78.00% on KinFaceW-II database.

Yan et al. [40] proposed a prototype discriminative feature-learning algorithm, where midlevel features representing decision values from support vector machine hyperplanes are used and a metric is learned that minimizes the distances between kin and maximizes neighboring nonkin samples. The algorithm reports an accuracy of 70.10% on KinFaceW-I and 77.00% on KinFaceW-II database. A new ensemble similarity learning metric was proposed by Zhou et al. [43] where a sparse bilinear similarity function was used to model the relative characteristics encoded in kin images. An ensemble of similarity models was employed to achieve strong generalization ability. A mean accuracy of 78.60% was reported on the KinFaceW-I data set and a mean accuracy of 75.70% was reported on the KinFaceW-II data set.

Several feature descriptors have been proposed for the problem of kinship verification. Zhou et al. [31] presented a new Spatial Pyramid Learning-based feature descriptor (SPLE) for the purpose of kinship verification. They used normalized absolute histogram distance and reported a performance accuracy of 67.75% for kinship verification on their in-house database. Kohli et al. [33] applied Weber faces for illumination correction and proposed self-similarity–based approach for handling kinship variations in facial images. They reported an accuracy of 75.20% accuracy on the IIITD-Kinship Database. Zhou et al. [35] also proposed a Gabor-based gradient-oriented pyramid for the problem of kinship verification. They used a support vector machine classifier to classify if a given pair of images was kin and reported a mean accuracy of 69.75% on their database. Guo and Wang [34] used DAISY descriptors along with a product of likelihood ratio for verifying kin and their algorithm yielded 75.00% accuracy. Liu et al. [41] proposed a novel inheritable Fisher vector feature, which maximizes the similarity between kin pairs while minimizing the similarity between nonkin pairs. They reported an accuracy of 73.45% on the KinFaceW-I database and 81.60% on the KinFaceW-II database.

Recently, there has been a shift in machine-learning algorithms to employ deep learning frameworks if there are large amounts of training data available. Dehgan et al. [39] used gated autoencoders to fuse generated features with a discriminative neural layer at the end to delineate parent–offspring relationships. The authors reported an accuracy of 74.50% on the KinFaceW-I database and 82.20% on the KinFaceW-II database. Robinson et al. [42] used the learned VGG convolutional neural network for the problem of kinship verification. Additionally, they also released the largest kinship database, FIW database, that contains images of more than 1000 families. They showed that the combination of VGG features alongside PCA gives the best performance of 66.90% on the FIW database.

6.1.1 Research problem

The direct comparison of algorithms proposed in the literature is challenging because of the lack of standardized experimental protocol in classifying kin. At the same time, a small number of images in publicly available data sets led to difficulties in training deep learning models. The study of kinship verification can be divided into two parts: (1) pairwise kinship verification, which involves analyzing face pairs of an individual with their family members, and (2) self-kinship, which is kinship verification with age-separated images of the same individual (or age-invariant face verification). This chapter presents in-depth analysis of performance of deep learning algorithms in the Kinship Verification via Representation Learning (KVRL) framework proposed by Kohli et al. [44]. KVRL relies on representation learning of faces, and the learned representations are used to model the kinship characteristics in a facial image. These learned face representations are then used for classifying kin and nonkin pairs and for self-kinship verification (age-invariant face verification of the same individual). Efficacy of deep belief networks (DBNs) and Stacked Denoising Autoencoders (SDAEs) in the KVRL framework is demonstrated using five kinship data sets.

6.2 Deep Learning Algorithms for Kinship Verification Using Representation Learning Framework

Deep learning has been successfully applied to model representations in natural language processing [45–48], speech recognition [49], image segmentation [50], and object recognition [16,51]. These algorithms learn the deep features in an abstract manner by using a network of nonlinear transformations in a greedy layer-wise mechanism [53,54]. Among many deep learning algorithms, SDAEs and DBNs are two popular techniques. This section describes these deep learning techniques, including how they are hierarchically structured in a KVRL framework to learn the representations of faces for kinship verification.

6.2.1 Stacked denoising autoencoder

Let $\mathbf{x} \in R^{\alpha}$ represent the input feature vector. An autoencoder [55] maps the input vector to a reduced feature $\mathbf{y}$ using a deterministic encoder function f_{θ} as given in Equation 6.1.

$$\mathbf{y} = f_{\theta}(\mathbf{x}) = \sigma(W\mathbf{x} + b) \tag{6.1}$$

where:

$\theta = \{W, b\}$ represents the weight and bias to be learned
σ represents the activation function used in the multilayer neural network

This feature $\mathbf{y}$ is then used by a decoder function to reconstruct the input as shown in Equation 6.2.

$$\hat{\mathbf{x}} = f_{\theta'}(\mathbf{y}) = \sigma(W'\mathbf{y} + b') \tag{6.2}$$

where:

$\theta' = \{W', b'\}$ is the approximate weight and bias
σ is the sigmoid activation function and
$\hat{\mathbf{x}}$ represents the probabilistic approximation of $\mathbf{x}$ obtained from $\mathbf{y}$

The autoencoder minimizes the reconstruction error between $\mathbf{x}$ and $\hat{\mathbf{x}}$.

$$\arg \min_{\theta} ||\mathbf{x} - \hat{\mathbf{x}}||^2 \tag{6.3}$$

The autoencoders are further optimized by introducing sparsity to activate few hidden units during training. A sparsity constraint [56] on the hidden units of the autoencoder ensures that a sparse representation is obtained according to the optimization objective function given in Equation 6.4.

$$min \left(\sum_{i=1}^{m} ||\mathbf{x_i} - \hat{\mathbf{x}_i}||^2 + \beta \sum_{j=1}^{n} KL(\rho||\hat{\rho}_j) \right) \tag{6.4}$$

where:

m is the input size of the data
n is the number of hidden nodes
ρ is the sparsity constant
β is the weight for sparsity penalty term

KL(.) is the Kullback–Leibler divergence metric given as,

$$KL(\rho||\hat{\rho}_j) = \rho \, \log\frac{\rho}{\hat{\rho}_j} + (1-\rho) \, \log\left(\frac{1-\rho}{1-\hat{\rho}_j}\right) \tag{6.5}$$

Here, $\hat{\rho}_j$ is the average activation of the hidden units in the autoencoder during the training phase. The algorithm aims to optimize the value of $\hat{\rho}_j$ close to ρ, which will make the penalty term, $KL(\rho||\hat{\rho}_j) = 0$ in the objective function. For learning robust and useful higher-level representations, denoising autoencoders are used. Denoising autoencoders are introduced to circumvent learning of identity mapping where the input vector $\mathbf{x}$ is first corrupted to $\bar{\mathbf{x}}$ and trained to get the reconstructed vector $\hat{\mathbf{x}}$. In effect, these autoencoders are called sparse denoising autoencoders.

6.2.2   Deep belief networks

A Deep belief network (DBN) is a graphical model that consists of stacked restricted Boltzmann machines (RBMs), which are trained greedily layer by layer [51]. An RBM represents a bipartite graph where one set of nodes is the visible layer and the other set of nodes is the hidden layer. A DBN models the joint distribution between the observed vector $\mathbf{x}$ and $\mathbf{n}$ hidden layers $(\mathbf{h})$ as follows:

$$P(\mathbf{x}, h^1, \ldots, h^n) = \left(\prod_{j=0}^{n-2} P(h^j|h^{j+1}) \right) P(h^{n-1}, h^n) \tag{6.6}$$

where:

> $\mathbf{x} = \mathbf{h}^0, P(h^k|h^{k+1})$ is a conditional distribution for the visible units conditioned on the hidden units of the RBM at level $k+1$
> $P(h^{n-1}, h^n)$ is the visible-hidden joint distribution in the top-level RBM

In their recent work, Hinton et al. [57], proposed *dropout* training as a successful way to prevent over-fitting and an alternate method for regularization in the network. This inhibits the complex coadaptation between the hidden nodes by randomly dropping out few neurons during the training phase. It can be thought of as a sampling process from a larger network to create random subnetworks with the aim of achieving good generalization capability. Let f denote the activation function for the nth layer, and $\mathbf{W}, \mathbf{b}$ be the weights and biases for the layer, $*$ denotes the element-wise multiplication, and $\mathbf{m}$ is a binary mask with entries drawn *i.i.d.* from *Bernoulli* $(1-r)$ indicating which activations are not dropped out. Then the forward propagation to compute the activation $\mathbf{y_n}$ of nth layer of the architecture can be calculated as,

$$y_n = f\left(\frac{1}{1-r} y_{n-1} * \mathbf{m}\mathbf{W} + \mathbf{b} \right) \tag{6.7}$$

6.2.3   Kinship verification via representation learning framework

The KVRL framework proposed in [44] comprises two stages and is shown in Figure 6.3. In the first stage, the representations of each facial region are

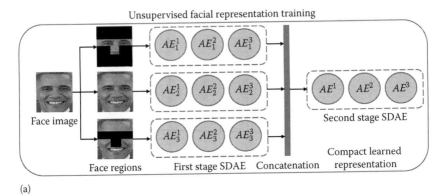

(a)

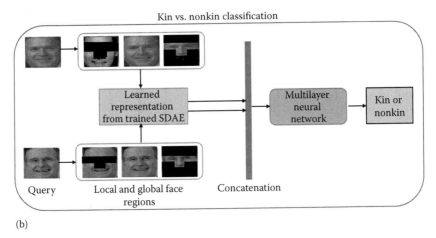

(b)

FIGURE 6.3
Two-stage KVRL with three-level SDAE approach: (a) Local and global facial representation training using the SDAE framework and (b) Supervised kin versus nonkin classification using learned representations from Figure 6.3a.

learned. These individually learned representations are combined to form a compact representation of the whole face in the second stage. Finally, a multilayer neural network is trained using these reduced feature representations of the whole face for supervised classification of kin and nonkin. The two stages of the KVRL framework–with SDAEs and DBNs are described.

KVRL-SDAE: A representation of a face image is learned by stacking sparse denoising autoencoders in two stages and learning in a greedy layer-wise manner. Stacking autoencoders reduces the classification error [58] and thus learns better representations than a single autoencoder. The complete face image is taken to model the global features while T-region and not-T region are chosen to model the local features as reported in [44]. These regions are also shown in Figure 6.3. Therefore, in the first stage of the KVRL framework, one SDAE is trained for each facial region so that each stacked autoencoder can learn a representation of the given face region. Each region is resized to a standard $M \times N$ image and converted to a $1 \times MN$ vector, which is provided as an input to a three-layer stacked sparse denoising autoencoder. Each SDAE is trained using a stochastic gradient descent followed by fine-tuning using backpropagation.

$\mathbf{AE_j^i}$ represents an autoencoder that maps the input vector to a reduced feature vector. Here, **i** denotes the layer to which the autoencoder belongs, and **j** represents the face region on which the autoencoder is trained. Therefore, a stacked autoencoder can be represented as $[AE_1^1, AE_1^2, AE_1^3]$, indicating that the stacked autoencoders are trained on face region 1 using three individual autoencoders. Once the stacked autoencoders are trained for each face region, the output from the last layer of each SDAE $[AE_1^3, AE_2^3, AE_3^3]$ are concatenated and given to the second-stage stacked autoencoder. This represents the combination of higher-level features learned from local and global facial regions. The SDAE in the second stage represents dimensionality reduction of the input feature vector and is used for training the classifier.

KVRL-DBN: Similar to the KVRL-SDAE approach, a representation of a face image is also learned by stacking RBMs and learning greedily layer by layer to form a DBN. Each facial region is resized to a standard M × N image and converted to a 1 × MN vector, which is provided as an input to a three-layer RBM. Let $\mathbf{RBM_j^i}$ represent an RBM, where **i** denotes the layer to which the RBM belongs, and **j** represents the face region on which the RBM is trained. In the first stage, a stacked RBM is trained and the output from $[RBM_1^3, RBM_2^3, RBM_3^3]$ is concatenated and given to the second-stage DBN.

Dropout is introduced in the second stage to learn the reduced features. The advantage of applying *dropout* is that different configurations of neural networks are trained based on the larger architecture and by employing shared weights for the nodes that are not dropped out. After learning the individual features in the first stage, the objective is to create a combined vector that encodes this information in a reduced manner. In kinship verification, the positive class has to encode large variation among the two pairs unlike face recognition (where the subjects are the same). Therefore, the aim is to create a representation of faces that generalizes well. By introducing *dropout* in this approach, efficient generalization is obtained that emulates sparse representations to mitigate any possible overfitting.

6.2.3.1 Kinship verification via supervised training of extracted features/kinship classification

The number of images in currently available kinship data sets are limited and cannot be used directly to train the deep learning algorithms. Therefore, a separate database is required to train the model employed in the KVRL framework. Once the KVRL-SDAE and KVRL-DBN algorithms are trained, representations of face images are extracted from them. This model provides a compact representation of the combined local and global features representing the whole face image. For a pair of kin images, the features are concatenated to form an input vector for the supervised classification. A three-layer feed-forward neural network is trained for binary classification of kinship.

6.3 Experimental Evaluation

6.3.1 Kinship data sets

Five existing kinship databases are used for performance evaluation of the KVRL framework and comparative analysis. For experimental evaluations, the following kinship databases are used (shown in Figure 6.4):

1. *Cornell kinship* [28]: It consists of 286 images pertaining to 143 subject pairs. The facial images in this database are frontal pose and have neutral expression.

Cornell Kin
- Subjects: 286
- Images: 286
- Kin relations: Father–Daughter, Father–Son, Mother–Daughter, Mother–Son
- Multiple images: No

UB KinFace
- Subjects: 400
- Images: 600
- Kin relations: Father–Daughter, Father–Son, Mother–Daughter, Mother–Son
- Multiple images: Yes (of parents)

WVU Kinship
- Subjects: 226
- Images: 906
- Kin relations: Father–Daughter, Father–Son, Mother–Daughter, Mother–Son, Brother–Brother, Sister–Sister, Brother–Sister
- Multiple images: Yes

KinFace–I
- Subjects: 1066
- Images: 1066
- Kin relations: Father–Daughter, Father–Son, Mother–Daughter, Mother–Son
- Multiple images: No

KinFace–II
- Subjects: 2000
- Images: 2000
- Kin relations: Father–Daughter, Father–Son, Mother–Daughter, Mother–Son
- Multiple images: No

FIGURE 6.4
Characteristics of kinship databases used for experimental evaluation.

2. *KinFaceW-I* [37]: This database consists of 1066 images corresponding to 533 kin pairs. It has 156 Father–Son, 134 Father–Daughter, 116 Mother–Son, and 127 Mother–Daughter kin pair images.

3. *KinFaceW-II* [37]: This database has been created such that images belonging to the kin pair subjects are acquired from the same photograph. It consists of 1000 kin pair images with an equal number of images belonging to the four kinship relationships: Father–Son, Father–Daughter, Mother–Son, and Mother–Daughter.

4. *UB KinFace* [29]: This database consists of 200 groups consisting of 600 images. Each group has one image of the child and one image belonging to the corresponding parent when they were young and when they were old. The database has 91 Father–Son, 79 Father–Daughter, 15 Mother–Son, and 21 Mother–Daughter kin pair images.

5. *WVU kinship* [44]: All the aforementioned kinship data sets contain only one image per kin pair and hence are not suitable for face recognition experiments. The **WVU kinship database** [44] was developed for both kinship-verification research and incorporates kinship scores to improve the performance of face recognition. The WVU kinship data set consists of 113 pairs of individuals from Caucasian and Indian ethnicity. The data set has four images per person, which introduces intra-class variations for a specific kin pair. It consists of the following kin relations:

   1. Brother–Brother (BB): 22 pairs,
   2. Brother–Sister (BS): 9 pairs,
   3. Sister–Sister (SS): 13 pairs,
   4. Mother–Daughter (MD): 13 pairs,
   5. Mother–Son (MS): 8 pairs,
   6. Father–Son (FS): 34 pairs, and
   7. Father–Daughter (FD): 14 pairs.

These multiple images per kin pair also include variations in pose, illumination, and occlusion.

6.3.2 Architectural details of the deep learning KVRL framework

The results with two variants of the KVRL framework, namely KVRL-SDAE and KVRL-DBN, are presented. Training the SDAE and DBN algorithms to learn a representation of faces for kinship requires a large number of face images. For this purpose, around 600,000 face images are used. These images are obtained by combining existing face databases such as [59] and [60]. For detecting faces, all the images are aligned using an affine transformation and a Viola-Jones face detection algorithm [61] is used. The face regions namely: full face, T region, and not-T region are extracted from each face image. Each facial region is resized to a 32×32 image and a 1024 vector is given as input to individual SDAE or RBM deep learning algorithms in the first stage. For every individual SDAE, three different autoencoders are stacked together and all of them are learned in a greedy layer-wise fashion where each level receives the representation of the output from the previous level. The number of nodes in the SDAE learned for the three facial regions in the first stage is same $[AE_j^1, AE_j^2, AE_j^3] = [1024, 512, \text{and } 256]$. An output vector of size 256 is obtained from the third level of each SDAE. These outputs are the representations learned from the global and local face regions and are concatenated to form a vector of size 768. This vector is given to a SDAE in the second stage to reduce the dimensionality. The final vector has a size of 192 and is the learned representation of the *face* image.

Similarly, three different RBMs are stacked together and all of them are learned one layer at a time. The performance of the KVRL-DBN algorithm is also tested using the top three facial regions from the human study. In the first stage, the number of nodes in $[RBM_1^1, RBM_1^2, RBM_1^3]$ are 1024, 512, and 512, respectively. An output vector of size 512 is obtained from the third level of each DBN and is concatenated to form a vector of size 1536. A compact representation is learned from the DBN in the second stage and is used for training the classifier. In the second stage of the DBN, the size of the three layers are 1536, 768, and 384, respectively. *Dropout* is introduced in the second stage with probability of 0.5.

6.3.3 Experimental protocol

The performance of the KVRL framework is evaluated on the same experiment protocol as described in [38], where five-fold cross validation for kin classification is performed by keeping the number of pairs in all kin relations to be roughly equal in all folds. This is done to make the experiments directly comparable even though the list of negative pairs included may vary. Random negative pairs for kinship are generated ensuring no overlap between images used for training and testing.

6.3.4 Performance results on five existing kinship databases

The kinship-verification accuracy results obtained using experiments conducted on the Cornell kinship database, KinFaceW-I database, KinFaceW-II database, UB KinFace database, and WVU kinship database are summarized in Tables 6.2 through 6.5. From the results, SDAE and DBN algorithms in the KVRL framework show better performance than the current state-of-the-art kinship-verification results on all the databases. Comparative analysis is also performed with algorithms such as the multiview neighborhood repulsed metric learning (MNRML) [37], discriminative multimetric learning (DMML) [38], and discriminative model [39].

TABLE 6.2

Kinship-verification accuracy (%) on Cornell kinship data set with FS, FD, MS, and MD kin relations

Method	Father–Son	Father–Daughter	Mother–Son	Mother–Daughter
MNRML [37]	74.5	68.8	77.2	65.8
DMML [38]	76	70.5	77.5	71.0
KVRL-SDAE	85.0	80.0	85.0	75.0
KVRL-DBN	88.3	80.0	90.0	72.5

TABLE 6.3

Kinship-verification accuracy (%) on KinFaceW-I and KinFaceW-II databases with FS, FD, MS, and MD kin relations

(a) KinFaceW-I Data set				
Method	Father–Son	Father–Daughter	Mother–Son	Mother–Daughter
MRNML [37]	72.5	66.5	66.2	72.0
DML [38]	74.5	69.5	69.5	75.5
Discriminative Model [39]	76.4	72.5	71.9	77.3
KVRL-SDAE	95.5	88.8	87.1	96.9
KVRL-DBN	96.2	89.6	87.9	97.6

(b) KinFaceW-II Data set				
Method	Father–Son	Father–Daughter	Mother–Son	Mother–Daughter
MNRML [37]	76.9	74.3	77.4	77.6
DML [38]	78.5	76.5	78.5	79.5
Discriminative Model [39]	83.9	76.7	83.4	84.8
KVRL-SDAE	94.0	89.2	93.6	94.0
KVRL-DBN	94.8	90.8	94.8	95.6

TABLE 6.4

Kinship-verification accuracy (%) on child-young parent and child-old parent sets of UB kinship data set

Method	Child–Young Parents	Child–Old Parents
MNRML [37]	66.5	65.5
DMML [38]	74.5	70.0
KVRL-SDAE	85.9	84.8
KVRL-DBN	88.5	88.0

TABLE 6.5

Kinship-verification accuracy (%) on WVU kinship data set on FS, FD, MS, MD, BB, BS, and SS kin relations

Method	FS	FD	MS	MD	BB	BS	SS
KVRL-SDAE	80.9	76.1	74.2	80.7	81.6	76.5	80.3
KVRL-DBN	85.9	79.3	76.0	84.8	85.0	79.9	85.7

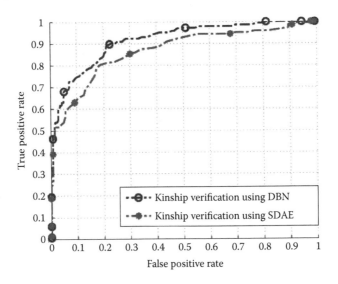

FIGURE 6.5

ROC curves for kinship verification with KVRL-DBN and KVRL-SDAE on Cornell kinship data set.

6.3.4.1   Cornell kinship data set

Table 6.2 and the receiver operating characteristic (ROC) curve in Figure 6.5 summarize the results of the kinship verification on the Cornell kinship data set [28]. It is observed that KVRL-DBN yields the highest kinship verification for the four kin relationships in the database. It outperforms existing MNRML [37] and DMML [38] techniques for kinship verification. In this database, kin pairs belonging to Mother–Son relation are correctly detected with highest accuracy of 90.0% by the KVRL-DBN framework. Similarly, the KVRL-SDAE framework detects Mother–Son kin pairs with 85.0% kinship-verification accuracy.

6.3.4.2   KinFaceW-I and KinFaceW-II databases

The results with KinFaceW-I and KinFaceW-II [37] databases are shown in Table 6.3 and the ROC curve is shown in Figure 6.6. The KVRL-SDAE and KVRL-DBN framework outperform the existing MRNML [37], DML [38], and

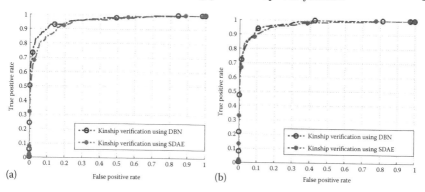

(a)   (b)

FIGURE 6.6

ROC curves for kinship verification with KVRL-DBN and KVRL-SDAE on (a) KinFaceW-I database and (b) KinFaceW-II database.

Discriminative Model [39] techniques for kinship verification on KinFaceW-I and KinFaceW-II databases. KVRL-DBN yields the highest kinship verification of 96.9% for Mother–Daughter relation on the KinFaceW-I database. For the KinFaceW-II database, the KVRL-DBN achieves 94.0% accuracy for Father–Son and Mother–Daughter relations. A general trend appears for the KinFaceW-I and KinFaceW-II databases where the images of same-gender kin perform better than different-gender kin images. Thus, Father–Son and Mother–Daughter kinship relations have a higher kinship verification accuracy than Father–Daughter and Mother–Son.

6.3.4.3 UB KinFace database

UB KinFace database [29] consists of groups of images, which includes images of children, young parents, and old parents. The database is built on the hypothesis that images of parents when they were young are more similar to images of children, as compared to images of parents when they are older. Results of comparative analysis of kinship-verification performance on the two sets of this database are reported in Table 6.4 and Figure 6.7. It is observed that the kinship-verification performance is better in child–young parent kin pair (Set 1) as compared to child–old parent kin pair (Set 2) where there is a significant age gap between the kin pairs.

6.3.4.4 WVU kinship database

WVU kinship database [44] consists of seven kin relations: Father–Son, Father–Daughter, Mother–Son, Mother–Daughter, Brother–Brother, Brother–Sister, and Sister–Sister. The kinship-verification performance of the KVRL-SDAE and KVRL-DBN frameworks on WVU kinship database is summarized in Table 6.5 and Figure 6.8. The performance of the KVRL-DBN framework is higher than the KVRL-SDAE framework for all seven kin relations. Similar to KinFaceW-I and KinFaceW-II databases, kin pairs

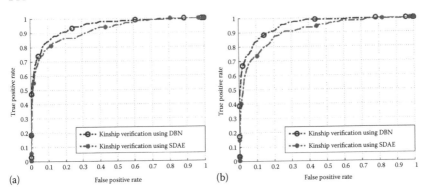

FIGURE 6.7
ROC curves for kinship verification with KVRL-DBN and KVRL-SDAE on Set 1 and Set 2 of UB KinFace database: (a) KinFaceW-I database and (b) KinFaceW-II database: (a) UB-young parent and young child database and (b) UB-old parent and young child database.

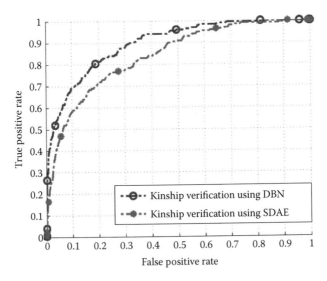

FIGURE 6.8
ROC curves for kinship verification with KVRL-DBN and KVRL-SDAE on WVU kinship database.

belonging to same-gender have greater kinship-verification accuracy as compared to different-gender kin pairs.

From all these experiments, it is observed that the best performance is obtained with hierarchical two-stage DBN (KVRL-DBN). One reason for such a performance can be that the DBNs' hidden layers are able to better model kinship. Being a probabilistic generative model, it may be learning the subtle similarities that occur in local regions among kin pairs. This reinforces that an unsupervised representation learning model using both global and local face regions yields better performance for kinship verification. We also observe an improvement in kinship-verification performance when *dropout* is used in the second stage of the model.

6.4 Self-Kinship Problem (Age-Invariant Face Verification)

From an anthropological point of view, kinship information is based on the degree of genetic similarity between two individuals. Thus, it is logical to expect that an ideal kinship-verification algorithm will give perfect kinship-verification accuracy if the input pair of face images belong to the same individual. This pair of images belonging to the same individual can also be separated by age gap. Therefore, the classical age-invariant face-verification research problem [62,63] can be considered under the umbrella of kinship verification. This scenario is identified as the problem of *self-kinship* where the objective is to identify the age-progressed images of the same individual as kin. Figure 6.9 illustrates the problem of self-kinship where the age-progressed face images of an individual can be verified using kinship-verification algorithms.

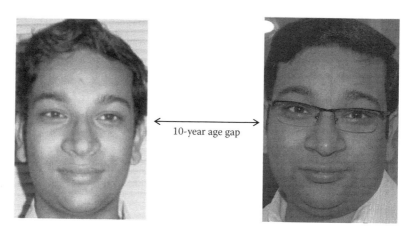

10-year age gap

FIGURE 6.9

An example of age-progressed pair of images of an individual. This is also an example of self-kinship because these age-separated face images can be verified using kinship verification algorithms.

6.4.1   Self-kinship experimental results

To validate the efficacy of the KVRL framework in this situation, the experiment is performed on two databases: FG-Net database [64] and UB KinFace data set [29] using the KVRL-SDAE and KVRL-DBN algorithms. The FG-Net database consists of 1002 images from 82 subjects with an average of 12 images per subject. The database consists of 5,808 positive inter-class samples and 12,000 negative inter-class samples. Three-fold cross-validation is performed similar to Lu et al. [37] where each subject appears in the training or testing set exclusively. The UB database also consists of images of parents when they were young and when they were old. The database consists of 200 images of individuals in each of the aforementioned scenarios. A three-fold cross validation is conducted with each pair present either in the training or testing data set.

Figure 6.10 shows the results obtained from the self-kinship experiment. An equal error rate (EER) of 14.14% is observed on the UB data set, whereas an EER of 16.45% is observed on the FG-Net Database using SDAE in the KVRL framework. A further reduction in EER of 10.95% on the UB data set and 15.09% on the FG-Net Database is observed when using the DBN model in the KVRL framework. This result is considerably better than the 22.5% observed by the current state-of-the-art algorithm for kinship verification: MNRML [37]. These algorithms outperform the previous best algorithm in the age-invariant face-verification (self-kinship) experiment for the FG-Net Database [65] and performs equally good on the UB KinFace database [29].

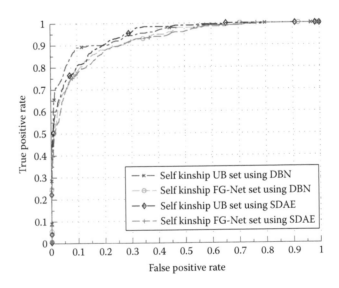

FIGURE 6.10
ROC curves demonstrating the performance of self-kinship (age-invariant face verification) using the KVRL algorithm.

6.5 Summary and Future Research

Kinship verification has been explored in anthropological, neuroscience, and computer-vision research domains. Kinship-verification research using facial images has real-world applications such as locating missing relatives and validating kinship claims during asylum-seeking process. Kinship verification can be delineated as two distinct, but related problems: (1) kinship verification between different individuals and (2) self-kinship, which characterizes age-invariant face verification associated with the same individual. Different approaches have been proposed for kinship verification such as metric learning, features-based solutions, and deep learning algorithms. In this chapter, we examine the efficacy of two deep learning algorithms: DBNs and SDAEs in the KVRL framework [44]. KVRL is a two-stage hierarchical representation learning framework that uses the trained deep learning representations of faces to calculate a kinship similarity score. We demonstrate that these kinship verification approaches outperform recently reported results on five different kinship data sets publicly available. The results of self-kinship using the KVRL framework are reported and outperforms the current state-of-the-art kinship algorithms using the FG-Net and UB KinFace databases.

The current research focuses on lineal kin relationships such as parents, children, and grandparents. However, there is no research that exists when the traditional kin relationship is broadened to include collateral extended family such as uncle, aunts, cousins, nephews, and nieces. This kin verification can become even more complex when the genetic link among kin may be weak, especially when families include half-brothers, half-sisters, or step-parents. Additionally, the majority of the techniques proposed for kinship verification have been developed for facial images. There is a dearth of research for kinship verification in videos. Kinship verification in unconstrained videos can be highly beneficial for surveillance purposes. Such systems can be used at border control to prevent illegal child trafficking by validating relationships between parents and their children using surveillance videos. Addressing such research scenarios will be challenging and will need new approaches to study how information learned in one domain, that is well understood, can be represented or transformed to another domain to accurately identify such kin relationships that may not be easily accomplished even through DNA testing.

References

1. J. Carsten, *Cultures of Relatedness: New Approaches to the Study of Kinship.* Cambridge University Press, 2000 [Online]. Available: https://books.google.com/books?id=HDJrqc-U-y8C

2. L. H. Morgan, A conjectural solution of the origin of the classificatory system of relationship, *Proceedings of the American Academy of Arts and Sciences*, 7, 436–477, 1868.

3. C. Lévi-Strauss, *The Elementary Structures of Kinship:*, ser. Beacon Paperback. no. BP 340. Beacon Press, 1969 [Online]. Available: https://books.google.com/books?id=4QhYYQi6CoMC

4. K. Finkler, C. Skrzynia, and J. P. Evans, The new genetics and its consequences for family, kinship, medicine and medical genetics, *Social Science & Medicine*, 57(3), 403–412, 2003.

5. G. Murdock, *Social Structure*. New York, Macmillan, 1949.

6. C. M. Berman and B. Chapais, *Kinship and Behavior in Primates*. Oxford, UK, Oxford University Press, 2004.

7. L. Taylor and R. Sussman, A preliminary study of kinship and social organization in a semi-free-ranging group of Lemur catta, *International Journal of Primatology*, 6, 601–614, 1985.

8. D. L. Pastor, The quality of mother–infant attachment and its relationship to toddlers' initial sociability with peers, *Developmental Psychology*, 17(3), 326, 1981.

9. K. Kelly, A. Slade, and J. F. Grienenberger, Maternal reflective functioning, mother–infant affective communication, and infant attachment: Exploring the link between mental states and observed caregiving behavior in the intergenerational transmission of attachment, *Attachment & Human Development*, 7(3), 299–311, 2005.

10. M. Noriuchi, Y. Kikuchi, and A. Senoo, The functional neuroanatomy of maternal love: Mother's response to infant's attachment behaviors, *Biological Psychiatry*, 63(4), 415–423, 2008.

11. P. Sinha, B. Balas, Y. Ostrovsky, and R. Russell, Face recognition by humans: Nineteen results all computer vision researchers should know about, *Proceedings of the IEEE*, 94(11), 1948–1962, 2006.

12. L. Hogben, The genetic analysis of familial traits, *Journal of Genetics*, 25(2), 211–240, 1932.

13. M. Daly and M. I. Wilson, Whom are newborn babies said to resemble? *Ethology and Sociobiology*, 3(2), 69–78, 1982.

14. N. J. Christenfeld and E. A. Hill, Whose baby are you? *Nature*, 378(6558), 669, 1995.

15. S. Bredart and R. M. French, Do babies resemble their fathers more than their mothers? A failure to replicate Christenfeld and Hill (1995), *Evolution and Human Behavior*, 2, 129–135, 1999.

16. P. Bressan and M. F. Martello, Talis pater, talis filius: perceived resemblance and the belief in genetic relatedness, *Psychological Science*, 13(3), 213–218, 2002.

17. R. Burch and G. Gallup, Perceptions of paternal resemblance predict family violence, *Evolutionary Human Behavior*, 21(6), 429–435, 2000.

18. S. M. Platek, J. P. Keenan, G. G. Gallup, and F. B. Mohamed, Where am I? The neurological correlates of self and other, *Brain Research: Cognitive Brain Research*, 19(2), 114–122, 2004.

19. L. T. Maloney and M. F. Dal Martello, Kin recognition and the perceived facial similarity of children, *Journal of Vision*, 6(10), 1047–1056, 2006.

20. M. F. Dal Martello and L. T. Maloney, Where are kin recognition signals in the human face? *Journal of Vision*, 6(12), 1356–1366, 2006.

21. M. F. Dal Martello and L. T. Maloney, Lateralization of kin recognition signals in the human face, *Journal of Vision*, 10(8), 1–10, 2010.

22. G. Kaminski, S. Dridi, C. Graff, and E. Gentaz, Human ability to detect kinship in strangers' faces: Effects of the degree of relatedness, *The Royal Society Biological Sciences*, 276, 3193–3200, 2009.

23. D. McLain, D. Setters, M. P. Moulton, and A. E. Pratt, Ascription of resemblance of newborns by parents and nonrelatives, *Evolution and Human Behavior*, 21(1), 11–23, 2000.

24. R. Oda, A. Matsumoto-Oda, and O. Kurashima, Effects of belief in genetic relatedness on resemblance judgments by Japanese raters, *Evolution and Human Behavior*, 26(5), 441–450, 2005.

25. D. Lieberman, J. Tooby, and L. Cosmides, The architecture of human kin detection. *Nature*, 445(7129), 727–31, 2007.

26. G. Kaminski, F. Ravary, C. Graff, and E. Gentaz, Firstborns' disadvantage in kinship detection, *Psychological Science*, 21(12), 1746–1750, 2010.

27. S. M. Platek and S. M. Kemp, Is family special to the brain? An event-related fMRI study of familiar, familial, and self-face recognition, *Neuropsychologia*, 47(3), 849–858, 2009.

28. R. Fang, K. D. Tang, N. Snavely, and T. Chen, Towards computational models of kinship verification, in *IEEE International Conference on Image Processing*, 2010, pp. 1577–1580.

29. S. Xia, M. Shao, and Y. Fu, Kinship verification through transfer learning, in *International Joint Conference on Artificial Intelligence*, Menlo Park, CA, AAAI Press, 2011, pp. 2539–2544.

30. M. Shao, S. Xia, and Y. Fu, Genealogical face recognition based on UB KinFace database, in *IEEE Computer Vision and Pattern Recognition Workshops*, June 2011, pp. 60–65.

31. X. Zhou, J. Hu, J. Lu, Y. Shang, and Y. Guan, Kinship verification from facial images under uncontrolled conditions, in *ACM Multimedia*, New York, ACM, 2011, pp. 953–956.

32. S. Xia, M. Shao, J. Luo, and Y. Fu, Understanding kin relationships in a photo, *IEEE Transactions on Multimedia*, 14(4), 1046–1056, 2012.

33. N. Kohli, R. Singh, and M. Vatsa, Self-similarity representation of weber faces for kinship classification, in *IEEE International Conference on Biometrics: Theory, Applications and Systems*, 2012, pp. 245–250.

34. G. Guo and X. Wang, Kinship measurement on salient facial features, *IEEE Transactions on Instrumentation and Measurement*, 61(8), 2322–2325, 2012.

35. X. Zhou, J. Lu, J. Hu, and Y. Shang, Gabor-based gradient orientation pyramid for kinship verification under uncontrolled environments, in *ACM Multimedia*, New York, ACM, 2012, pp. 725–728.

36. H. Dibeklioglu, A. Salah, and T. Gevers, Like father, like son: Facial expression dynamics for kinship verification, in *IEEE International Conference on Computer Vision*, December 2013, pp. 1497–1504.

37. J. Lu, X. Zhou, Y.-P. Tan, Y. Shang, and J. Zhou, Neighborhood repulsed metric learning for kinship verification, *IEEE Transactions on Pattern Analysis and Machine Intelligence*, 36(2), 331–345, 2014.

38. H. Yan, J. Lu, W. Deng, and X. Zhou, Discriminative multimetric learning for kinship verification, *IEEE Transactions on Information Forensics and Security*, 9(7), 1169–1178, 2014.

39. A. Dehghan, E. G. Ortiz, R. Villegas, and M. Shah, Who do I look like? Determining parent-offspring resemblance via gated autoencoders, in *IEEE Computer Vision and Pattern Recognition*, 2014, pp. 1757–1764.

40. H. Yan, J. Lu, and X. Zhou, Prototype-based discriminative feature learning for kinship verification, *IEEE Transactions on Cybernetics*, PP (99), 1–1, 2014.

41. Q. Liu, A. Puthenputhussery, and C. Liu, Inheritable fisher vector feature for kinship verification, in *IEEE International Conference on Biometrics: Theory, Applications and Systems*, 2015, pp. 1–6.

42. J. P. Robinson, M. Shao, Y. Wu, and Y. Fu, Families in the wild (FIW): Large-scale kinship image database and benchmarks, in *ACM Multimedia*, New York, ACM, 2016, pp. 242–246.

43. X. Zhou, Y. Shang, H. Yan, and G. Guo, Ensemble similarity learning for kinship verification from facial images in the wild, *Information Fusion*, 32 (Part B), 40–48, 2016.

44. N. Kohli, M. Vatsa, R. Singh, A. Noore, and A. Majumdar, Hierarchical representation learning for kinship verification, *IEEE Transactions on Image Processing*, 26, 289–302, 2017.

45. G. E. Hinton, Learning distributed representations of concepts, in *Conference of the Cognitive Science Society*, Vol. 1, Mahwah, NJ, Lawrence Erlbaum Associates, 1986.

46. Y. Bengio, R. Ducharme, P. Vincent, and C. Jauvin, A neural probabilistic language model, *Journal of Machine Learning Research*, 3, 1137–1155, 2003.

47. A. Bordes, X. Glorot, J. Weston, and Y. Bengio, Joint learning of words and meaning representations for open-text semantic parsing, in *International Conference on Artificial Intelligence and Statistics*, Amhurst, MA, PMLR, 2012, pp. 127–135.

48. X. Glorot, A. Bordes, and Y. Bengio, Domain adaptation for large-scale sentiment classification: A deep learning approach, in *International Conference on Machine Learning*, Madison, WI, Omnipress, 2011, pp. 513–520.

49. G. Dahl, A. R. Mohamed, and G. E. Hinton, Phone recognition with the mean-covariance restricted boltzmann machine, in *Advances in Neural Information Processing Systems*, Red Hook, NY, Curran Associates Inc, 2010, pp. 469–477.

50. G. Carneiro, J. Nascimento, and A. Freitas, The segmentation of the left ventricle of the heart from ultrasound data using deep learning architectures and derivative-based search methods, in *IEEE International Conference on Image Processing*, 21(3), 968–982, 2012.

51. G. Hinton, S. Osindero, and Y.-W. Teh, A fast learning algorithm for deep belief nets, *Neural Computation*, 18(7), 1527–1554, Cambridge, MA, MIT Press, 2006.

52. A. Krizhevsky, I. Sutskever, and G. E. Hinton, ImageNet classification with deep convolutional neural networks, in *Advances in Neural Information Processing Systems*, Red Hook, NY, Curran Associates Inc, 2012, pp. 1097–1105.

53. Y. Bengio, Learning deep architectures for AI, *Foundations and trends in Machine Learning*, 2(1), 1–127, 2009.

54. Y. Bengio, A. Courville, and P. Vincent, Representation learning: A review and new perspectives, *IEEE Transactions on Pattern Analysis and Machine Intelligence*, 35(8), 1798–1828, 2013.

55. D. E. Rumelhart, G. E. Hinton, and R. J. Williams, Learning internal representations by error propagation, *Parallel Distributed Processing: Explorations in the Microstructure of Cognition*, 1, 318–362, 1986.

56. A. Ng, Sparse autoencoders, 2011. Available: http://web.stanford.edu/class/cs294a/sparseAutoencoder.pdf/

57. N. Srivastava, G. Hinton, A. Krizhevsky, I. Sutskever, and R. Salakhutdinov, Dropout: A simple way to prevent neural networks from overfitting, *Journal of Machine Learning Research*, 15(1), 1929–1958, 2014.

58. P. Vincent, H. Larochelle, I. Lajoie, Y. Bengio, and P.-A. Manzagol, Stacked denoising autoencoders: Learning useful representations in a deep network with a local denoising criterion, *Journal of Machine Learning Research*, 11, 3371–3408, 2010.

59. L. Wolf, T. Hassner, and I. Maoz, Face recognition in unconstrained videos with matched background similarity, in *IEEE Computer Vision and Pattern Recognition*, 2011, pp. 529–534.

60. R. Gross, I. Matthews, J. Cohn, T. Kanade, and S. Baker, Multi-PIE, *Image and Vision Computing*, 28(5), 807–813, 2010.

61. P. Viola and M. J. Jones, Robust real-time face detection, *International Journal of Computer Vision*, 57(2), 137–154, 2004.

62. D. Yadav, M. Vatsa, R. Singh, and M. Tistarelli, Bacteria foraging fusion for face recognition across age progression, in *IEEE Computer Vision and Pattern Recognition Workshops*, 2013, pp. 173–179.

63. D. Yadav, R. Singh, M. Vatsa, and A. Noore, Recognizing age-separated face images: Humans and machines, *PLOS ONE*, 9(12), 1–22, 12 2014.

64. A. Lanitis, Comparative evaluation of automatic age progression methodologies, *EURASIP Journal on Advances in Signal Processing*, 2008, (1), 1–10, 2008.

65. A. Lanitis, Comparative evaluation of automatic age-progression methodologies, *EURASIP Journal on Advances in Signal Processing*, 2008, 101:1–101:10 doi:10.1155/2008/239480

7

What's Hiding in My Deep Features?

Ethan M. Rudd, Manuel Günther, Akshay R. Dhamija,
Faris A. Kateb, and Terrance E. Boult

CONTENTS

7.1   Introduction 153
7.2   Related Work 156
7.3   Analysis of Face Networks 158
          7.3.1   Attribute prediction experiments 159
          7.3.2   Pose-prediction experiments 161
7.4   Toward an Invariant Representation 163
          7.4.1   Preliminary evaluation on rotated MNIST 163
          7.4.2   Proposed architectures to enhance invariance 165
                      7.4.2.1   The μ-LeNet 165
                      7.4.2.2   The PAIR-LeNet 166
          7.4.3   Analysis and visualization of the proposed
                      representations 167
7.5   Conclusions and Future Work 171
Acknowledgment 172
References 172

7.1 Introduction

Effective recognition hinges on the ability to map inputs to their respective class labels despite exogenous input variations. For example, the associated identity label from a face-recognition system should not change because of variations in pose, illumination, expression, partial occlusion, or accessories. Deep neural networks have been hailed for their ability to deliver state-of-the-art recognition performance, even under noticeable input variations. Although raw outputs from end-to-end networks are commonly used in idealized benchmark settings (e.g., on the ImageNet classification challenge [1]), realistic tasks seldom use actual network outputs because training a full network is time-consuming, and classes in the application at hand are often not so rigidly defined as in benchmark settings. Moreover, fusing additional information through back-propagation is sometimes nontrivial.

Particularly in biometrics applications, a network that is trained on many examples for a specific modality is commonly used as a feature extractor by taking feature representations from lower layers of the network. During enrollment, samples from the gallery are submitted to the network, and the output feature vectors are used for template construction. These templates may be constructed in a variety of ways (e.g., as sets of raw feature vectors or aggregations thereof [2,3]). At match time, probe samples are compared to the gallery by taking the cosine distance to gallery templates [3]. Critically, operational constraints often require fast enrollment, which precludes training an end-to-end network and significantly reduces the likelihood that samples in the gallery will be from the same classes as samples in the training set; many face-recognition protocols even require that identities in training and enrollment sets have no overlap.

Unlike the outputs of an end-to-end network, in which learned classifications exhibit invariances to exogenous factors in the input (they have to, assuming reasonable recognition accuracy), "there is no concrete reasoning provided [in the literature] on the invariance properties of the representations out of the fully connected layers," [4] nor is there any intuitive reason to assume that deep features prior to the final fully connected layer will exhibit such invariances because typical loss functions have no explicit constraint to enforce such invariance. Although some exogenous properties of input samples are presumably attenuated by the network because these higher-level abstractions still offer quite good recognition performance, there is still significant motivation to discern what type of exogenous information resides in these features.

As a primary motivation, note that lack of invariance is not necessarily a bad thing: exogenous information preserved from the input is partially what allows deep representations to generalize well to other tasks. For example, popular approaches to facial attribute classification use features derived from networks trained on identification tasks [5]. A truly invariant representation would preclude information about attributes that are unrelated to identity from being learned in such a manner (e.g., Smiling); yet until recently, this was the state-of-the-art facial attribute classification approach and is still widely used. The new state-of-the-art [6,7] leverages attribute data directly, precisely for this reason. More generally, many learning approaches to task transfer and domain adaptation often make the implicit assumption that a feature space derived from a different task or a different domain will work well for the task or domain at hand, provided that enough data have been used to derive the feature space, with the actual transfer learning conducted via classifiers within this feature space. We show that such an assumption may or may not be true depending on the information content of the feature space.

A secondary motivation for understanding the nature of exogenous information embedded in deep features—namely, the desire to create more invariant representations—becomes important when a truncated network is used to enroll novel samples not present in the training set, specifically if those samples constitute novel identities. The original end-to-end network is optimized so that classes constituted by the original training samples separate well; even if the underlying deep features are not invariant to exogenous factors, the last layer will combine these deep features into an overall classification. Because we ignore the last layer during enrollment, variations resulting from exogenous factors could end in confusion between classes for newly enrolled samples. Thus, there is reason to explore whether we can attain more invariant deep feature spaces because this could increase performance scalability of applied machine learning systems, especially for biometric recognition.

In this chapter, we conduct a formal exploration of the invariant and non-invariant properties of deep feature spaces. Although there has been related research conducted in the machine-learning and computer-vision communities in areas of domain-adaptation and task-transfer learning, there has been little direct concentration on this topic, in part we surmise, because of the already impressive recognition rates and representational performance that deep neural networks provide. We analyze two deep representations. The first, geared toward realistic applications, is the combined output of multiple face-recognition networks [8,9], which is an approach that achieved state-of-the-art accuracy on the IJB-A data set [10]. As indicated in Figure 7.1,

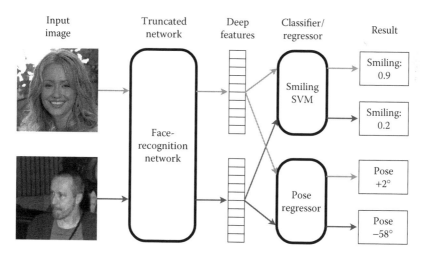

FIGURE 7.1

Classifying image properties from deep features. Nonidentity-related properties such as smiling or pose variations should not affect a good face-recognition system's output. So, people expect that deep features extracted from truncated face-recognition networks are invariant to these properties. But they are not. We demonstrate this lack of invariance by using simple classifiers or regressors to predict these image properties from the deep features of an identification network with high reliability.

we demonstrate that from this representation we can not only accurately predict pose, but we can also predict facial attributes, despite an added triplet-loss metric learning phase atop these representations. With respect to attribute classification, we find that classifiers trained in this deep feature space—which is purely identity derived—achieve near state-of-the-art performance on identity-related attributes such as gender and, surprisingly, they are also able to achieve impressively high performance on nonidentity-related attributes (e.g., Smiling), which an invariant representation would have down-weighted or pooled out. The second deep representation that we employ is the canonical LeNet/MNIST architecture [11], with which we attempt several different training procedures to enforce invariant representations. Our analysis demonstrates that we are indeed able to extract a more invariant feature space with little accuracy loss, but many noninvariances still remain.

7.2   Related Work

Face biometric systems have seen remarkable performance improvements across many tasks since the advent of deep convolutional neural networks. Taigman et al. [12] pioneered the application of modern deep convolutional neural networks to face-recognition tasks, with DeepFace, the first network to reach near-human verification performance on the Labeled Faces in the Wild (LFW) benchmark [13]. In their work, they used an external image preprocessing to frontalize images and trained their network on a private data set of 4.4 million images of more than 4000 identities. Later, Oxford's Visual Geometry Group (VGG) publicly released a face-recognition network [2] that omits the frontalization step, while training the network with a relatively small data set containing 95% frontal and 5% profile faces. Parkhi et al. [2] also implemented a triplet-loss embedding and demonstrated comparable performance to [12] on LFW despite the lower amount of training data. Lately, the IJB-A data set and challenge [10], which contains more profile faces, was proposed. Chen et al. [3] trained two networks on a small-scale private data set containing more profile faces than the DeepFace and VGG training sets. Using a combination of these two networks and a triplet-loss embedding that was optimized for comparing features with the dot product, they achieved the current state-of-the-art results on the IJB-A challenge. The combination of these deep features is the basis for our analysis in Section 7.3.

Facial-attribute classification using deep neural network was pioneered by Liu et al. [5]. The authors collected a large-scale data set (CelebA) of more than 200,000 images, which they labeled with 40 different facial attributes. They trained a series of two localization networks (LNets) and one attribute classification network (ANet). The ANet was pretrained for a face identification task and fine-tuned using the training partition of CelebA attribute data. Finally, they trained individual support vector machines (SVMs) atop the learned deep features of the penultimate layer of their ANet to perform final attribute prediction. Wang et al. [14] pretrained a network using data that they collected themselves via ego-centric vision cameras and augmented that data set with ground-truth weather and geo-location information. They then fine-tuned it on the CelebA training set. Although previous approaches that advanced the state-of-the-art on CelebA relied on augmented training data sets and separate classifiers trained atop deep features, Rudd et al. [6] recently advanced the state-of-the-art beyond these by using an end-to-end network trained only on the CelebA training set, but with a multi-task objective, optimizing with respect to all attributes simultaneously. Their Mixed Objective Optimization Network (MOON) is based on the VGG topology and also introduces a balancing technique to compensate for the high bias for some attributes in CelebA. Finally, Günther et al. [7] extended the approach in [6] to accommodate unaligned input face images using the Alignment Free Facial Attribute Classification Technique (AFFACT), which is able to classify facial attributes using only the detected bounding boxes (i.e., without alignment). This network provides the current state-of-the-art on the CelebA benchmark using no ground-truth landmark locations from the test images. In this chapter, we investigate a network that is a clone of the AFFACT network, which was trained using the same balancing method presented in [6].

The use of deep learned representations across visual tasks, including the aforementioned face biometric, can be traced back to the seminal work of Donahue et al. [15], in which the authors used a truncated version of AlexNet [16] to arrive at a Deep Convolutional Activation Feature (DeCAF) for generic visual recognition. Several efforts to remove/adapt variations in features that transfer across domains or tasks have been conducted, including some that are similar in nature to our research [4,17,18]. Li et al. [17] introduced a multiscale algorithm that pools across domains in an attempt to achieve invariance to out-of-plane rotations for object-recognition tasks. Mopuri and Babu [4] formulated a compact image descriptor for semantic search and content-based image retrieval applications with the aim of achieving scale, rotation, and object placement invariance by pooling deep features from object proposals. Tzeng et al. [18] formulated a semi-supervised domain transfer approach that, during fine-tuning to the target domain, uses a new loss-function that combines standard softmax loss with a "soft label" distillation (cf. [19]) to the softmax output mean vector and a domain confusion loss for domain alignment, which iteratively aims to first optimize a classifier that best separates domains and then optimizes the representation to degrade this classifier's performance. The approach by Tseng et al. [18] is similar to one of our methods in Section 7.4, but the goal is different; their approach aims to transfer domains within and end-to-end deep network, whereas ours aims to obtain an invariant representation for training lighter-weight classifiers on new samples from an unspecified target domain.

The use of pretrained deep representations is not new to face biometrics, but investigating the content of these deep features has only recently attained interest. Particularly, Parde et al. [20] investigated how well properties of deep features can predict nonidentity-related image properties. They examined images from the IJB-A data set [10] and concluded in the abstract of [20] that "DCNN features contain surprisingly accurate information about yaw and pitch of a face." Additionally, they revealed that it is not possible to determine individual elements of the deep-feature vector that contained the pose information, but that pose is encoded in a different set of deep features for each identity. Our work builds on theirs by investigating pose issues in much greater detail as well as exploring even more generic invariance for attribute-related information content across identity-trained deep representations.

7.3   Analysis of Face Networks

A typical example of a face image-processing network is a face-recognition network [2,3]. These networks are usually trained on large data sets using millions of images of thousands of identities, with the objective of minimizing negative log likelihood error under a softmax hypothesis function. Choice of training set is often made to capture wide variations both within and between identities, so training sets usually contain images with a wide variety of image resolutions, facial expressions, occlusions, and face poses. The resulting network is generally able to classify the correct training identities independently of the presence of these traits in the images.

Contrary to many closed-set image classification tasks, one characteristic of face-recognition systems in practice is that the identities in the training and test sets differ. As previously mentioned, this partially stems from the fact that training a deep neural network is computationally expensive, but in deployment settings, novel identities must be enrolled frequently. Hence, the softmax output for a given test image provides little value because it saturates with respect to training identities. Although one could train a secondary classifier atop softmax outputs, this is typically not done because the saturating effects remove potentially useful information. Instead, common practice is to use the output of the presoftmax layer of the network as a feature vector to represent an input face image. We will refer to vectors within this vector space as *deep features*.

To compute the similarity between two faces, the deep features are compared using some sort of distance function or classifier (e.g., Euclidean distance [2], cosine distance [3], and Siamese networks [12]). To increase classification accuracy, a metric-learned embedding is added (e.g., triplet loss [2,9] or joint Bayesian [12]), which projects the deep features into a space that is trained to increase the similarity of deep features from the same identity, while decreasing the similarity of deep features extracted from different identities. Because of the enormous boost in face-recognition performance that these sorts of systems have provided, a prevailing, albeit disputed and factually ungrounded, belief is that the deep features are mostly independent of image parameters that are not required for face recognition. In this section, however, we show that we are able to classify nonidentity-related facial attributes as well as face pose from the deep features—both before and after triplet-loss embedding—demonstrating that deep features maintain significant information in their representation about image parameters that are independent from identity.

7.3.1 Attribute prediction experiments

To investigate the independence of deep features from nonidentity-related facial attributes, we performed experiments on the CelebA data set [5], which contains around 200,000 face images, each of which is hand-annotated with 40 binary facial attributes. We obtained the deep features (832-dimensional before and 256-dimensional after triplet-loss embedding) and the according hyperface annotations from Chen et al. [3] for all images of the CelebA data set. Using the deep features from the CelebA training set, we trained 40 linear SVMs—one for each attribute—and optimized the C parameter for each attribute individually using the CelebA validation set. Because of the large imbalance of many of the attributes within the CelebA data set (cf. [6]; for example approximately 98% of the images lack the Bald attribute), we trained all SVM classifiers to automatically balance between positive and negative labels. On the test set, we predicted all 40 attributes and compared them with the ground-truth labels from the data set. We split the errors into false-positives and false-negatives, where false-positives in this case correspond to an attribute labeled as absent, but predicted to be present.

To get an idea how well the attributes can be predicted, we compare predictions with a version of the state-of-the-art AFFACT network [7] that was trained using the attribute balancing proposed [6]. The results of this experiment are shown in Figure 7.2, where we have split the 40 attributes into identity-related, -correlated, and -independent. For identity-dependent attributes such as Male, Pointy Nose, and Bald, we can see that the prediction from the deep features results in approximately the same error as AFFACT's. Hence, these attributes are well-contained in the deep features, despite the fact that the network was not trained to predict these attributes. Exceptions are attributes like Narrow Eyes and Oval Face. These attributes have a high overall prediction error, and we hypothesize that they may have been difficult for data set providers to consistently label.

Identity-correlated attributes (e.g., hair style and hair color) are usually stable, but might change more or less frequently. These attributes are generally more difficult to predict from the deep features. Although predictions from the deep features are still better than random, the associated errors are considerably higher than those corresponding to classifications made by the AFFACT network. Note that the features after the triplet-loss embedding

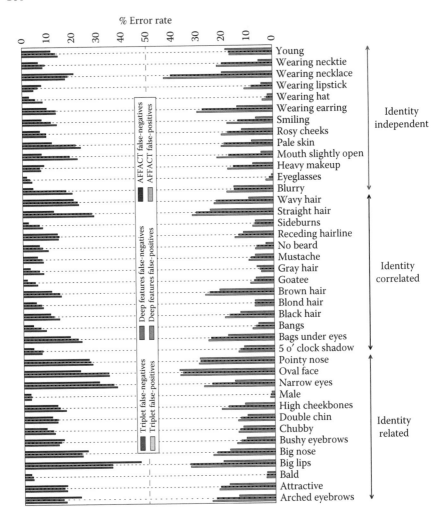

FIGURE 7.2

Attribute prediction from deep features. The error in attribute prediction from the deep features before and after triplet-loss embedding is displayed, split into false-negatives (attributes labeled as present, but predicted as absent) on the left and false-positives (absent attributes predicted as present) on the right. For comparison, the AFFACT network results show the state-of-the-art attribute prediction. Attributes are loosely grouped into identity-related, identity-correlated, and identity-independent subsets.

predict the identity-correlated attributes worse than before the embedding. Interestingly, for some attributes like Gray Hair or Sideburns we can observe a difference in false-positives and false-negatives. Although we can predict the *absence* (low false-positives) of these attributes similarly or even better than

AFFACT, the prediction of the *presence* (low false-negatives) of them is reduced for the deep features.

Finally, the identity-independent attributes such as `Smiling`, `Blurry`, and `Wearing Necklace` are far more difficult to predict from the deep features than from the AFFACT network, but the classification is still noticeably better than random. This suggests that identity-unrelated attribute information is still contained within the deep features because otherwise both the false-positive and false-negative error rates should be 50%. Hence, the prediction capability of deep features for those attributes is reduced when training the network to classify identity, and even more reduced after triplet-loss embedding, but some attributes like `Wearing Hat` or `Eyeglasses` can still be predicted with very high accuracy. This ultimately means that although some nonidentity-related information corresponding to some attributes is attenuated during network training and triplet-loss embedding, other nonidentity-related information is preserved, and we do not arrive at a feature space that is truly independent of nonidentity-related attribute information.

7.3.2 Pose-prediction experiments

When the pose of a face changes from frontal to full profile, its visual appearance alters dramatically. Two faces of different identities in the same pose are generally more similar than the face of the same identity in different poses. Hence, the network needs to learn a representation that is able to differentiate poses from identities, a task that has been shown to be difficult for non-network–based algorithms [21]. When training the network using softmax, the last layer can combine different elements of the deep features to obtain a representation that is independent of pose. In practice this means that the deep features may very well contain the pose information, and it is, thus, possible to predict the pose from the deep features.

We performed another experiment on the CelebA data set, in which we attempt to predict pose from deep features, using the same splits in training, validation, and test sets. Because the employed networks were trained with using horizontally flipped images, there is no way to differentiate between positive and negative yaw angles, and hence, we used the absolute yaw angle as target. Because the CelebA data set does not provide the pose information, we took the yaw angles automatically estimated by the state-of-the-art hyperface algorithm [3] as target values for the pose prediction. Note that the hyperface yaw angle estimates are relatively precise for close-to-frontal images, but they become unreliable for larger yaw angles greater than 45 degrees.

To determine if pose information is generally contained in deep features—not just the deep features of Chen et al. [3]—we extracted the penultimate layer from two more networks: the VGG face network [2] (4096-dimensional, layer `FC7`, post-ReLU) used for identity recognition and the AFFACT network [7] (1000-dimensional) that was trained for facial-attribute prediction. Despite the different tasks that the networks are trained for, intuition suggests that

TABLE 7.1

Predicting pose from deep features. Results are given for the pose-prediction experiments on the CelebA test set, using yaw angles automatically extracted with the hyperface algorithm as target values. Shown are the average absolute difference between predicted and hyperface yaw angles in degrees. The count of images in the according hyperface yaw range (shown left) are given in the rightmost column

Yaw	AFFACT	VGG	Deep feat.	Triplet-loss	Count
0–15	6.1	7.7	8.1	10.6	14,255
15–30	4.6	6.3	5.8	7.1	4,527
30–45	5.6	7.2	5.7	6.7	872
>45	9.1	11.4	10.3	13.9	268
Total	5.8	7.4	7.5	9.7	19,922

all of the networks should have learned their tasks independently of face pose because changes in pose do not change the target labels. For feature extraction, images of the CelebA data set were cropped according to the detected hyperface bounding box* and scaled to resolution 224×224 (cf. [7]), which happens to be the input resolution for both VGG and AFFACT.

For each type of deep feature, we trained a linear regression model using the CelebA training set. Using this model, we predicted the yaw angle contained inside the according deep features and compared it with the hyperface yaw angle. The results given in Table 7.1 display the average distance between the two values in degree. Interestingly, a global trend is that the yaw angle in half-profile pose (from 15 to 45 degrees) could be predicted with the highest precision, whereas close-to-frontal pose angles seem to be a little more difficult. This suggests that poses up to 15 degrees do not make a large difference for feature extraction, possibly because of the over-representation of data in this range. On closer examination of the deep feature types, the outputs of the penultimate layer of the AFFACT network seem to be least stable to pose variations—potentially because how the network was trained—although the deep features from Chen et al. [3] and VGG [2] have higher prediction errors, which are once more superseded by the triplet-loss-projected versions of [3]. Interestingly, we observe the general trend that deep features with more elements (4096 for VGG and 1000 for AFFACT) contain more yaw information than shorter vectors (832 before and 256 after triplet-loss embedding). Given that the average pose-prediction errors are generally below 10 degrees, we can conclude that the yaw angle (and we assume that the same is true for pitch and roll angles) can still be predicted from all kinds of deep features, and

---

*Both the VGG and AFFACT networks have shown to be stable to different scales and rotation angles. Proper alignment of the face is, hence, not necessary for either of the two networks.

hence, the deep features are not invariant to pose. Finally, choosing nonlinear basis functions could almost certainly enhance pose-prediction accuracy, but the fact that we are able to do so well without them already demonstrates the presence of noticeable pose information within the deep features of all three networks.

7.4 Toward an Invariant Representation

In this section, we explore the problem of formulating an invariant representation. We perform this exploration using perturbations to the canonical handwritten digit-classification data set MNIST [11], augmenting the familiar LeNet topology that is shipped as an example with the Caffe framework [22].

7.4.1 Preliminary evaluation on rotated MNIST

Because we are interested in investigating the deep-feature representation, we do not use the final softmax output of the network during evaluation. This is in contrast to common practice for the MNIST data set. Instead, we train the network using a softmax loss to learn the representation, but then remove the final layer (`softmax` and `ip2`) of the network to investigate the output of the penultimate `ip1` layer. This is analogous to using the penultimate representation for enrollment and recognition in a face-recognition network.

The task that we aim to accomplish with this set of experiments is to evaluate the invariance of the `ip1` layer to rotations on the inputs and to explore how to enforce such invariance. As a baseline approach, we artificially rotated the MNIST images using seven different angles in total: $-45°$, $-30°$, $-15°$, $0°$, $15°$, $30°$, and $45°$. We limited our rotations to this range to avoid legitimate confusions of digits induced by rotation (e.g., 9 and 6). Using the default training images, augmented with each rotation, we trained the standard LeNet network with the default solver until convergence was attained on a similarly augmented form of the validation set. Using the softmax output, we obtained a final classification accuracy of 98.98% on the augmented form of the test set. This result indicates that the network was able to learn to classify rotated images successfully.

Using the learned representation, we truncate the network after the `ip1` layer's `ReLU` activation function, extracting 500-dimensional feature vectors for all images rotated at all angles. To represent each of the classes, we simply average the `ip1` feature vectors (which we call the Mean Activation Vector [MAV]) of all training images of a class—an approach similar to that commonly used in face pipelines [2]. We ran a simple evaluation on the `ip1` features of the test set by computing the cosine similarity to all 10 MAVs; if the correct class had highest similarity, we consider the sample to be classified correctly. Using this approach, the total classification accuracy dropped to 89.90%, which is a noticeable decline from using softmax classifications. A more detailed comparison between the two models is given in Figure 7.4.

Momentarily ignoring the decline in accuracy, a more fundamental question arises: has training across rotations led the representation to become invariant to rotation angle (i.e., is there noticeable information regarding the representation embedded within ip1)? As a first step in answering this question, we train a linear regressor for each label in the training set and attempt to predict rotation angle from the extracted ip1 features. As the regression target, we use the known rotation angle. On the test set, we classified the ip1 features with the regressor of the corresponding label. The mean and the standard deviation—averaged over all 10 labels—are shown in light-gray color color in Figure 7.3. Even though the original images have an inherent rotation angle—people slant hand-written digits differently—we can reasonably predict the angle with a standard error of around 15 degrees.

Noting our ability to predict pose angle with reasonable accuracy, we then computed an MAV of ip1 features of each label for each angle separately from the training set. At test time, we estimated the angle of the test ip1 feature and computed the similarity to all 10 MAVs with that angle. Using this approach, our average classification accuracy increased to 95.44%, cf. Figure 7.4. Hence, exploiting information from exogenous input features contained in the ip1 representation actually allows us to improve classification accuracy.

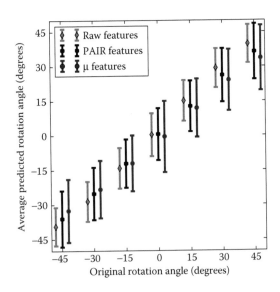

FIGURE 7.3

Angle prediction. This figure shows the average results of predicting the angles used to rotate the MNIST test images from the ip1 features of LeNet.

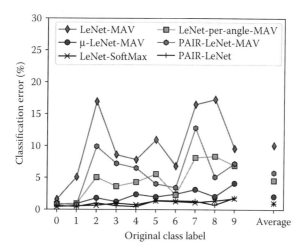

FIGURE 7.4

Digit-classification errors. Classification errors are given for several techniques, individually for each digit and as an average. Techniques that use the Mean Activation Vector (MAV) are shown with fill markers, whereas softmax-approaches are given in black and only for comparison. For LeNet-MAV, the MAV on the original LeNet is used, evaluation is performed as the class with the lowest cosine distance. For LeNet-per-angle-MAV, one MAV per angle is computed on the training set, and the angle is estimated for test images, and evaluation is performed using only the MAVs for this angle. The μ-LeNet was trained using the corresponding MAV as target, evaluation is similar to LeNet-MAV. The PAIR-LeNet was trained using pairs of images, evaluation is similar to LeNet-MAV.

7.4.2 Proposed architectures to enhance invariance

In this subsection, we modify LeNet in two distinct ways, for which intuition suggests that it would lead to a representation that is more invariant to exogenous input variations. The first such architecture, μ-LeNet, uses a distillation-like approach by regressing to the mean `ip1` vector across each class at the `ip1` layer of the truncated LeNet. The second architecture, PAIR-LeNet, introduces a Siamese-like topology to encourage two distinct inputs to have the same representation.

7.4.2.1 The μ-LeNet

Distillation, introduced by Hinton et al. [19], was designed for the purpose of "knowledge transfer" between end-to-end networks, generally of different topologies. The procedure involves softening the output softmax distribution of the network to be distilled and using the softened distribution outputs as soft labels for the distillation network. Using this as motivation, we use a related, but different approach, in which we aim to achieve a more invariant representation. Namely, using the MAV of the `ip1` output of a trained LeNet, we train a new μ-LeNet `ip1` representation by using the MAVs as regression targets. To stabilize and speed up the learning process, we also performed a min-max normalization on the feature space from 0 to 1 as a preprocessing step.

Using cosine distance with respect to the MAV, our recognition rate was 97.18%, which was, noticeably better than using the `ip1` layer under the original MNIST topology, cf. Figure 7.4. On the other hand, although a decrease in angle classification success is noticeable, this decrease is very slight, as shown in dark-gray color in Figure 7.3. The μ-LeNet also took significantly longer time to train than the typical LeNet topology.

7.4.2.2   The PAIR-LeNet

The μ-LeNet blatantly attempts to force input images to their respective class's mean `ip1` representation output from a trained LeNet. A different approach is to consider pairs of images at a time and optimize their representations to be as identical as possible. We refer to this approach as PAIR because it uses image pairs with Perturbations to Advance Invariant Representations. The idea is depicted in Figure 7.5 and is similar to a Siamese network topology. We trained the network on the rotated MNIST training set using randomized pairs of rotations until convergence was attained on the validation set. The classification rate of the end-to-end network on the augmented test set was 99.02%. Using cosine distance with respect to each label's MAV, we

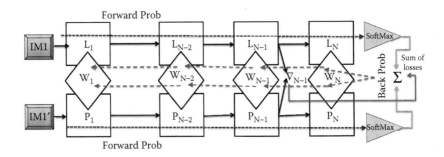

FIGURE 7.5
PAIR network topology. A generalized schematic of the Perturbations to Advance Invariant Representations (PAIR) architecture. Pairs of training images are presented to a network with different losses, but shared weights. Conventional softmax loss is used for both of the input images, then Euclidean loss between the activation vectors of the two images is added. This aims to force a similar representation between image pairs of the same class.

obtained a recognition rate of 94.21% in the derived `ip1` feature space, cf. Figure 7.4. As shown in black color in Figure 7.3, angle-classification success decreased, but this decrease was only slight. This suggests that, although the topology learns a representation that is useful for recognition, information about rotation angle still resides in the representation.

7.4.3 Analysis and visualization of the proposed representations

Our MNIST experiments in the previous section demonstrate that we are able to obtain marginal improvements with respect to rotation invariance, but contrary to our expectations, we were suprised how small the improvements are. In this section, we empirically analyze the feature spaces learned by our proposed approaches to better understand their properties.

Consider an idealized invariant representation designed to characterize the 10 digit classes from MNIST. One characteristic that we would expect the feature space representation to have is a rank no greater than 10. Thus, after subtracting the mean feature vector of the data matrix and performing singular value decomposition, the vast majority of the variance should be explained by the first singular vectors with the rest accounting for only minor noise.

In Figure 7.6, we plot the scree diagram for each of our proposed approaches using `ip1` layer vectors extracted from the rotated MNIST test set as the data matrix for the singular value decomposition. From analyzing the scree diagrams, we see that the μ-LeNet representation is approximately rank-10, with little variance explained by subsequent components. This indicates not only that the network has converged well on the training set, but it also indicates that the training set seems to have similar characteristics to the test set. Because the test set is basically balanced in terms of class labels, the sharp degradation in singular values suggests either that their respective singular vectors either do not represent the underlying digits, but rather constituent parts of the underlying digits, or some digits are more difficult to discriminate than others and require more variance to do so.

As a follow-up analysis, we performed heat-map visualizations, depicting activations across the rotated test set. These heat maps are shown in Figure 7.7. In each of the figures, digit classes are sorted in ascending order (0 through 9) from left to right. Within each digit class, poses are sorted in ascending order ($-45°$ to $45°$). For all three of the feature spaces, we can discern 10 modes corresponding to each of the digits, but the Raw-LeNet's feature space is far more scattered. Moreover, within each of the 10 modes, we see a trend of amplification or attenuation moving from right to left. This suggests a noticeable dependence on pose in the Raw-LeNet representation. The μ-LeNet in Figure 7.6(b) exhibits far more stability in the relative value of feature vector elements within a mode. The PAIR-LeNet representation in Figure 7.6(c) fluctuates for a given element within a mode more than the

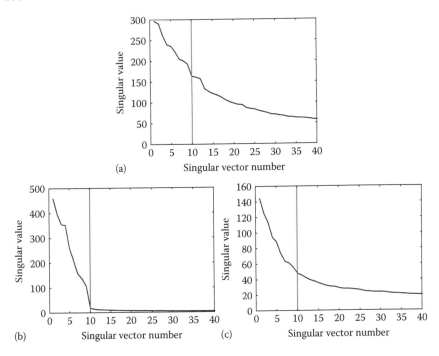

FIGURE 7.6

Screen diagrams for derived LeNet feature spaces: (a) Raw LeNet, (b) μ-LeNet, and (c) PAIR-LeNet. These screen plots depict the variance (singular value) explained by the first 40 of 500 singular vectors from the test set data matrices. Light-gray colored lines reflect the max number of singular vectors expected under an ideal representation. All feature vectors were min-max normalized to $[-1, 1]$ for visual comparison.

μ-LeNet, resulting in noisier looking horizontal lines, but there is little visual evidence of clear-cut dependence on pose. Especially for the μ-LeNet, the fact that we see discernible horizontal line segments within each mode of the data matrix suggests that remaining pose-related information is present in low-magnitude noise.

Separate colorbars are shown for each plot in Figure 7.7 to get good relative visualization of the data matrices. Note that the Raw-LeNet's features are somewhat greater in intensity than the PAIR-LeNet. The μ-LeNet's features are noticeably smaller in magnitude than either Raw-LeNet or μ-LeNet because of normalization required to get the distillation-like training to converge. Note that the PAIR-LeNet has a much sparser feature space than either of the other two architectures. We hypothesize that this is a result of the PAIR architecture's shared weights.

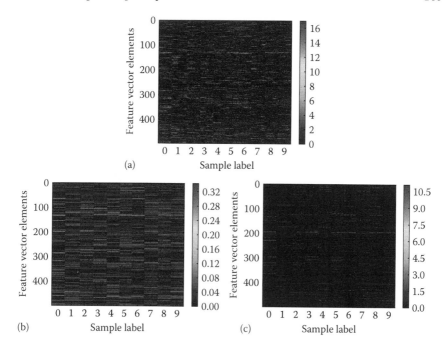

FIGURE 7.7
Heat maps for derived LeNet feature spaces: (a) Raw LeNet, (b) μ-LeNet, and
(c) PAIR LeNet. These deep features were taken from the `ip1` layer of the
network after the ReLU activation. Rows represent feature vector elements,
while columns correspond to samples from the rotated MNIST test set.

Another interesting question is: to what degree are individual feature vector elements associated with a particular class. To this end, we attempt to
block-diagonalize each of the data matrices by performing row operations so
as to maximize contrast. Specifically, given a $M \times N$ data matrix, where M
is the feature dimension and N is the number of samples with known labels,
we iterate through the M rows. For each unique label l, we assign error for
the ith feature vector element and the lth label as:

$$E_{il} = \sum_{j=1}^{N} \frac{D_{ij}(1 - I(y_j, l))}{D_{ij} I(y_j, l)}, \tag{7.1}$$

where $I(\cdot)$ is an indicator function that yields 1 if y_j is equal to l and 0 otherwise. Picking a cluster associated with the value of l for which Equation (7.1)
is minimized and assigning the ith row of the matrix to that cluster for all rows
$(i = 1, \ldots, M)$, then vertically stacking the clusters from minimum to maximum l value yields the optimal block-diagonalization by the error measure in
Equation (7.1).

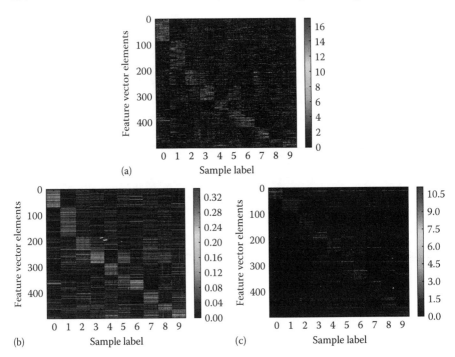

FIGURE 7.8
Diagonalizations of the data matrices of LeNet: (a) Raw-LeNet, (b) μ-LeNet, and (c) PAIR-LeNet. Like Figure 7.7, these deep features were taken from the `ip1` layer of the network after the ReLU activation. Rows represent feature vector elements, while columns correspond to samples from the rotated MNIST test set. In contrast to Figure 7.7 which displays rows in order, rows were reordered to best diagonalize the data matrix as per Equation (7.1).

The respective block-diagonalizations are shown in Figure 7.8. Although we can see a clear-cut diagonal structure in all three of the plots, the Raw-LeNet is far noisier than either the μ-LeNet or the PAIR-LeNet. The block-diagnol structure of the PAIR-LeNet is faint because of sparsity, but has the fewest high-intensity off-block diagonal elements. However, the few off-block diagonal elements are highly saturated. Although the μ-LeNet architecture has a strong block diagonal, we see that certain feature vector elements are easily confused between classes (e.g., those with high response for 3 and 5). Interestingly, far more features have high response for 1 than any other class, even though classes are relatively well distributed. In all three cases, however, although there is clear-cut block-diagonal structure, the strong co-occurrences of certain elements between classes suggest that enforcing saturation for a given class requires higher-level abstractions.

7.5   Conclusions and Future Work

The research and evaluations that we have conducted in this chapter suggest not only that one should expect the presence of noninvariances in deep feature spaces derived via common objective functions, but also that attempting to attenuate such noninvariances and simultaneously maintain recognition performance is a challenging task, even with objective functions explicitly designed to do so.

Generally, our analysis of the performance of face-attribute prediction from face-identity–derived deep features suggests what we would expect: that identity-related attributes tend to be more readily discriminated than identity-correlated attributes, which are in turn more easily discriminated than identity-independent attributes. However, the fact that second-stage classifiers were able to recognize all attributes with noticeably better than random performance dictates that there is still some information about those attributes contained in the deep-feature representations, and consequently, that the deep features are sensitive to variations in these attributes. Also, the fact that some identity-independent attributes (e.g., `Wearing Hat` or `Wearing Lipstick`) were easily recognized suggests that—perhaps fundamentally—the representational capacity to best recognize an identity necessarily carries information that is highly relevant to recognizing the presence or absence of these nonidentity-related attributes.

With respect to predicting the pose of a face, we find that across several network topologies trained for different tasks and on different data sets, pose information can be accurately recovered, even with simple linear regressors. We hypothesize that the reason the AFFACT and VGG network feature spaces offered most readily predictable pose information may have something to do with their lack of sensitivity to alignment, training across jittered data, and the generally high dimensionality of the feature space. Future research to ascertain which factors lead to pose discrimination could involve deepening these networks and reducing dimensionality by bottlenecking the output. In this chapter, we have not addressed how noninvariance changes as a result of network depth, which is a relevant topic since adding layers to the network literally adds layers of abstraction.

Both topology changes that we introduced to LeNet—the μ-LeNet and the PAIR-LeNet—exhibited some characteristics that we would expect from an invariant representation, but did not contribute noticeably to rotation-invariance because our ability to recognize rotation barely diminished. Perhaps variations resulting from rotation (and pose) are difficult to attenuate. This would again be an interesting problem on which to explore the effects of deeper representations. Another interesting experiment would be to analyze the effects of using similar architectures to attenuate nonpose-related exogenous input variations.

Although the features extracted from the μ-LeNet still included information about pose, the scree diagram in Figure 7.6(b) suggests that this information is contained only in the part that is not varying much (i.e. it is not expressed strongly). Hence, as shown in Figure 7.4, this network achieved the best cosine-based classification accuracy. However, because the MAVs are identical between training and test set, this result is surely biased. Translating this experiment to face recognition, the MAV would be computed over training set identities, whereas evaluation would be performed on test set identities, which are different. In future work, we will investigate a μ-network approach in such a face recognition setting.

For the PAIR-LeNet, we chose to reduce representational differences only between pairs of images. Although this reduced pair-wise distances in the feature space, it did not work well to reduce class-wise variance. An interesting extension to the PAIR network would be to use several images of one class or identity at the same time and try to reduce the variance over the batch. Perhaps, with large enough batch size and images of homogeneous label per batch randomly selected across the training set, we could further reduce variance in the underlying deep feature space.

Acknowledgment

This research is based on work supported in part by NSF IIS-1320956 and in part by the Office of the Director of National Intelligence (ODNI), Intelligence Advanced Research Projects Activity (IARPA), via IARPA R&D Contract No. 2014-14071600012. The views and conclusions contained herein are those of the authors and should not be interpreted as necessarily representing the official policies or endorsements, either expressed or implied, of the ODNI, IARPA, or the U.S. Government. The U.S. Government is authorized to reproduce and distribute reprints for governmental purposes notwithstanding any copyright annotation thereon.

References

1. O. Russakovsky, J. Deng, H. Su, J. Krause, S. Satheesh, S. Ma, Z. Huang, A. Karpathy, A. Khosla, M. Bernstein, A. C. Berg, and L. Fei-Fei. ImageNet large scale visual recognition challenge. *International Journal of Computer Vision (IJCV)*, 115(3):211–252, 2015.

2. O. M. Parkhi, A. Vedaldi, and A. Zisserman. Deep face recognition. *British Machine Vision Conference (BMVC)*, 1(3):6, 2015.

3. J.-C. Chen, R. Ranjan, A. Kumar, C.-H. Chen, V. M. Patel, and R. Chellappa. An end-to-end system for unconstrained face verification with deep convolutional neural networks. In *International Conference on Computer Vision (ICCV) Workshop*, pp. 360–368, Piscataway, NJ, IEEE, 2015.

4. K. R. Mopuri and R. Venkatesh Babu. Object level deep feature pooling for compact image representation. In *Conference on Computer Vision and Pattern Recognition (CVPR) Workshops*, pp. 62–70, Piscataway, NJ, IEEE, 2015.

5. Z. Liu, P. Luo, X. Wang, and X. Tang. Deep learning face attributes in the wild. In *International Conference on Computer Vision (ICCV)*, pp. 3730–3738. IEEE, 2015.

6. E. M. Rudd, M. Günther, and T. E. Boult. MOON: A mixed objective optimization network for the recognition of facial attributes. In *European Conference on Computer Vision (ECCV)*, Amsterdam, 2016.

7. M. Günther, A. Rozsa, and T. E. Boult. AFFACT - Alignment free facial attribute classification technique. In International Joint Conference on Biometrics (IJCB). Springer, 2017.

8. J.-C. Chen, V. M. Patel, and R. Chellappa. Unconstrained face verification using deep CNN features. In *Winter Conference on Applications of Computer Vision (WACV)*, pp. 1–9. IEEE, 2016.

9. S. Sankaranarayanan, A. Alavi, C. D. Castillo, and R. Chellappa. Triplet probabilistic embedding for face verification and clustering. In *Biometrics Theory, Applications and Systems (BTAS)*. IEEE, 2016.

10. B. F. Klare, B. Klein, E. Taborsky, A. Blanton, J. Cheney, K. Allen, P. Grother, A. Mah, and A. K. Jain. Pushing the frontiers of unconstrained face detection and recognition: IARPA Janus benchmark A. In *Conference on Computer Vision and Pattern Recognition (CVPR)*. IEEE, June 2015.

11. Y. LeCun, L. Bottou, Y. Bengio, and P. Haffner. Gradient-based learning applied to document recognition. *Proceedings of the IEEE*, 86(11):2278–2324, 1998.

12. Y. Taigman, M. Yang, M. A. Ranzato, and L. Wolf. Deepface: Closing the gap to human-level performance in face verification. In *Conference on Computer Vision and Pattern Recognition (CVPR)*. IEEE, June 2014.

13. G. B. Huang, M. Ramesh, T. Berg, and E. Learned-Miller. Labeled faces in the wild: A database for studying face recognition in unconstrained environments. Technical Report 07-49, University of Massachusetts, Amherst, MA, October 2007.

14. J. Wang, Y. Cheng, and R. S. Feris. Walk and learn: Facial attribute representation learning from egocentric video and contextual data. In *Conference on Computer Vision and Pattern Recognition (CVPR)*. IEEE, 2016.

15. J. Donahue, Y. Jia, O. Vinyals, J. Hoffman, N. Zhang, E. Tzeng, and T. Darrell. DeCAF: A deep convolutional activation feature for generic visual recognition. In *International Conference on Machine Learning (ICML)*, pp. 647–655, 2014.

16. A. Krizhevsky, I. Sutskever, and G. E. Hinton. Imagenet classification with deep convolutional neural networks. In *Advances in Neural Information Processing Systems*, pp. 1097–1105, Red Hook, NY, Curran Associates, 2012.

17. C. Li, A. Reiter, and G. D. Hager. Beyond spatial pooling: Fine-grained representation learning in multiple domains. In *Conference on Computer Vision and Pattern Recognition (CVPR)*, pp. 4913–4922. IEEE, 2015.

18. E. Tzeng, J. Hoffman, T. Darrell, and K. Saenko. Simultaneous deep transfer across domains and tasks. In *International Conference on Computer Vision (ICCV)*, pp. 4068–4076, Piscataway, NJ, IEEE, 2015.

19. G. Hinton, O. Vinyals, and J. Dean. Distilling the knowledge in a neural network. *arXiv preprint arXiv:1503.02531*, 2015.

20. C. J. Parde, C. Castillo, M. Q. Hill, Y. I. Colon, S. Sankaranarayanan, J.-C. Chen, and A. J. O'Toole. Deep convolutional neural network features and the original image. In *International Conference on Automatic Face and Gesture Recognition (FG)*. IEEE, 2017.

21. M. Günther, L. El Shafey, and S. Marcel. Face recognition in challenging environments: An experimental and reproducible research survey. In T. Bourlai (Ed.), *Face Recognition Across the Imaging Spectrum*, 1st ed. Springer, Cham, Switzerland, 2016.

22. Y. Jia, E. Shelhamer, J. Donahue, S. Karayev, J. Long, R. Girshick, S. Guadarrama, and T. Darrell. Caffe: Convolutional architecture for fast feature embedding. In *ACM International Conference on Multimedia (ACMMM)*. ACM, 2014.

8

Stacked Correlation Filters

Jonathon M. Smereka, Vishnu Naresh Boddeti, and
B. V. K. Vijaya Kumar

CONTENTS

8.1   Introduction 175
8.2   Correlation Filter Background 178
8.3   Stacked Correlation Filters 178
    8.3.1   Initial correlation output (Layer 0) 180
    8.3.2   Stacked layers 182
    8.3.3   Correlation output refinement 183
8.4   Experiments and Analysis 185
8.5   Summary 191
References 192

8.1   Introduction

A correlation filter (CF) is a spatial-frequency array (equivalently, a template in the image domain) designed from a set of training patterns to discriminate between similar (authentic) and nonsimilar (impostor) match pairs. The CF design goal is to produce a correlation output displaying a sharp peak at the location of the best match from an authentic comparison and no such peak for an impostor comparison. CFs have been proven to be useful in challenging image-matching applications such as object alignment [1], detection [2–4], and tracking [5,6]. In this chapter, we will discuss the use of CFs for biometric recognition, when in verification scenarios there is limited training data available to represent pattern distortions.

A biometric-recognition system functions by matching a new unknown sample, referred to as a *probe*, with separate stored template(s), referred to as the *gallery*. In identification scenarios (essentially asking the question, "Who is this person?"), a probe sample is compared against a set of stored gallery templates, and the biometric system assigns the probe the same identity as the template resulting in the highest score. Under verification scenarios (asking the question, "Is this person who he or she claims to be?"), a single probe sample is compared against the gallery corresponding to the claimed identity with a preset threshold determining whether the probe is authentic or impostor (also referred to as 1:1 matching). There are a large number of CFs [2,7–9] that have been previously shown to perform well in biometric-recognition applications like face [10], iris [11], periocular [12], fingerprint [13], and palm-print [14] recognition.

The matching challenge is noticeably more difficult when only a single image is available for the gallery template, for example, as in real-world applications such as when matching crime-scene face images to face images in surveillance videos and in several NIST biometric competitions [15–17] designed to mimic such real-world scenarios. CFs can implicitly and efficiently leverage shifted versions of an image as negative training samples. Therefore CFs are better suited for the 1:1 matching problem in comparison to other classifiers like support vector machines and random forests, which are designed to discriminate between two or more classes. However, in challenging matching scenarios (e.g., because of the presence of in-plane deformations, occlusions, etc.), an authentic correlation output may be difficult to distinguish from an impostor correlation output as shown in Figure 8.1. This failure occurs because of lack of training data or discriminative content between the probe and gallery. This problem is not new or unique to biometrics, and usual efforts to address it include varying features, changing the method of recognizing a peak (e.g., peak-to-sidelobe ratio, peak-to-correlation-energy, etc.), and filter design. For example, the CF response of a trained maximum average correlation height (MACH) [18] filter is highly dependent on the mean training image, which may not always be a good representation of the authentic class. Accordingly, variants of the MACH design were introduced as a possible solution [8,19–23], such as the extended MACH (EMACH) [20,24] that includes a parameter to weight the bias from low-frequency components represented by the mean training image.

In much of the traditional design and usage of CFs, the previous work has focused specific attention on improving the discrimination ability of the resulting correlation outputs with the remainder (after the peak height or location are extracted) of the correlation shape being discarded. In this chapter, we will discuss and demonstrate a novel technique, known as stacked correlation filters (SCFs), for improving the effectiveness of CFs by using the insight that the expected shape of a correlation output for an authentic match pair is supposed to be noticeably different from the CF output shape for an impostor pair. Moreover, the process of identifying an authentic correlation shape can be used to refine the correlation outputs after the initial matching for improved discrimination.

The rest of this chapter is organized as follows. Section 8.2 provides a brief review of CF design and Section 8.3 details an overview of the SCF architecture. An examination of the effectiveness of the SCF method is provided in Section 8.4 through experimentation and analysis, and Section 8.5 provides a summary.

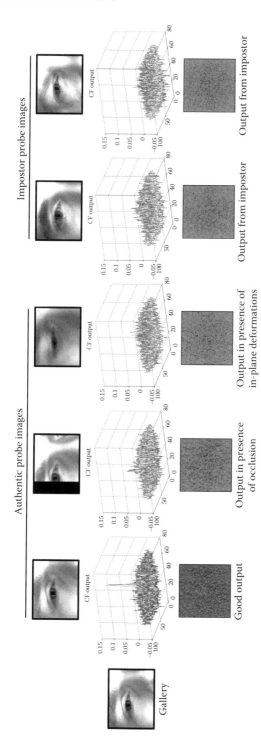

FIGURE 8.1

Example correlation outputs when comparing a gallery image to several different authentic probe images. A less discernible correlation output is obtained under difficult match scenarios (occlusion and in-plane deformations), whereas a good output (sharp peak at the location of best match) is obtained from the match pair with few distortions.

8.2   Correlation Filter Background

As previously discussed, CFs represent a family of classifiers generally designed for high-localization performance and can even be built with a single training image. However, because CFs are well explained in previous publications [25,26], we provide only a brief summary.

The main idea behind CF design is to control the shape of the cross-correlation output between the training image and the template (loosely called a *filter*) by minimizing the mean square error between the actual cross-correlation output and the ideal desired correlation output for an authentic (or impostor) input. For an authentic pair, the correlation output response, denoted as $\mathbf{C}$ (or $\mathbf{C}_i$ if the aim is to obtain a collection of CF outputs; e.g., for the i−th patch comparison of N total patches) from the probe image and gallery template should exhibit a peak at the location of the best match. For an impostor pair comparison, the CF output should exhibit no such peak. Each gallery template is designed to achieve said behavior on training data, which ideally extends to testing data from the same user.

The trained template, $\mathbf{H}$, is compared to a probe image, $\mathbf{Y}$, by obtaining the cross-correlation as a function of the relative shift between the template and the query. For computational efficiency, this is computed in the spatial frequency domain (u,v):

$$\hat{\mathbf{C}}(u, v) = \hat{\mathbf{H}}(u, v)\hat{\mathbf{Y}}^*(u, v) \qquad (8.1)$$

where:

    ^indicates the two-dimensional (2D) discrete Fourier transform (DFT)

    $\hat{\mathbf{Y}}(u, v)$ is the 2D DFT of the probe image pattern

    $\hat{\mathbf{H}}(u, v)$ is the CF (i.e., 2D DFT of the trained template)

    $\hat{\mathbf{C}}(u, v)$ is the 2D DFT of the correlation output $\mathbf{C}(x, y)$ with superscript

    $*$ denoting the complex conjugate.

The correlation output response peak value is used to indicate how similar the probe image is to the gallery template. In addition, the location of the peak can be used as an indication of the location of the trained pattern within the probe image. Thus, CFs can simultaneously locate and detect a biometric signature.

8.3   Stacked Correlation Filters

Conceptually, CFs are regressors, which map the image features to a specified output. Under challenging conditions like 1:1 matching of images with significant appearance variability (e.g., because of illumination, pose changes, etc.),

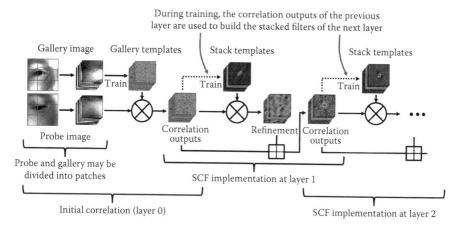

During training, the correlation outputs of the previous layer are used to build the stacked filters of the next layer

Gallery image   Gallery templates   Stack templates   Stack templates

Train   Train   Train

Probe image

Correlation outputs   Refinement   Correlation outputs

Probe and gallery may be divided into patches

SCF implementation at layer 1

Initial correlation (layer 0)

SCF implementation at layer 2

FIGURE 8.2
SFC overview: Operating first on the outputs from an initial matching stage, additional sets of CFs are consecutively layered with each set designed to refine the previous layer's output to the desired ideal output.

a single CF may be insufficient to deal with the wide range of variations in the image appearance (such as displayed in Figure 8.1). SCFs address this problem by using a layered approach to refine the correlation outputs from the initial matching and provide better discrimination between authentic and impostor pairs. The "stack" is built as a set of sequential CFs, the first layer being applied to the output after correlating the image features, and the subsequent layers applied to the refined outputs of the previous layer, as in Figure 8.2. The additional degrees of freedom allow SCFs to better deal with image variability.

Operating first on the outputs from an initial CF (referred to as *layer 0*), additional sets of CFs are consecutively layered with each set designed to refine the output from the previous layer to the ideal desired correlation output. The intuition is that because correlation outputs for matching pairs have an expected shape, an additional CF can be trained to recognize said shape and use that result to improve the final output. The resulting SCF architecture thus simplifies the matching process to a series of sequential predictions with the output of the previous layer feeding into the next. The approach can be applied to a single CF output; however, as we will discuss, there can be a more significant improvement by dividing the image into a set of patches such as shown in Figure 8.2.

The use of sequential predictions (feeding the output of predictors from a previous stage to the next) has been revisited many times in the literature. In Cohen and Carvelho et al. [27] and Daumé et al. [28], sequential prediction is applied to natural language processing tasks, whereas in Viola and Jones [29], a face-detection system was developed consisting of a cascaded series of classifiers. More recently the inference machines architecture [30,31] was proposed that reduces structured prediction tasks, traditionally solved using probabilistic graphical models, to a sequence of simple machine learning subproblems. Within biometrics, sequential predictions have been applied to perform score fusion [32,33]. SCFs operate on a similar intuition (iteratively applying weak classifiers to improve the final output), but offer a novel approach in both biometric recognition as well as in CF application.

8.3.1   Initial correlation output (Layer 0)

The initial matching between the probe and gallery images is contained within what is referred to as layer 0 of the SCF architecture. Distinct and separate from the proposed model, this layer relates the extracted features from the probe and gallery images to provide the initial correlation outputs that are fed to the SCFs.

Considering that each image (or patch region) only provides limited training information for the specific user in a 1:1 matching scenario, several feature sets (K total) are extracted to build a more robust template for the first stage of recognition (i.e., each pixel or a group of pixels is represented by a vector of features). We design one CF per feature channel such that the correlation output is a sum of individual outputs as depicted in Figure 8.3. Contrary to building K-independent CFs, we design all K CFs *jointly* to ensure the output satisfies our design criteria. The joint CF design is posed as the following optimization problem:

$$\min_{\mathbf{h}_1,\ldots,\mathbf{h}_K} \frac{1}{r} \sum_{j=1}^{r} \left\| \sum_{k=1}^{K} \mathbf{z}_k^j \otimes \mathbf{h}_k - \mathbf{g}_j \right\|_2^2 + \lambda \sum_{k=1}^{K} \|\mathbf{h}_k\|_2^2$$

$$s.t. \sum_{k=1}^{K} \mathbf{h}_k^T \mathbf{z}_k^j = u_j \quad j = 1, 2, \cdots, r \tag{8.2}$$

where:

$\otimes$ denotes the vectorized 2D correlation operator of the implied 2D arrays represented by its constituent vectors

$\mathbf{z}_k^j$ is vectorized representation of the k-th feature dimension of image $\mathbf{X}_j$ (total of r images)

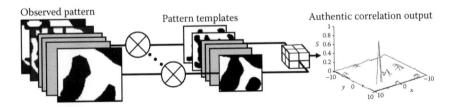

FIGURE 8.3
The outputs of individual feature channels for an image patch are aggregated to produce the final correlation output, which would have a sharp peak at the target location.

$\mathbf{h}_k$ is vectorized representation of the k–th correlation filter

$\mathbf{g}_j$ is vectorized representation of the desired correlation output for the j–th image

u_j is the constrained inner-product output, and $\lambda > 0$ is used to provide regularization from additive noise.

The terms in Equation 8.2 aim to minimize the mean square error between the actual and desired correlation output for each training image while constraining the filter to produce a large value (i.e., $u_j = 1$) given an authentic correlation and small value (i.e., $u_j = 0$) otherwise.

The following closed form expressions for the CF, $\hat{\mathbf{H}} = \left[\hat{\mathbf{h}}_1^T, \dots, \hat{\mathbf{h}}_K^T\right]^T$ (where $\hat{\mathbf{H}} \in \mathbb{C}^{KM \times 1}$, K feature sets, and M is the dimensionality of each feature set) and can be derived by posing the optimization problem in the frequency domain provided $\mathbf{g}_j$ is a delta function (see Rodriqeuz [25] and Kumar et al. [34] for more details):

$$\hat{\mathbf{H}} = \hat{\mathbf{T}}^{-1}\hat{\mathbf{Z}}\left(\hat{\mathbf{Z}}^{\dagger}\hat{\mathbf{T}}^{-1}\hat{\mathbf{Z}}\right)^{-1}\mathbf{u} \tag{8.3}$$

where $\dagger$ denotes the conjugate transpose and $\hat{\mathbf{Z}} = \left[\hat{\mathbf{z}}^1, \hat{\mathbf{z}}^2, \dots, \hat{\mathbf{z}}^r\right]$ is the training matrix composed of concatenated vectors $\hat{\mathbf{z}}_k^j$ from the r training images of K feature sets (i.e., $\hat{\mathbf{z}}^j = \left[(\hat{\mathbf{z}}_1^j)^T, \dots, (\hat{\mathbf{z}}_K^j)^T\right]^T$, where $\hat{\mathbf{z}}_k^j \in \mathbb{C}^M$ and $\hat{\mathbf{Z}} \in \mathbb{C}^{KM \times r}$), and $\hat{\mathbf{T}} = \lambda\mathbf{I} + \hat{\mathbf{A}}$ where $\mathbf{I}$ is the identity matrix to provide regularization to noise and $\hat{\mathbf{A}}$ is the power spectrum of the extracted features (average energy on the diagonal and cross-power spectrum on the off diagonals):

$$\hat{\mathbf{A}} = \begin{bmatrix} \frac{1}{r}\sum_j \hat{\mathbf{Z}}_1^{(j)\dagger}\hat{\mathbf{Z}}_1^{(j)} & \cdots & \frac{1}{r}\sum_j \hat{\mathbf{Z}}_1^{(j)\dagger}\hat{\mathbf{Z}}_K^{(j)} \\ \vdots & \ddots & \vdots \\ \frac{1}{r}\sum_j \hat{\mathbf{Z}}_K^{(j)\dagger}\hat{\mathbf{Z}}_1^{(j)} & \cdots & \frac{1}{r}\sum_j \hat{\mathbf{Z}}_K^{(j)\dagger}\hat{\mathbf{Z}}_K^{(j)} \end{bmatrix} \tag{8.4}$$

where $\hat{\mathbf{Z}}^j \in \mathbb{C}^{M \times M}$ is a diagonal matrix with $\hat{\mathbf{z}}^j$ along the diagonal.

Determining the template in Equation 8.3 requires the inverse of a rather large but sparse matrix $\hat{\mathbf{T}} \in \mathbb{C}^{KM \times KM}$ with a special block structure where each block is a diagonal matrix. By leveraging this structure, we recursively invert this matrix blockwise using the schur complement [35], which can be computed efficiently because the blocks are diagonal matrices:

$$\hat{\mathbf{T}}^{-1} = \begin{bmatrix} \mathbf{D} & \mathbf{E} \\ \mathbf{F} & \mathbf{G} \end{bmatrix}^{-1} = \begin{bmatrix} \mathbf{J} & \mathbf{K} \\ \mathbf{L} & \mathbf{M} \end{bmatrix} \tag{8.5}$$

where $\mathbf{J} = (\mathbf{D} - \mathbf{EG}^{-1}\mathbf{F})^{-1}$ and $\mathbf{K}, \mathbf{L}$, and $\mathbf{M}$ are functions of $\mathbf{J}, \mathbf{E}, \mathbf{F}$, and $\mathbf{G}$. Because $\mathbf{E}, \mathbf{F}$, and $\mathbf{G}$ are diagonal matrices, the matrix product $\mathbf{EG}^{-1}\mathbf{F}$ can be computed easily. Further, $\mathbf{J}$ has the same block structure as $\hat{\mathbf{T}}$ and can be inverted recursively using the same procedure.

8.3.2   Stacked layers

The SCFs are trained using only the correlation outputs and corresponding similarity or dis-similarity labels per match pair from the previous layer. In the case in which the image is divided into multiple patches, the SCF for the next layer is designed as a multichannel CF [1,4] with the correlation outputs of the patches (N total) from the previous layer constituting the features for each channel. At each layer l, the filter design problem is posed as (using τ training correlation outputs from the previous layer):

$$\min_{\mathbf{w}_1^l,\ldots,\mathbf{w}_N^l} \frac{1}{\tau} \sum_{i=1}^{\tau} \left\| \sum_{j=1}^{N} (\mathbf{c}_i^{l-1})_j \otimes \mathbf{w}_j^l - \mathbf{g}_i \right\|_2^2 + \Lambda^l \sum_{j=1}^{N} \left\| \mathbf{w}_j^l \right\|_2^2 \qquad (8.6)$$

where:

   $(\mathbf{c}_i^{l-1})_j$ is the vectorized representation of the j−th patch correlation output of the i−th image pair from the previous layer

   $\mathbf{w}_j^l$ is vectorized representation of the j−th SCF at layer l

   $\mathbf{g}_i$ is the vectorized representation of the desired correlation output

   Λ^l is used to provide regularization from additive noise.

Unlike the CFs used for the initial outputs (layer 0), the SCFs are *unconstrained*, that is, the result of an inner product between the input and template is not forced to a specific value.

The optimization problem in Equation 8.6 can be solved for the l−th SCF $\hat{\mathbf{W}}^l = \left[(\hat{\mathbf{w}}_1^l)^T, \cdots, (\hat{\mathbf{w}}_N^l)^T \right]^T$ (with $\hat{\mathbf{W}}^l \in \mathbb{C}^{NR \times 1}$, N is the number of patches, and R is the dimensionality of each patch), in the frequency domain (see [1] for more details):

$$\hat{\mathbf{W}}^l = \hat{\mathbf{S}}^{-1}\hat{\mathbf{Q}} \qquad (8.7)$$

where $\hat{\mathbf{S}} = \Lambda^l \mathbf{I} + \hat{\mathbf{B}}$ trades-off regularization to noise ($\mathbf{I}$ is the identity matrix) and the cross-power spectrum of the correlation outputs $\hat{\mathbf{B}}$:

$$\hat{\mathbf{B}} = \frac{1}{\tau} \begin{bmatrix} \sum_i (\hat{\mathbf{C}}_i^{(l-1)*})_1 (\hat{\mathbf{C}}_i^{(l-1)})_1 & \cdots & \sum_i (\hat{\mathbf{C}}_i^{(l-1)*})_1 (\hat{\mathbf{C}}_i^{(l-1)})_N \\ \vdots & \ddots & \vdots \\ \sum_i (\hat{\mathbf{C}}_i^{(l-1)*})_N (\hat{\mathbf{C}}_i^{(l-1)})_1 & \cdots & \sum_i (\hat{\mathbf{C}}_i^{(l-1)*})_N (\hat{\mathbf{C}}_1^{(i)})_N \end{bmatrix}, \qquad (8.8)$$

and $\hat{\mathbf{Q}}$ is the weighted average (or simple average when $\mathbf{g}_i$ represents a delta function as in our experiments) of the correlation outputs from the previous layer $(l$–$1)$:

$$\hat{\mathbf{Q}} = \begin{bmatrix} \frac{1}{\tau} \sum_{i=1}^{\tau} (\hat{\mathbf{C}}_i^{(l-1)*})_1 \hat{\mathbf{g}}_i \\ \vdots \\ \frac{1}{\tau} \sum_{i=1}^{\tau} (\hat{\mathbf{C}}_i^{(l-1)*})_N \hat{\mathbf{g}}_i \end{bmatrix}, \qquad (8.9)$$

and $(\hat{\mathbf{C}}_i^{(l-1)})_j$ is a diagonal matrix with the vectorized 2D DFT representation of $(\hat{\mathbf{c}}_i^{(l-1)})_j$ along the diagonal.

Equation 8.9 shows that the design and suitability of the filter relies on the average of the inputs, which, as previously discussed, for nonregular patterns may produce a bloblike result. However, because correlation planes are used as the inputs, which have an expected shape (authentic comparisons should contain a sharp peak at the location of best match), the SCFs will not necessarily suffer from this drawback.

Similar to the constrained filter used for the initial CF outputs (layer 0), the filter design for the SCFs requires computing the inverse of a rather large, but sparse matrix $\hat{\mathbf{S}} \in \mathbb{C}^{NR \times NR}$. As in the constrained case, the matrix has a block structure in which each block is a diagonal matrix, thus can also be computed efficiently as previously described.

Finally, the SCF method is computationally efficient because the application of all the filters in the SCF stack can be done in the frequency domain thereby requiring the Fourier transform only on the image features and a final inverse Fourier transform at the end of the stack. The SCF architecture requires computing the DFT of the K extracted features of dimensionality R for a complexity of $\mathcal{O}(KR \log(R))$. Applying the CFs in the frequency domain consists of single element-wise matrix multiplications for a complexity of $\mathcal{O}(KR)$ at each layer. Adding SCF layers only increases the cost linearly with each layer ($\mathcal{O}(LKR)$ where L is the number of layers). Thus, the overall cost (including the DFT cost) is $\mathcal{O}(KR(L + \log(R)))$.

8.3.3 Correlation output refinement

Because correlation is linear, as in other multilayer classifiers, a nonlinear operation is implemented to separate the layers (without the nonlinearity, the "stack" is equivalent to a single filter). In our design, we considered the following nonlinear operations.

- *Peak correlation energy (PCE)*: $f(\mathbf{C}) = \frac{\mathbf{C} - \mu_{\mathbf{C}}}{\sigma_{\mathbf{C}}}$, where $\mu_{\mathbf{C}}$ is the mean and $\sigma_{\mathbf{C}}$ is the standard deviation of that correlation output

- *Hyperbolic tangent (Tanh)*: $f(\mathbf{C}) = \tanh \mathbf{C}$

- *Sigmoid function (Sig)*: $f(\mathbf{C}) = \frac{1}{1 + e^{-\mathbf{C}}}$

- *Rectified linear unit (ReLU)*: $f(\mathbf{C}) = \max(0, \mathbf{C})$,

where the nonlinear operation, $f(\cdot)$, is applied to the output(s) of the SCFs when correlated with the previous layer's outputs. Recall that the purpose of the SCFs at each layer is to refine the previous layer's correlation output (i.e., to sharpen existing peaks for authentic comparisons while flattening the

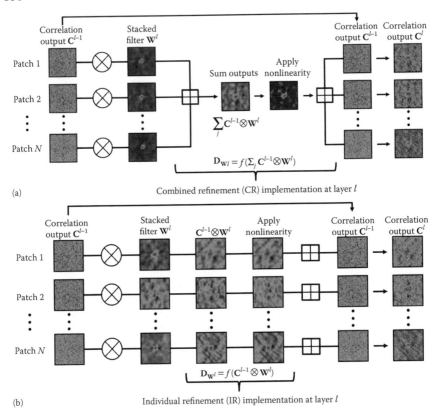

FIGURE 8.4
Visual representation of each method for computing the refinement to be
added back to the previous layer's outputs when applying a set of SCFs:
(a) Combined refinement and (b) Individual refinement.

outputs for impostor comparisons). To this end, we developed two refinement
methods shown in Figure 8.4 for determining the SCF outputs:

- *Combined Refinement* (*CR*): The outputs from the previous layer are correlated with the SCFs. The resulting correlation outputs are added, and
  a nonlinear function is applied to the sum. This result is added back to
  refine the outputs from the previous layer as shown in Figure 8.4a:

$$E(\mathbf{C}) = f\left(\sum_{j=1}^{N} \mathbf{C}_j\right). \tag{8.10}$$

- *Individual Refinement* (*IR*): The outputs from the previous layer are correlated with the SCFs. A nonlinear function is applied to each correlation

output, and the result is added back to refine the output from the previous layer as shown in Figure 8.4b:

$$E(\mathbf{C}) = f(\mathbf{C}_j). \tag{8.11}$$

An overview of the process used to train and test the SCFs for a given refinement method and nonlinear function is displayed by Algorithms 8.1 and 8.2, respectively.

Algorithm 8.1 SCFs: Training

Require: Image Pairs $\{\mathbf{X}_i, \mathbf{Y}_i\}_{i=1}^{\tau}$
1: **for** $i = 1$ to τ **do**
2:      Learn Gallery CF $\mathbf{H}_i$ using $\mathbf{X}_i$
3:      $\mathbf{C}_i^0 = f(\mathbf{H}_i \otimes \mathbf{Y}_i)$
4: **end for**
5: **for** $l = 1$ to L **do**
6:      Train SCF $\mathbf{W}^l$ using $\{\mathbf{C}_i^{l-1}\}_{i=1}^{\tau}$
7:      **for** $i = 1$ to τ **do**
8:          $\mathbf{D}_{\mathbf{W}_i^l} = E(\mathbf{C}_i^{l-1} \otimes \mathbf{W}^l)$
9:          $\mathbf{C}_i^l = \mathbf{D}_{\mathbf{W}_i^l} + \mathbf{C}_i^{l-1}$
10:      **end for**
11: **end for**

Algorithm 8.2 SCFs: Testing

Require: Image Pair $\{\mathbf{X}, \mathbf{Y}\}$
1: Learn Gallery CF $\mathbf{H}$ using $\mathbf{X}$
2: $\mathbf{C}^0 = f(\mathbf{H} \otimes \mathbf{Y})$
3: **for** $l = 1$ to L **do**
4:      $\mathbf{D}_{\mathbf{W}^l} = E(\mathbf{C}^{l-1} \otimes \mathbf{W}^l)$
5:      $\mathbf{C}^l = \mathbf{D}_{\mathbf{W}^l} + \mathbf{C}^{l-1}$
6: **end for**
7: Patch Score $P = \max \mathbf{C}^L$
8: Final Score $S = \sum_i P_i$

8.4 Experiments and Analysis

As we will show, achieving optimal performance by manually encoding a single layer or set of layers to a specific refinement method and nonlinearity is a nontrivial task. Thus, during training we searched over each combination and determined the best selection with cross-validation, a procedure we designated as *Dynamic Refinement* (DYN). The result allowed the SCF architecture to actively adjust to the quality of the outputs of the previous layer.

We briefly examine each SCF configuration and the benefits of the DYN approach using the Extended Yale B face data set (YaleB+) [36] in a 1:1 image-to-image matching scenario (excluding self-comparisons) with fivefold cross validation. As a measure of overall system performance, we report equal error rates (EERs) and verification rates (VRs) from the scores obtained by concatenating the associated folds.* Finally, we preprocess the images by a simple histogram normalization to compensate for drastic variations in illumination, resize each to 128×128 pixels for computational efficiency, and set $\lambda = 10^{-5}$ when training the CFs at layer 0 (parameter which is used in CF design to trade-off output resistance to noise and average correlation energy) with Λ^l (trade-off parameter used in the SCFs) being found via cross-validation at each layer l.

Composed of 2414 frontal-face images from 38 subjects, the images from the YaleB+ face data set capture 9 lighting directions and 64 illumination conditions for each user (thus there are $\sim$64 samples per subject). As shown in Figure 8.5 the data set images are well aligned (cropped and normalized to 192×168 pixels) because of careful acquisition under a "dome camera" setting, but exhibit severe illumination variations. Traditionally the data set is divided into five subsets according to the angle between the light source direction and the central camera axis ($12°$, $25°$, $50°$, $77°$, $90°$) where the recognition challenge increases as the angle increases. However, for the presented experiments in this chapter all of the images were treated equally to eliminate any bias.

When training the SCFs, we followed the procedure described by Algorithm 8.1. However, to prevent overfitting at each layer, a new randomized

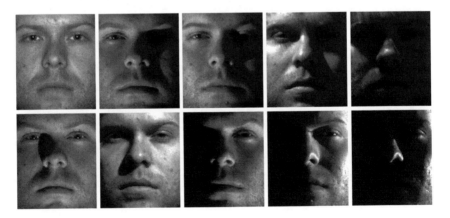

FIGURE 8.5
Sample intraclass images from the YaleB+ face database.

---

*We don't include rank-1 identification rates since performance is always >99%.

subset of the available match pairs are selected to be used for training. For example, within the first fold there are 1931 images available for training, and thus ~3.7 million possible match pairs including self comparisons (i.e., 1931×1931). When training the first layer we randomly select 23,000 (20,000 authentic, 3,000 impostor) of the available match pairs to build the SCFs. Then when training the next layer, we apply the SCFs from the first layer to a *new* random subset of another 23,000 match pairs, and so on. Significantly fewer outputs from impostor match pairs are employed during training because they are largely just noise (with the exception of hard impostors, as we will discuss later), so we do not need to include many examples.

Figure 8.6 shows the resulting EERs (right y-axis, dashed lines) and VRs (left y-axis, solid lines) from running the match scenario in which three SCF layers are trained and implemented for varying patch configurations: 1×1, 2×2 (four total patches), 3×3, ..., 6×6 (36 total patches). Best results for the CR method (94.62% VR, 1.84% EER) are obtained using the first layer of a 5×5 patch configuration and ReLU nonlinearity. The best results for the individual refinement (IR) method are obtained using the first (3.78% EER) and second layers (85.18% VR) of a 3×3 patch configuration and sigmoid nonlinearity.

From the plots in Figure 8.6, we first notice that there is not a single patch configuration or nonlinearity that consistently outperforms the others. Nonetheless, some relationships do emerge when focusing on each refinement method individually. For the CR method, employing more patches produces better performance in most of the experiments. This is because, by taking the sum of the set, patches with poor performance can be strengthened by those with better performance. Thus, adding patches produces a larger response. Accordingly, the IR method needs fewer patches to ensure that each perform similarly because of no specific mechanism being in place for adjusting poor-performing patches, and thus, fewer patches limit the disparity. Implemented individually, the deficiencies become more apparent; however, the conflicting nature will be shown to be beneficial when adjusting the model at each layer.

The experiments on the CR and IR methods also reveal what is referred to as the "hard imposter" phenomenon. Figure 8.7 shows an example occurrence in which the impostor score distribution ultimately separates into multiple modes. Not being limited to either refinement method (CR and IR), the problem appears when a set of outputs causing false peaks in impostor correlation planes are refined or sharpened similar to authentic comparisons. Continuing to iterate with each layer only further perpetuates the problem and pushes more impostor scores closer to the authentic distribution (i.e., causing more false-positives and thereby decreasing the verification performance, but not necessarily affecting the rank-1 identification rate because a large number of authentic scores are well above the EER and VR score thresholds). This is mitigated by cross-validating over-refinement and nonlinearity for each layer.

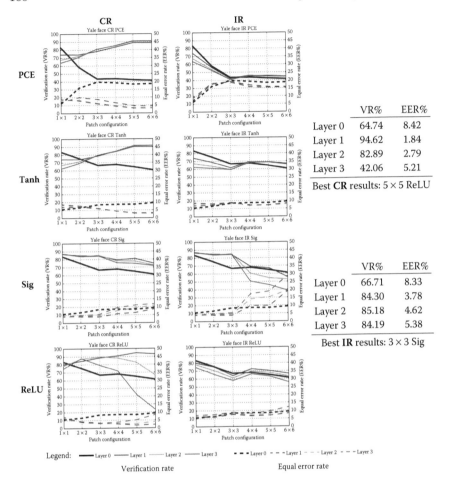

FIGURE 8.6
Performance of the CR and IR methods over varying patch comparisons and nonlinearities on the YaleB+ face data set. For each patch configuration (*x*-axis), each graph shows VR on the left *y*-axis (solid lines) and EER on the right *y*-axis (dashed lines). The last column displays two tables with the best results from each refinement method over all patch configurations. The tables illustrate numerically that aggressive refinement can quickly cause performance to degrade.

Figure 8.8 contains the distribution of nonlinearities and refinements resulting from searching over each nonlinearity and refinement during training, whereas Figure 8.9 shows the corresponding performance. Best results are obtained at the second layer of a 6 × 6 patch configuration (92.11% VR, 2.52% EER). Despite the best individual performance being achieved from using a single nonlinearity and refinement method (CR with ReLU on 5 × 5 patches),

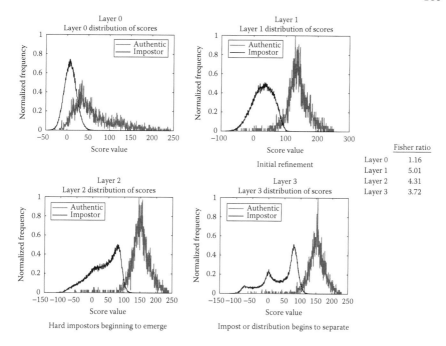

FIGURE 8.7

Example of how hard impostors can negatively effect the SCF output (can occur in both CR and IR methods). As shown, the impostor score distribution splits into multiple modes—easier impostors being separated from harder impostors, which have false peaks sharpened—causing the higher layers to perform poorly. The neighboring table shows the Fisher's ratio at each layer.

there is a significant improvement of the stability of the model (across all patches) when using the DYN. Progress is no longer limited to the first or second layer, with quick degradation thereafter, but instead improves more gradually across several layers. Thus, rather than needing to empirically test each refinement method and nonlinearity, we can now largely ensure that improvement will occur as long the two images are divided into patches for matching.*

By examining the histograms in Figure 8.8 we notice that there is little correlation with regard to when one refinement or nonlinearity is better than another (aside from the 1×1 patch configuration where the IR and CR methods are equivalent). Rather, what does stand out from the histograms is that the folds make similar choices (e.g., based on these results it is unlikely that the nonlinear operation implemented layer 2 of one fold will differ from that of another fold). However, we do notice that the Tanh and Sig operations are rarely used after the initial correlation (layer 0).

---

*Similar to the CR method, the DYN method works best with more patches.

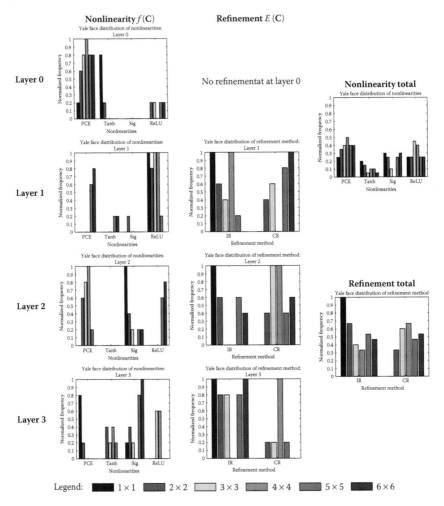

FIGURE 8.8
Distribution across nonlinearities (left column) and refinement methods (right column) used at each layer when searching over each during cross-validation (referred to as *dynamic refinement*) for each patch configuration on the YaleB+.

Finally, we examine the computational cost of applying each refinement method. Computation times (in milliseconds) from the IR and CR methods are shown in the table that follows (using a 4×4 patch configuration). Each time comparison is based on matching a random assortment of 1000 authentic and 9000 impostor comparisons (generating a 100×100 score matrix) and averaging the result. These tests were completed (using a single thread) on a 64-bit laptop with a 2.67 GHz dual core Intel i7-620M CPU and 8 GB of RAM

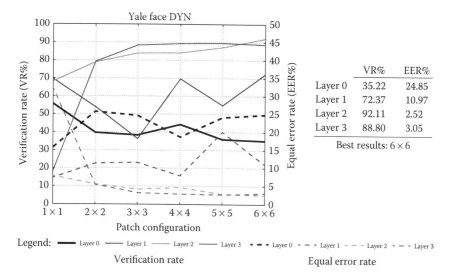

FIGURE 8.9
Performance of the dynamic refinement (DYN) method on YaleB+.

		Layer 1	
	Layer 0	**CR**	**IR**
Authentic	18.179 ms	29.295 ms	26.539 ms
Impostor	18.070 ms	28.924 ms	26.846 ms

The table shows the average computation time required for authentic and impostor match pairs after applying one SCF layer. As expected, the computational cost of the IR and CR methods is equivalent.

8.5 Summary

CFs are designed to specify a desired output for authentic and impostor matches and are widely used in many biometric applications. In this chapter we discussed SCFs, a fundamentally new CF paradigm in which instead of a single CF, we use a cascaded stack of filters to achieve the desired CF outputs. The included experiments demonstrate the effectiveness of SCFs for biometric verification, achieving substantial performance gains over a single CF under 1:1 image matching scenarios.

Although this chapter discussed the use of CFs for this purpose, the idea behind the process is not so limited. Accordingly, future efforts could investigate different classifiers and training configurations for improved performance (e.g., most impostor planes are essentially noise), thus discriminative methods may have difficulty trading off against the lack of structure, instead possibly requiring to mine hard impostors during training. One benefit of the SCF approach discussed in this chapter is that adding SCF layers only increases the computational cost linearly with each layer, accordingly, a challenge of any future research will be to demonstrate that any increased complexity is worth the increased computation or is limited to filter training.

References

1. V. N. Boddeti, T. Kanade, and B. V. K. Vijaya Kumar. Correlation filters for object alignment. In *IEEE International Conference on Computer Vision and Pattern Recognition*, pp. 2291–2298, June 2013.

2. A. Rodriguez, V. N. Boddeti, B. V. K. Vijaya Kumar, and A. Mahalanobis. Maximum margin correlation filter: A new approach for localization and classification. *IEEE Trans. on Image Processing*, 22(2):631–643, 2012.

3. J. F. Henriques, J. Carreira, R. Caseiro, and J. Batista. Beyond hard negative mining: Efficient detector learning via block-circulant decomposition. In *2013 IEEE International Conference on Computer Vision*, pp. 2760–2767, December 2013.

4. V. N. Boddeti and B. V. K. Vijaya Kumar. Maximum margin vector correlation filter. *ArXiv e-prints*, abs/1404.6031, 2014.

5. J. F. Henriques, R. Caseiro, P. Martins, and J. Batista. Exploiting the circulant structure of tracking-by-detection with kernels. In *European Conference on Computer Vision*, pp. 702–715. Springer-Verlag, October 2012.

6. D. S. Bolme, J. R. Beveridge, B. A. Draper, and Y. M. Lui. Visual object tracking using adaptive correlation filters. In *IEEE Conference on Computer Vision and Pattern Recognition*, pp. 2544–2550, June 2010.

7. J. F. Henriques, R. Caseiro, P. Martins, and J. Batista. High-speed tracking with kernelized correlation filters. *IEEE Transactions on Pattern Analysis and Machine Intelligence*, 37(3):583–596, 2015.

8. M. D. Rodriguez, J. Ahmed, and M. Shah. Action MACH a spatio-temporal maximum average correlation height filter for action recognition. In *IEEE International Conference on Computer Vision and Pattern Recognition*, pp. 1–8, June 2008.

9. H. Kiani, T. Sim, and S. Lucey. Multi-channel correlation filters for human action recognition. In *IEEE International Conference on Image Processing*, pp. 1485–1489, October 2014.

10. C. K. Ng, M. Savvides, and P. K. Khosla. Real-time face verification system on a cell-phone using advanced correlation filters. In *IEEE Workshop on Automatic Identification Advanced Technologies*, pp. 57–62, October 2005.

11. M. Zhang, Z. Sun, and T. Tan. Perturbation-enhanced feature correlation filter for robust iris recognition. *Biometrics, IET*, 1(1):37–45, 2012.

12. J. M. Smereka, V. N. Boddeti, and B. V. K. Vijaya Kumar. Probabilistic deformation models for challenging periocular image verification. *IEEE Transactions on Information Forensics and Security*, 10(9):1875–1890, 2015.

13. A. Meraoumia, S. Chitroub, and A. Bouridane. Multimodal biometric person recognition system based on fingerprint & finger-knuckle-print using correlation filter classifier. In *IEEE International Conference on Communications*, pp. 820–824, June 2012.

14. P. H. Hennings-Yeomans, B. V. K. Vijaya Kumar, and M. Savvides. Palmprint classification using multiple advanced correlation filters and palm-specific segmentation. *IEEE Transactions on Information Forensics and Security*, 2(3):613–622, 2007.

15. G. W. Quinn and P. J. Grother. Performance of face recognition algorithms on compressed images. Technical Report NISTIR 7830, National Institute of Standards and Technology (NIST), December 2011.

16. P. J. Phillips, W. T. Scruggs, A. J. O'Toole, P. J. Flynn, K. W. Bowyer, C. L. Schott, and M. Sharpe. FRVT 2006 and ICE 2006 large-scale results. Technical report, National Institute of Standards and Technology (NIST), IEEE, March 2007.

17. P. J. Phillips, J. R. Beveridge, B. A. Draper, G. Givens, A. J. O'Toole, D. S. Bolme, J. Dunlop, Y. M. Lui, H. Sahibzada, and S. Weimer. An introduction to the good, the bad, & the ugly face recognition challenge problem. In *IEEE International Conference on Automatic Face Gesture Recognition and Workshops*, pp. 346–353, March 2011.

18. A. Mahalanobis, B. V. K. Vijaya Kumar, S. Song, S. R. F. Sims, and J. F. Epperson. Unconstrained correlation filters. *Applied Optics*, 33(17): 3751–3759, 1994.

19. A. Mahalanobis and H. Singh. Application of correlation filters for texture recognition. *Applied Optics*, 33(11):2173–2179, 1994.

20. M. Alkanhal, B. V. K. Vijaya Kumar, and A. Mahalanobis. Improving the false alarm capabilities of the maximum average correlation height correlation filter. *Optical Engineering*, 39:1133–1141, 2000.

21. P. K. Banerjee, J. K. Chandra, and A. K. Datta. Feature based optimal trade-off parameter selection of frequency domain correlation filter for real time face authentication. In *International Conference on Communication, Computing & Security*, pp. 295–300. New York, ACM, February 2011.

22. B. V. K. Vijaya Kumar and M. Alkanhal. Eigen-extended maximum average correlation height (EEMACH) filters for automatic target recognition. In *Proceedings of the SPIE - Automatic Target Recognition XI*, Vol. 4379, pp. 424–431, Bellingham, WA, SPIE, October 2001.

23. B. V. K. Vijaya Kumar, A. Mahalanobis, and D. W. Carlson. Optimal trade-off synthetic discriminant function filters for arbitrary devices. *Optical Letters*, 19(19):1556–1558, 1994.

24. Y. Li, Z. Wang, and H. Zeng. Correlation filter: An accurate approach to detect and locate low contrast character strings in complex table environment. *IEEE Transactions on Pattern Analysis and Machine Intelligence*, 26(12):1639–1644, 2004.

25. B. V. K. Vijaya Kumar, A. Mahalanobis, and R. Juday. *Correlation Pattern Recognition*. New York: Cambridge University Press, 2005.

26. V. N. Boddeti and B. V. K. Vijaya Kumar. Extended depth of field iris recognition with correlation filters. In *IEEE International Conference on Biometrics: Theory, Applications and Systems*, pp. 1–8, September 2008.

27. W. W. Cohen and V. R. Carvalho. Stacked sequential learning. In *International Joint Conference on Artificial Intelligence*, pp. 671–676, San Francisco, CA, Morgan Kaufmann Publishers Inc., August 2005.

28. H. Daumé Iii, J. Langford, and D. Marcu. Search-based structured prediction. *Machine learning*, 75(3):297–325, 2009.

29. P. Viola and M. Jones. Rapid object detection using a boosted cascade of simple features. In *IEEE Conference on Computer Vision and Pattern Recognition*, pp. I-511–I-518, Piscataway, NJ, IEEE, December 2001.

30. D. Munoz, J. A. Bagnell, and M. Hebert. Stacked hierarchical labeling. In *European Conference on Computer Vision*, pp. 57–70, Verlag Berlin, Heidelberg, Springer, September 2010.

31. S. Ross, D. Munoz, M. Hebert, and J. A. Bagnell. Learning message-passing inference machines for structured prediction. In *IEEE Conference on Computer Vision and Pattern Recognition*, pp. 2737–2744, June 2011.

32. T. Murakami and K. Takahashi. Accuracy improvement with high confidence in biometric identification using multihypothesis sequential probability ratio test. In *IEEE International Workshop on Information Forensics and Security*, pp. 67–70, December 2009.

33. V. P. Nallagatla and V. Chandran. Sequential decision fusion for controlled detection errors. In *International Conference on Information Fusion*, pp. 1–8, Piscataway, NJ, IEEE, July 2010.

34. A Rodriguez. *Maximum Margin Correlation Filters*. PhD thesis, Carnegie Mellon University, Pittsburgh, PA, 2012.

35. D. V. Ouellette. Schur complements and statistics. *Linear Algebra and its Applications*, 36:187–295, 1981.

36. K. Lee, J. Ho, and D. Kriegman. Acquiring linear subspaces for face recognition under variable lighting. *IEEE Transactions on Pattern Analysis and Machine Intelligence*, 27(5):684–698, 2005.

9

Learning Representations for Unconstrained Fingerprint Recognition

Aakarsh Malhotra, Anush Sankaran, Mayank Vatsa, and Richa Singh

CONTENTS

9.1   Introduction 197
9.2   Deep Learning for Fingerprint: A Review 203
      9.2.1   Constrained fingerprint recognition 203
      9.2.2   Unconstrained fingerprint recognition 207
9.3   Latent Fingerprint Minutiae Extraction Using Group Sparse
      Autoencoder 209
      9.3.1   Experimental protocol 212
      9.3.2   Results and analysis 213
9.4   Deep Learning for Fingerphoto Recognition 215
      9.4.1   Fingerphoto matching using deep ScatNet 215
      9.4.2   Experimental protocol 217
      9.4.3   Results and analysis 218
           9.4.3.1   Expt1: Fingerphoto to fingerphoto
                    matching 218
           9.4.3.2   Expt2: Fingerphoto to live-scan matching 218
9.5   Summary 220
References   221

9.1   Introduction

For many decades, fingerprints have been a popular and successful biometric for person identification. They are gradually becoming the universal identifiers for a multitude of applications such as e-commerce, law enforcement, forensics, and banking. A recent survey shows that the market of fingerprint biometrics in 2016 was worth USD \$4.1 billion, accounting to 91% of the overall biometrics market [1]. Supported by the recent advancements in technology and data-handling capacity, automated fingerprint-recognition systems are extensively used in many civil and law enforcement applications, such as access-control systems, financial-transaction systems, and cross-border security at immigrations, where it is required to capture fingerprint images in a semi-controlled or uncontrolled manner.

With variety in requirements and applications, fingerprint literature comprises multiple kinds of fingerprints captured using different capture mechanisms [2]: (1) inked fingerprints, obtained by applying a specialized ink on finger skin and the ridge characteristics are captured using a fingerprint card, (2) live-scan fingerprints captured using optical, thermal, and capacitive sensors, (3) latent fingerprints captured using offline capture mechanisms, and (4) fingerphotos, captured using cameras from mobile devices such as cell phone or web camera. The type of fingerprint captured and tested varies with the variation in the application. Traditional civil and commercial applications of fingerprint recognition involves capturing the fingerprint ridge-valley structure in a controlled environment using live scanners to provide access control. Law enforcement and forensic applications involve matching latent fingerprints lifted from a crime scene to be matched with live scans or tenprint cards captured in a controlled environment. Recently, popular banking and access-control applications require logging with fingerprints captured from sensors in mobile phones, that are either capacitive or optical cameras.

Figure 9.1 shows multiple fingerprint images from the same finger captured during the same session using different controlled and uncontrolled capture mechanisms. Figure 9.1a–d shows controlled fingerprints captured using ink and live-scanner mechanisms. Figure 9.1e shows a latent fingerprint lifted from a ceramic surface, and Figure 9.1f contains a fingerphoto image captured using a smartphone. It can be observed that fingerprint data/information content visually differs based on the capture mechanism. Inked fingerprint or live-scan fingerprints are captured in a highly controlled environment producing high clarity, continuous ridge-valley patterns with very little or no background

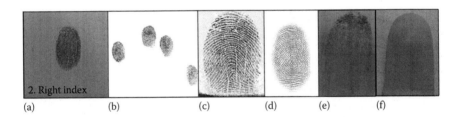

(a)   (b)   (c)   (d)   (e)   (f)

FIGURE 9.1
Images of the right index finger of a subject captured during the same session using different controlled and uncontrolled capture mechanisms: (a) inked fingerprint, (b)–(d) live-scan fingerprints: (b) CrossMatch sensor, (c) Secugen Hamster IV sensor, (d) Lumidigm multispectral sensor, (e) latent fingerprint lifted using black powder dusting method, and (f) fingerphoto image captured using smartphone camera.

variations. However, as we move into more uncontrolled fingerprint capture environments such as latent fingerprints or contact-less capture of fingerphoto images, the obtained ridge-valley pattern is very different from the traditional capture mechanisms.

The focus of this chapter is on unconstrained fingerprint recognition, that is, processing latent fingerprints and fingerphotos. Latent fingerprints are unintentional reproductions of fingerprints by transfer of materials from the skin to the surface in contact. The secretions in the surface of the skin such as sweat, amino acids, proteins, and natural secretions when come in contact with the surface, a friction ridge impression of skin is deposited on the surface. These impressions depend on the nature of the skin and the nature of the surface and often are not easily visible directly to human eyes unless operated by some external recovering technique such as powdering or chemical fuming [4]. Latent fingerprints are extensively used in forensic applications as common evidences in crime scene applications. Fingerphoto is a contact-less imaging of the finger ridge impression using a camera. A common application of fingerphoto includes use of present-day smartphone device or any other handheld electronic device to capture a photo of the frontal region of the finger. Because of the contact-less nature of the capture, the ridge-valley contrast obtained in a fingerphoto image will be highly different from a fingerprint image captured using a live-scan capture device. In such fingerprints, the ridge flow becomes highly discontinuous, a lot of background noise is introduced during capture, and only a partial fingerprint is obtained, while the rest is either lost or smudged during capture. Figures 9.2 and 9.3 show different capture variations and their corresponding challenges such as partial information, poor quality, distortions, and variations including background and illumination. This demonstrates that the variations in sensors and acquisition environment introduce a wide range of both intra- and interclass variability in the captured fingerprint in terms of resolution, orientation, sensor noise, and skin conditions. Extensive research has been undertaken in fingerprints captured using inked and live-scan methods [2,5–7]. However, these underpinned challenges and capture variations limit the use of existing automated live-scan matching algorithms for matching latent fingerprints and fingerphoto images.

The pipeline of an automated fingerprint identification system (AFIS) can be divided into a set of sequential stages: (1) region of interest (ROI) segmentation, (2) ridge quality assessment and enhancement, (3) feature extraction to uniquely represent the fingerprint ridge pattern, and (4) feature matching. As demonstrated previously, the most significant impact of varying acquisition environments and devices is on the feature extraction stage, which plays a pivotal role into the overall pipeline. Traditionally, fingerprint features can be categorized into three types: level-1, level-2, and level-3 features (Figure 9.4). Core and delta points comprise level-1 features, which are used for indexing large data sets, whereas dots, pores, and incipient ridges comprise level-3 features and are used for matching high-resolution fingerprint images [10]. Minutiae (level-2 features) are the most widely and commonly accepted features in

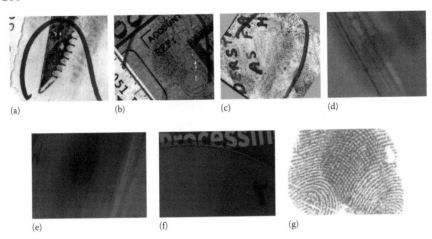

(a) (b) (c) (d)

(e) (f) (g)

FIGURE 9.2
Illustration of different challenges in latent fingerprint recognition: (a) partial information, (b) similar/complex background, (c) orientation variation, (d) slant surface/nonlinear distortions, (e) poor quality or blur, (f) illumination variation, and (g) overlapping fingerprints. (From CASIA-Fingerprint V5. *Chinese Academy of Sciences Institute of Automation (CASIA) Fingerprint Image Database Version 5.0.*)

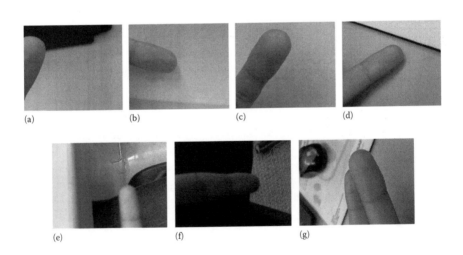

(a) (b) (c) (d)

(e) (f) (g)

FIGURE 9.3
Illustration of different challenges in fingerphoto recognition: (a) partial information, (b) similar/complex background, (c) pose variation, (d) slant skin surface, (e) poor quality or blur, (f) illumination variation, and (g) multiple fingers.

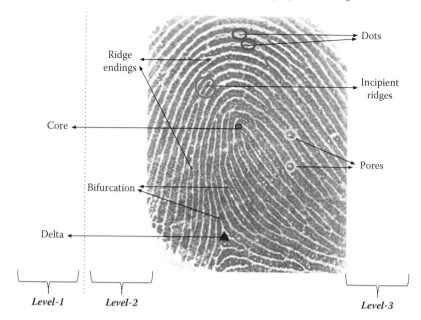

FIGURE 9.4
An illustration of traditional fingerprint features consisting of level-1, level-2, and level-3 features.

representing fingerprints. However, the noisy capture of fingerprints in uncontrolled environments affects the reliability of minutiae extraction. As shown in Figure 9.5, one of the state-of-the-art minutia extraction algorithms yields a high percentage of spurious minutiae [9,11,12].

Puertas et al. [13] studied the performance of manual and automated minutiae extraction in 2010. They also compared minutiae extraction from latent fingerprints with slap and rolled fingerprints. To analyze these experiments, they created a database of 50 latent fingerprints, 100 plain and rolled fingerprints, along with a 2.5 million extended gallery of tenprint cards. One of the major findings of their research is that despite a higher average number of minutia being detected in latent fingerprints by automated commercial-of-the-shelf system (COTS) (31.2) compared to 25.2 by manual marking, the system performed lower with 48.0% rank-1 identification compared to 72.0% for manual marking. This is because of the presence of many false minutia points detected in the automated minutiae extraction phase. The study performed by Puertas et al. [13] shows the need for a good automated minutiae extractor for latent fingerprints. The automated minutiae extractor should not only detect the true minutiae present in the latent fingerprints (or unconstrained fingerprints), but also reduce the number of spurious minutiae detected.

In literature, researchers have proposed several hand-crafted enhancements to minutia extraction and representation for unconstrained fingerprints.

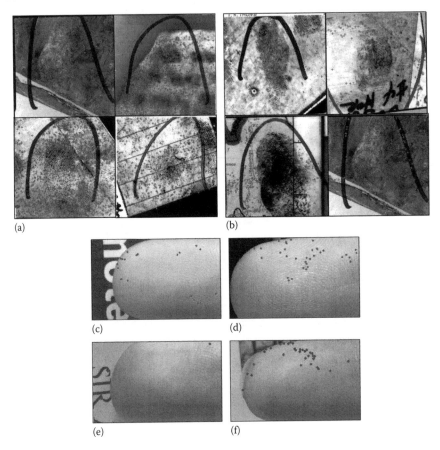

(a)   (b)

(c)   (d)

(e)   (f)

FIGURE 9.5
Sample images showing spurious minutiae extraction using VeriFinger 6.0 SDK in (a)–(b) latent fingerprints (from NIST SD-27 [8]) and (c)–(f) fingerphoto (from IIITD Fingerphoto database [9]).

In 2002, Lee et al. [14] proposed the usage of ridge frequency along with location, type, and orientation for each minutia. Mueller and Sanchez-Riello [15] used a hybrid approach of minutiae and skeleton data of the preprocessed fingerphoto image for fingerphoto matching. In 2010, Cappelli et al. [16] introduced Minutia Cylinder Code (MCC) as a robust and local descriptor of minutia. Many researchers have further used MCC descriptors for feature extraction in latent fingerprints [17–20]. In 2010, Jain and Feng [21] suggested the use of a collection of hand-crafted features including fingerprint skeleton, singular points, level-3 features, ridge flow, wavelength, and quality map as an extended feature set for latent fingerprint matching. Apart from minutiae-based features, researchers have also focused on other hand-crafted features such as Gabor features [22,23], local binary pattern, and Speeded

Up Robust Features (SURF) [24]. However, most of those approaches are not robust to large variations in unconstrained or poor-quality fingerprints and large data sets. Therefore, it is important to design a learning-based feature representation algorithm that uses different kinds of fingerprint samples from the training data to learn discriminatory feature representations that can tolerate noise and variations in data distribution.

In this chapter, we focus on two important and challenging applications of uncontrolled fingerprint recognition: (1) a forensic application of recognizing latent fingerprints and (2) a commercial application of matching fingerphoto images. We emphasize the importance of building "lights-out" automated matching for latent fingerprints and fingerphoto images using representation learning (specifically, deep learning) algorithms, without human intervention. We conduct a detailed survey of exiting approaches in uncontrolled fingerprint comparison and present state-of-the-art approaches for latent fingerprint and fingerphoto matching.

9.2 Deep Learning for Fingerprint: A Review

The literature for representation learning in fingerprint recognition can be segregated into constrained and unconstrained fingerprint recognition.

9.2.1 Constrained fingerprint recognition

As shown in Table 9.1, the earliest work for fingerprint recognition using the network-learning–based approach is proposed by Baldi and Chauvin [25] in 1993. Initially, the CCD camera-captured fingerprint image is subjected to preprocessing, segmentation, alignment, and normalization. Once these steps are performed both for gallery and probe images, the aligned pair of fingerprints are matched. A central 65×65 region is selected in gallery image and a 105×105 probable region is found in the probe image. Within this 105×105 patch of the probe image, the appropriate region corresponding to 65×65 regions of the gallery image is found by correlation. These 65×65 regions are compressed and downscaled to 32×32 regions. Two different filters of size 7×7 with a stride of 2 are convolved in this 32×32 region. The convolutions are followed by a sigmoid activation. The model is trained on 300 fingerprints and the testing is performed using 4650 fingerprint pairs. The network is trained using gradient descent on cross-entropy error. On the test pairs, authors reported an error rate of 0.5%.

Sagar and Alex [26] in their initial work in 1995 used fuzzy logic to model dark (ridges) and bright (valleys) levels of fingerprints to extract minutiae. As a further extension, in 1999, they used a combination of fuzzy logic system and neural networks for minutiae detection. A filter bank of 32 filters, each

TABLE 9.1

Summarizing the literature on the application of deep learning techniques for fingerprint recognition in constrained acquisition environment

Research	Application	Architecture	Database
Baldi and Chauvin [25]	Representation of the central patch of a fingerprint for matching	[Conv-1: two (7×7) filters, Sigmoid Activation]	*Train:* 300 fingerprints, *Test:* 4650 fingerprint pairs (In-house)
Sagar and Alex [26]	Minutiae prediction for a fingerprint pixel by modeling dark and bright patterns	[Fuzzification engine, $64\times32\times32\times2$ nodes]	—
Sahasrabudhe and Namboodiri [27]	Reconstruct noisy orientation field of fingerprints for better enhancement	[Patch-wise orientation field (Input), $f(\theta)$ and $s(\theta)$, 2 CRBF]	*Train:* 4580 good quality fingerprint images, *Test:* FVC 2002 DB3
Sahasrabudhe and Namboodiri [28]	Enhance noisy fingerprint using Convolutional Deep Belief Networks (CDBN)	[Fingerprint Image (Input), CDBN with Conv1 Probabilistic max Pool and Conv2, Gibbs sampling]	*Train:* FVC2000 DB1+FVC2002 DB1+NIST SD-4, *Test:* FVC2000 DB1+FVC2002 DB2+FVC2002 DB3
Wang et al. [29]	Orientation classification by creating a representation for fingerprint orientation field	3-layered sparse autoencoder [400, 100, 50 nodes]	NIST SD-4 (2000 *Train*, 400 *Test*)
Minaee and Wang [30]	ScatNet representation of the fingerprint for recognition	Mean and Variance of ScatNet features	PolyU fingerprint database
Jiang et al. [31]	Extract minutia by predicting minutia patches and location for fingerprints	JudgeNet and LocateNet: [Conv, Pool, Conv, Pool, Conv, Pool, Softmax]	*Test:* In-house 200 labeled fingerprints
Michelsanti et al. [32]	Classify orientation of fingerprint in four classes	VGG-F and VGG-S, replace FC with FC layer with four outputs, fine-tuned using fingerprints	NIST SD-4 (2000 *Train*, 2000 *Test*)

of size 1×5 in eight different orientations are considered to detect minutiae from every pixel. The average response of each filter is served as input to a fuzzification engine that converts this average into two values: dark and bright. These two values for the 32 blocks are fed into a neural network architecture: $64 \times 32 \times 32 \times 2$, which outputs if the pixel location is minutia or not. Compared to the vanilla fuzzy system, the use of neural network improved the true positive rate for minutia detection by 7% and the false-recognition rate decreased by 10%.

In 2015, Minaee and Wang [30] used Scattering Network (ScatNet) [33] for controlled fingerprint recognition. ScatNet features are translation and rotation invariant representation of a signal x, which are generated using a scattering network. The scattering network has an architecture similar to a convolutional neural network (CNN), however, its filters are predefined wavelet filters. Once ScatNet features are extracted, authors applied a principal component analysis (PCA) for dimensionality reduction and classification is performed using a multiclass support vector machine (SVM). The proposed algorithm is tested on the PolyU fingerprint database [34]. On a total of 1480 images, 50% train-test split is performed. Images are resized to 80×60 and using 200 PCA features, training of multiclass SVM is performed. On testing data, an equal error rate (EER) of 8.1% and an accuracy of 98% is reported. As compared to the previously reported results of minutia-based matching with EER of 17.68%, the reported EER of 8.1% is a significant improvement.

Recently in 2016, Jiang et al. [31] proposed a minutiae extraction technique for constrained fingerprint recognition using JudgeNet and LocateNet. Four kinds of labeled patches are created: (1) 45 patch: 45×45, (2) 45b patch: 45×45 patch with blurred exterior region, (3) 63 patch: 63×63 patches, and (4) 63b patch: 63×63 patches with blurred exterior region. To increase the available data, patches are generated in an overlapping manner (overlap $r = 9$ pixels). Also, to make the learning of the classifier rotation invariant and increase its training data, additional patches are augmented by rotating the generated patches by 90 degrees, 180 degrees, and 270 degrees and adding them to the training set. In the first step, a CNN-based architecture called JudgeNet identifies if the input patch has a minutiae or not. The architecture of JudgeNet is as follows:

$$[Conv\text{-}1 \quad Pool\text{-}1 \quad Conv\text{-}2 \quad Pool\text{-}2 \quad Conv\text{-}3 \quad Pool\text{-}3 \quad Softmax(2\ class)],$$

which uses an activation based on a rectified linear unit (ReLu) [35] for each of the convolution layers. Once the minutia patches are detected, a LocateNet is used for finding the minutia location in the central 27×27 region. Training is performed by creating nine nonoverlapping 9×9 blocks in the central 27×27 regions, each signifying a minutia location. The architecture of LocateNet is same as JudgeNet except for the last layer, which can be seen as follows:

$$[Conv\text{-}1 \quad Pool\text{-}1 \quad Conv\text{-}2 \quad Pool\text{-}2 \quad Conv\text{-}3 \quad Pool\text{-}3 \quad Softmax(9\ class)].$$

Further, spurious minutia are removed by taking a mean of the probabilities for each minutia location and keeping only those which are greater than 0.5. Such an approach is able to remove multiple spurious minutiae in nearby locations, which are generated because of overlapping patches. Using an in-house data set of 200 labeled fingerprint images, the authors reported the highest accuracy of 92.86% on 63b patches compared to 90.68%, 91.79%, and 91.76% in 45, 45b, and 63 patches, respectively. These results signify that a larger patch size with exterior region blurred is suitable for minutia detection using a deep CNN.

In literature, deep learning techniques are not only used to extract minutiae in controlled fingerprints, but they can also be used to enhance ridge patterns and reconstruct noisy orientation fields. In 2013, Sahasrabudhe and Namboodiri [27] proposed the use of unsupervised feature extraction using two continuous restricted Boltzmann machines (CRBMs). The horizontal and vertical gradients obtained from Sobel operator is given to two different orientation extraction functions. The output from each of these functions is used to train two CRBMs. The training is performed using 4580 good quality fingerprint images. For each hidden neuron, sigmoid activation along with a "noise control" parameter is used. The patch size used in the experiment is 60×60, which is downscaled to 10×10 and fed into the visible layer with 100 nodes. The number of hidden nodes used are 90. The testing is performed on the FVC 2002 DB3 database [36], and the outputs from both CRBMs are fused to get corrected orientation field, which is further used for enhancement using Gabor filters. The proposed approach detected fewer spurious minutiae (4.36%) and an improved true acceptance rate (TAR) of 49% at 10% false acceptance rate (FAR).

Sahasrabudhe and Namboodiri in 2014 extended their research [28] to enhance fingerprints by proposing convolutional deep belief network (CDBN)–based fingerprint enhancement. CDBNs are multiple convolutional restricted Boltzmann machines stacked together to form one deep network. The proposed architecture has two convolutional layers with a probabilistic max-pooling in between and is fed with the grayscale image. Training is performed using good quality hand-picked images from FVC 2000 DB1, FVC 2002 DB1, and NIST SD-4 database. Using this trained CDBN architecture, a representation of the input fingerprint image is obtained, which is reconstructed to get an enhanced fingerprint image. The testing is performed on FVC 2000 DB1 and DB2 [37] and FVC 2002 DB3 [36]. Compared to Gabor-based enhancement, an improved EER of 6.62%, 8.52%, and 23.95% is reported on the three data sets, respectively.

Wang et al. [29] used a sparse autoencoder [38] to classify fingerprints based on the orientation pattern (whorl, left loop, right loop, and arch). The sparse autoencoder is fed with an orientation field of the fingerprint image, and the autoencoder tries to reconstruct the input field. Features from the sparse autoencoder are fed into a multiclass softmax classifier where the input features x are classified to class y out of all of the k classes (in this case, $k = 4$). The sparse autoencoder model is trained using 2000 images and tested using 400 images from the NIST SD-4 database [39]. With three hidden layers with 400, 100, and 50 nodes, respectively, a classification accuracy of 91.9% is reported.

Recently in 2017, Michelsanti et al. [32] worked on classifying fingerprints based on the orientation: whorl (W), left loop (L), right loop (R), and arch (A). The arch orientation (A) is a combination of arch and tented arch from the NIST SD-4 because the database has a fewer number of samples for these two classes. Pretrained CNNs are used in the proposed algorithm (VGG-F [40] and VGG-S [40]) and are fine-tuned on a subset of fingerprint images taken from the NIST SD-4 database. For fine-tuning, the last fully connected layer of respective CNNs are removed and a fully connected layer with four nodes is added. To increase the number of samples while fine-tuning, data are augmented by adding left-right flipped images to the data set and interchanging the labels for the left and the right loop. While fine-tuning, adaptive local learning rate and momentum are used to speed up training. On the testing set, an accuracy of 95.05% using VGG-S is reported compared to 94.4% on VGG-F.

9.2.2   Unconstrained fingerprint recognition

Deep learning research in unconstrained fingerprint recognition has focused on only latent impressions (though limited) and research in fingerphoto matching has not used the capabilities of deep learning. Unlike constrained fingerphoto research, one of the primary challenges in latent fingerprint recognition is the lack of a large-scale labeled data set. Table 9.2 shows some of the popular data sets in literature for matching latent fingerprints. Most of the research papers in the literature have used these public data sets or have shown results on some private in-house data sets, impacting the reproducibility of the results.

To estimate orientation of ridge flow in latent fingerprints, Cao and Jain [47] trained a ConvNet to classify orientation field of a latent fingerprint in one of the 128 orientation patterns. These 128 orientations are learned from orientation patches obtained from the NIST SD-4 database, followed by training each of the 128 orientation classes of ConvNet by 10,000 fingerprint patches from the NIST SD-24 database [48]. The architecture for the trained ConvNet is as follows:

$$[Conv\text{-}1 \quad Pool\text{-}1 \quad Conv\text{-}2 \quad Pool\text{-}2 \quad Conv\text{-}3 \quad fc \quad output]$$

ReLu activations are used in each convolution layer followed by max-pooling layers. To introduce sparsity, dropout regularization is used after the fully connected fc layer.

For a probe latent fingerprint, a 160×160 patch from the foreground segmented latent fingerprint is fed into the trained ConvNet model. Orientation field is predicted and enhanced further using Gabor filtering. The experiments are performed on the NIST SD-27 [8] database. On the complete data set, the average root mean square deviation of 13.51 is reported, which is the lowest deviation in comparison to the other popular algorithms such as

TABLE 9.2

Characteristics of existing latent fingerprint databases

| Database | Number of | | Characteristics |
	Classes	Images	
NIST SD-27A [41]	258	258	Latent to rolled fingerprint matching. Contains 500 PPI and 1000 PPI exemplars. Manually annotated features are also available.
IIIT-D Latent Fingerprint [42]	150	1046	Latent fingerprint with mated 500 PPI and 1000 PPI exemplars, slap images of 500 PPI. Latent images are lifted using black powder dusting process and captured directly using a camera.
IIIT-D SLF [43]	300	1080	Simultaneous latent fingerprint with mated slap 500 PPI exemplars, two sessions of simultaneous latent fingerprint are lifted using black powder dusting.
MOLF Database [44]	1000	19,200	Dap, slap, latent and simultaneous latent fingerprints. Manually annotated features available.
WVU Latent Fingerprint [45]	449	449	500 PPI and 1000 PPI images of latent fingerprints with manually annotated features for latent to rolled fingerprint matching. However, database is not publicly available.
ELFT-EFS Public Challenge #2 [46]	1100	1100	500 PPI and 1000 PPI images in WSQ compressed format with manually annotated features. Database not publicly available.

LocalizedDict [49]. A rank-10 accuracy of 77.0% is obtained when latent fingerprints from NIST SD-27 database, which are enhanced using the proposed algorithm, are matched using a COTS system.

Sankaran et al. [50] proposed the first automated approach for minutia extraction from latent fingerprints using a stacked denoising sparse autoencoder (SDSAE). SDSAE learns a sparse complex nonlinear feature representation of the noisy input data. Once the latent fingerprint patch descriptor is learned, a softmax-based binary supervised classifier is learned to classify every image patch as a minutia or non-minutia patch. An autoencoder with an architecture of $[4096\ 1200\ 30\ 2\ 30\ 1200\ 4096]$ is trained using $20, 176, 436$ patches of size 64×64 obtained from a heterogenous data set of CASIA-FingerprintV5 [3], NIST SD-14 v2 [51], FingerPass [52], and MCYT [53]. On the NIST SD-27 database, minutiae-patch detection accuracy of 65.18% is obtained along with a rank-10 identification accuracy of 33.61% using NIST Biometric Imaging Software (NBIS) matcher.

Recently, Tang et al. [54] used fully FCN [55] and CNN instead of hand-crafted features for minutia detection. The proposed algorithm primarily has two steps: (1) generating candidates using a FCN on the input latent fingerprint image and (2) classifying candidates and estimating orientation. In addition to these two steps, each step also aims to get accurate coordinates of the minutia points with its multitask loss function. For generating candidates, a pretrained ZF-net [56], VGGNet [57], or residual-net [58] is used along with the classification and regression layer, as follows:

$$\left[\begin{array}{l} [ZF/VGG/Residual \\ \left\{ \begin{array}{l} (\text{Conv: } 256 \ (3 \times 3)\text{filters}) \ (\text{Conv: } 2 \ (1 \times 1)\text{filters}) : \text{Classification} \\ (\text{Conv: } 256 \ (3 \times 3)\text{filters}) \ (\text{Conv: } 2 \ (1 \times 1)\text{filters}) : \text{Location} \end{array} \right\} \end{array} \right]$$

To classify the proposals and to estimate the orientation, the pretrained models are connected to a pooling and softmax layer, as follows:

$$\left[ZF/VGG/Residual \ \text{pooling}(6 \times 6) \left\{ \begin{array}{l} \text{fc} \ \ \text{softmax}(2) : \text{Classification} \\ \text{fc} \ \ \text{fc}(3) : \text{Location and Orientation} \end{array} \right\} \right]$$

The network is trained on 129 latent prints from NIST SD-27 and an additional 4205 latent fingerprint images collected by China's police department from crime scenes. The database is tested on other 129 images of the NIST SD-27 database, and the precision and recall rate of 53.4% and 53.0%, respectively, are reported.

9.3   Latent Fingerprint Minutiae Extraction Using Group Sparse Autoencoder

In this section, we present a state-of-the-art approach for extracting minutiae from latent fingerprint images using group sparse autoencoder (GSAE) [59]. The primary idea is to use group sparse constraint in autoencoders so as to better distinguish between the minutia and nonminutia patches from latent fingerprints. It can be observed from Figure 9.6a that the local region around a minutia has a different ridge structure than a nonminutia patch. However, as shown in Figure 9.6b, latent fingerprint minutia patches lack a definite structure, making it challenging to learn meaningful information. Because of the nonuniform and uncertain variations in latent fingerprints, it has been challenging for researchers to define a model for extracting minutiae.

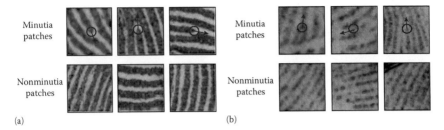

Minutia patches

Nonminutia patches

Minutia patches

Nonminutia patches

(a)   (b)

FIGURE 9.6

(a) High-quality fingerprint patches illustrating the difference in the ridge structure between minutia and nonminutia patches. (b) Local patches from latent fingerprints illustrating the lack of well-defined structures and noisy ridge patterns.

Human-engineered features such as gradient information and frequency-based information, provide limited performance because of the presence of background noise. Figure 9.7 illustrates the three main stages of the proposed GSAE-based latent fingerprint-minutia extraction algorithm.

1. *Pretraining*: In the first stage, lots of high-quality fingerprint image patches are used to pretrain a GSAE by preserving group sparsity, as follows,

$$J(W) = \arg\min_{W,U}[||X - U\phi(WX)||_2^2$$
$$+ \lambda(||WX_{hqm}||_{2,1} + ||WX_{hqnm}||_{2,1})] \qquad (9.1)$$

where, X_{hqm} and X_{hqnm} represent the high-quality fingerprint minutia and nonminutia patches. The regularization term ensures that the group sparsity is preserved within the minutia and nonminutia patches.

2. *Supervised fine-tuning*: In the second stage, labeled latent fingerprint image patches are used to fine-tune the GSAE. Further, a binary classifier (2ν-SVM) is trained using the extracted patches to differentiate between minutia patches and nonminutia patches.

3. *Testing*: In the third stage, the learned feature descriptor and the trained classifier are tested. An unknown latent fingerprint is divided into overlapping patches, a feature descriptor for each patch is extracted using the proposed fine-tuned GSAE algorithm, and then classified using the trained 2ν-SVM classifier.

For a latent fingerprint image for which minutia points need to be detected, the image is first divided into overlapping (16 pixels) patches of size 64×64. We have used 2ν-SVM [61] with radial basis function kernel for classification. 2ν-SVM is a "cost-sensitive" version of SVM that penalizes the training errors

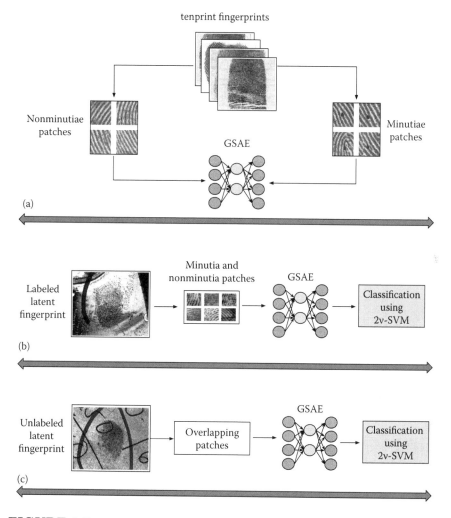

FIGURE 9.7
Architecture of the proposed minutiae-extraction algorithm using GSAE: (a) Pretraining, (b) supervised fine-tuning, and (c) testing.

of one class more than the other by assigning class specific weights to both the classes. This explicit penalty minimizes the false-negatives, while restricting the false-positives below a certain significance level. Hence, in the case of imbalanced class data or different cost of error, different importance can be given to the two types of errors, making sure that the majority class is not creating a bias. Further, in case of c-class classification problems, c different binary classifiers are created using the "one-vs-all" approach to train binary 2ν-SVMs. The primal form of 2ν-SVM optimization function [61] is given as

$$\min_{w,b,\psi,\rho} \frac{1}{2}||w||^2 - \nu\rho + \frac{\gamma}{n}\sum_{i\in I_+}\psi_i + \frac{1-\gamma}{n}\sum_{i\in I_-}\psi_i \qquad (9.2)$$

such that, (1) $y_i(k(w, x_i) + b) \geq \rho - \psi_i$, (2) $\psi_i \geq 0$, and (3) $\rho \geq 0$. Here, w is the decision boundary, x are the support vectors, y are the corresponding class labels, $k(w, x_i)$ is the kernel function, ψ_i are the slack variables, $\gamma \in \{0, 1\}$ is a parameter controlling the trade-off between false-positives and false-negatives, and $i = \{1, 2, \ldots, n\}$ for n support vectors.

9.3.1 Experimental protocol

GSAE requires training with a large number of minutia and nonminutia patches from the fingerprint images. These images are taken from a combined data set of CASIA-FingerprintV5 [3], NIST SD-14 v2 [51], FingerPass [52], and MCYT [53]. The description and properties of these data sets are summarized in Table 9.3. To make the feature learning supervised, minutiae are extracted from all the fingerprints using an open source minutia extractor *mindtct* of the NIST Biometric Imaging Software NBIS [62]. An image patch of size 64×64 ($w = 64$) is extracted with minutia at the center, thereby creating 10,088,218 minutia patches extracted from all the images. From every fingerprint, the same number of nonminutia patches and minutia patches are extracted to ensure same number of samples from both the classes. A stacked group sparse autoencoder is designed with the network layer sizes as {4096, 3000, 2000, 1000, 500}. With a mini-batch size of 10,000 image patches, we pretrained each layer for 40,000 epochs and then performed 10,000 epochs for fine-tuning. The architecture is trained with raw image intensities of these image patches (vector size 1×4096) as input. For evaluation, two publicly available latent fingerprint databases are used: NIST SD-27 and MOLF.

The primary objective is correctly extracting minutiae from latent fingerprint images. Therefore, the performance metric used in all these experiments is Correct Classification Accuracy (CCA), which denotes the ratio of correctly

TABLE 9.3
Summary of the composition and characteristics of the heterogeneous fingerprint database

Database	Capture type	No. of images	No. of minutiae
NIST SD-14 v2 [60]	Card print	54,000	8,188,221
CASIA-FingerprintV5 [3]	Optical	20,000	515,641
MCYT [53]	Optical, capacitive	24,000	571,713
FingerPass [52]	Optical, capacitive	34,560	812,643
Total		**132,560**	**10,088,218**

Note: This heterogeneous database is used as the pretraining data set for the proposed deep learning approach.

classified patches with the total number of patches. The overall accuracy is further split into class-specific classification accuracy: Minutia Detection Accuracy (MDA) and Nonminutia Detection Accuracy (NMDA). In terms of MDA and NMDA, although both the accuracies should be high, it is important to detect all the minutia patches accurately along with minimizing the occurrence of spurious minutia patches.

9.3.2   Results and analysis

The performance of the proposed approach is evaluated in terms of CCA, MDA, and NMDA on two different data sets, NIST SD-27 and MOLF, under four different experimental scenarios:

- Using VeriFinger [63], a popular commercial tool for fingerprints,
- Using the proposed architecture with only KLD,
- Using the proposed architecture with only GSAE, and
- Using the proposed architecture with KLD + GSAE.

We also compared the results with the current state-of-the-art algorithm proposed by Sankaran et al. [50]. Our proposed algorithm outperforms all the existing algorithms with a CCA of 95.37% and 90.74% on the NIST SD-27 and IIIT MOLF data set, respectively. The results on the NIST SD-27 and IIITD MOLF data set are summarized in Table 9.4, respectively. Further, in Table 9.5, we tabulate the approach and results of all the existing research works in literature. It is to be duly noted that in Table 9.5, the experimental protocols and data sets are different across the research works and thus cannot be directly compared and no strong conclusion can be drawn. However, it shows strong promises of deep learning toward automated latent fingerprint recognition.

TABLE 9.4

Classification results (%) obtained on the NIST SD-27 and MOLF latent fingerprint data sets

Database	Algorithm	Classifier	CCA	MDA	NMDA
NIST SD-27 [41]	VeriFinger	VeriFinger	90.33	20.41	96.80
	Sankaran et al. [50]	Softmax	46.80	65.18	41.21
	KLD	2ν-SVM	91.90	91.90	100
	GSAE	2ν-SVM	94.48	94.48	100
	KLD + GSAE	2ν-SVM	95.37	95.37	100
IIIT MOLF [44]	VeriFinger	VeriFinger	78.52	21.33	92.92
	KLD	2ν-SVM	59.25	84.17	52.97
	GSAE	2ν-SVM	90.14	90.44	90.07
	KLD + GSAE	2ν-SVM	90.74	90.63	90.37

CCA, Correct Classification Accuracy; MDA, Minutia Detection Accuracy; NMDA, Nonminutia Detection Accuracy.

TABLE 9.5
Performance comparison of different latent fingerprint minutia extraction algorithms using deep learning techniques

Paper	Problem	Algorithm	Train data	Test gallery	Test probe	Result
Cao and Jain [47]	Orientation estimation	7 layer ConvNet	1,280,000 patches	100,258 extended NIST SD-27	258 NIST SD-27	Average RMSD for orientation estimation: 13.51. Rank-1 identification accuracy: 75%.
Tang et al. [54]	Minutiae extraction	ZF/VGG/Residual Net + regression	129 NIST SD-27	258 NIST SD-27	129 NIST SD-27	F1-score for minutia extraction: 0.532. Rank-1 identification accuracy: 12%.
Sankaran et al. [50]	Minutiae extraction	Stacked Denoising Sparse Autoencoder	20,176,436 image patches	2,258 extended NIST SD-27	129 NIST SD-27	Minutiae detection accuracy: 65.18%. Rank-10 identification accuracy 34%.
Sankaran et al. [59]	Minutiae extraction	Group Sparse Autoencoder	20,176,436 image patches	258 NIST SD-27 + 2 million extended	129 NIST SD-27	Minutiae detection accuracy: 95.37%.
Sankaran et al. [59]	Minutiae extraction	Group Sparse Autoencoder	20,176,436 image patches	4,400 MOLF + 2 million extended	4,400 MOLF	Minutiae detection accuracy: 90.63%. Rank-10 identification accuracy 69.83%.

As shown in Table 9.4, the CCA for patches from the NIST SD-27 database is as high as 95% with the KLD + GSAE algorithm. It is approximately 5% higher compared to VeriFinger. However, the MDA of VeriFinger is about 20%, signifying that it rejects a lot of genuine minutia patches. Compared to Sankaran et al. [50], our proposed algorithm has an improvement of more than 30%. In regard to the accuracy of detection of nonminutiae patches, our algorithm outperforms both VeriFinger and Sankaran et al. [50] with 100% accuracy. The high accuracy of GSAE can be credited to 2v-SVM classification, which supports in making the false-positive error almost zero.

On the IIITD MOLF database, the accuracy is higher for the proposed algorithm KLD+GSAE algorithm. However, the accuracy of detection of non-minutiae patches is better by VeriFinger. The performance of the proposed algorithm on the MOLF database is not as good as on the NIST SD-27 because the number of testing data points on the MOLF database is colossal compared to the NIST SD-27 data set, and there are significant variations in the characteristics of the two databases.

9.4   Deep Learning for Fingerphoto Recognition

With increase in the usage of smartphones, the number of applications for various tasks such as banking, gaming, and Internet surfing has increased. Smartphones and other handheld devices have become one of the largest store houses of personal and critical information. Use of biometrics is a popular and successful mechanism to secure personal content and provide access control to handheld devices. As the name *handheld* devices suggests, the device is held in the hand most often, and hence the use of fingerprints to authenticate access seems rather intuitive. Though fingerprint authentication in itself is quite secure, it includes an overhead of the cost of an additional sensor and not all the smartphones have inbuilt fingerprint sensors. To overcome such a challenge, capturing the photo of the finger in a contactless fashion using the existing rear camera of a handheld device provides a viable alternative to fingerprints. Such contactless images of finger-ridge impression captured using a camera are called as *fingerphoto* [64]. However, the research in terms of fingerphoto authentication is still in its preliminary stages [11,24,65–68], and none of the existing works leverage the use of deep learning for extracting learned representation from fingerphoto images.

9.4.1   Fingerphoto matching using deep ScatNet

In this section, we present a novel authentication methodology for fingerphotos [9], which includes a novel feature representation based on Scattering Networks (ScatNet) [69]. The main steps of the algorithm are: (1) segmentation and enhancement, (2) ScatNet feature representation, and (3) feature matching. An adaptive skin-color–based segmentation is performed in the CMYK

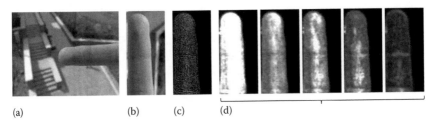

(a)   (b)   (c)   (d)

FIGURE 9.8
(a) Acquired image in natural outdoor environment, (b) segmented image after removing the background, (c) enhanced image to remove illumination noise, and (d) ScatNet feature representation for the enhanced image.

scale to segment the captured finger region from the noisy background. From the obtained region, an edge trace is performed on the boundary of the finger to extract an exact region of interest and a bounding box is made. The image is then enhanced to make ridge-valley information more distinctive and handle factors such as varying illumination and noise. Image is first converted to gray scale and median filtering is performed. In the next step, ridge-valley pattern is enhanced by sharpening, which involves subtracting Gaussian blurred image of $\sigma = 2$ from the original image. The output of segmentation and enhancement is shown in Figure 9.8.

For feature extraction, we want a representation that is not only rotation and translation invariant, but also preserves the high-frequency information contained in the fingerprint (ridges). We propose ScatNet-based feature representation [69], which has the property to encode and preserve high-frequency information and hence it is known to perform well on texture patterns. Let x be the input image signal in R^2 dimension. A representation that is locally affine invariant can be achieved by applying a low pass averaging filter as follows:

$$\phi_j(u) = 2^{-2j}\phi(2^{-j}u) \tag{9.3}$$

The representation obtained using such a filter is known as level zero ScatNet representation and is given as follows:

$$S_0x(u) = x \star \phi_j(u) \tag{9.4}$$

Though the representation is translation invariant, it loses high-frequency information as a result of averaging. To retain high-frequency information as well, a wavelet bank ψ is constructed by varying the scale and rotation parameter θ. The high frequency band pass wavelet filter is given as:

$$\psi_{\theta,j} = 2^{-2j}\psi(2^j r_\theta u) \tag{9.5}$$

and the wavelet modulus for high-frequency components is given as follows:

$$|x \star \psi_{\lambda 1}(u)| \tag{9.6}$$

Applying absolute function and taking the magnitude can be compared to the nonlinear pooling functions used in CNNs. To produce an affine invariant representation of these high-frequency components, a low pass filter is again applied to obtain first-order ScatNet features as follows:

$$S_1 x(u, \lambda_1) = |x \star \psi_{\lambda_1}(u)| \star \phi_j(u) \tag{9.7}$$

First-order ScatNet features are a concatenation of responses of all wavelets in filter bank ψ. The signal can be further convolved recursively by another set of high pass wavelet filter banks and averaged by an averaging filter to obtain higher order ScatNet features.

$$S_2 x(u, \lambda_1, \lambda_2) = ||x \star \psi_{\lambda_1}(u)| \star \psi_{\lambda_2}(u)| \star \phi_j(u) \tag{9.8}$$

The final feature representation for input signal x is the concatenation of all n-order coefficients (responses from all layers) and is given as $S = \{S_0, S_1, \ldots, S_n\}$. In our experiments, we used $n = 2$. Because these filters are predesigned, there is no extensive training required like a CNN. ScatNet features for a sample fingerphoto image from the ISPFDv1 database [9] are shown in Figure 9.8.

Given a $1 \times N$ length vectorized ScatNet representation for the probe (P) and the gallery (G), a supervised binary classifier $C : X \to Y$ is used to classify an input pair as a match or a nonmatch pair. The classifier learns to classify a pair of ScatNet feature representations input in the form $X = P - G$ as a match $(Y = 1)$ or a nonmatch $(Y = 0)$ pair. The classifier can be considered as a nonlinear mapping learned over the difference of the probe (P) and the gallery (G) representation as follows:

$$M(P, G) = \begin{cases} 1 & f_\theta(d_{L2}(P, G)) \geq t \\ 0 & f_\theta(d_{L2}(P, G)) < t \end{cases} \tag{9.9}$$

In our algorithm, we have used a random decision forest (RDF) as the nonlinear binary classifier [70,71]. The intuition behind the use of RDF for our problem is that RDF is known to perform well [72] in the presence of highly uncorrelated features.

9.4.2 Experimental protocol

Because the application of fingerphotos can range from device unlocking (gallery: fingerphotos) to banking applications (gallery: live-scan or inked prints), we have performed the following two experiments:

- *Expt1: fingerphoto to fingerphoto matching*

- *Expt2: fingerphoto to live-scan matching*

The experiments are performed using ISPFDv1 [9]. For both the scenarios, we match the gallery with all four fingerphoto subsets, namely, WI, WO, NI, and NO. Such a matching is performed because fingerphotos can be captured anywhere and under any illumination. For the experiments, the segmented images are resized to 400×840 after which the second level ScatNet features are extracted. To reduce the dimensionality of the extracted features and make the algorithm feasible for mobile platform, PCA is applied by preserving 99% Eigen energy. Principal components are then matched using the L2 distance metric or RDF classifier. We have also compared the performance of our proposed algorithm, ScatNet+PCA+RDF matching, with the state-of-the-art fingerprint-matching techniques, namely (1) CompCode [73] and (2) minutia-based matching (minutia from VeriFinger and matching using MCC descriptors [16]).

9.4.3   Results and analysis

9.4.3.1   Expt1: Fingerphoto to fingerphoto matching

In this experiment, WI is used as the gallery assuming that the user enrolls his or her fingerphoto in a controlled environment. The subset WI is matched with WO, NI, and NO independently. To perform a baseline experiment, we split the WI subset equally in two parts to match indoor with indoor images. Each of these results show the impact of changing environment during acquisition. Matching WI-WO demonstrates the impact of varying illumination; WI-NI matching illustrates the impact of changing backgrounds whereas WI-NO indicates the effect of varying background and illumination simultaneously. The results of these experiments are summarized in Tables 9.6 and 9.7.

Using ScatNet features, the proposed learning-based matching gives EER in the range of 3%–10%. The performance is much better compared to all the other matching approaches. The corresponding Receiver Operating Characteristic (ROC) graphs for ScatNet+PCA+RDF matching are shown in Figure 9.9.

9.4.3.2   Expt2: Fingerphoto to live-scan matching

In this experiment, LS (live-scan subset) is used as the gallery and the probe subsets are fingerphoto data subsets WI, WO, NI, and NO. To have

TABLE 9.6

EER (%) of the proposed algorithm for *fingerphoto-fingerphoto* matching

	Gallery	Probe	ScatNet+L2	ScatNet+PCA+RDF
Expt1	WI	WO	18.83	5.07
		NI	19.75	7.45
		NO	18.98	3.65
	WI/2	WI/2	28.42	6.00

TABLE 9.7

EER (%) of the proposed algorithm for *fingerphoto-fingerphoto* matching in comparison to matching based on CompCode and MCC descriptors

	Gallery	Probe	CompCode	MCC descriptors	ScatNet+PCA +RDF
Expt1	WI	WO	6.90	22.12	**5.07**
		NI	**5.02**	21.33	7.45
		NO	5.31	21.52	**3.65**
	WI/2	WI/2	6.61	37.25	**6.00**

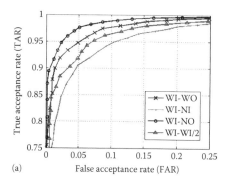 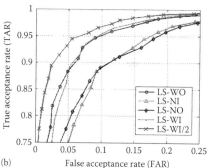

(a)   (b)

FIGURE 9.9

ROC for ScatNet+PCA+RDF based matching method for (a) *fingerphoto-to-fingerphoto* matching (Expt1) and (b) *live scan-to-fingerphoto* matching (Expt2).

TABLE 9.8

EER (%) of the proposed algorithm for *live scan-fingerphoto* matching

	Gallery	Probe	ScatNet+L2	ScatNet+PCA+RDF
Expt2	LS	WO	18.95	7.12
		NI	18.59	10.43
		NO	19.18	10.38
		WI	19.38	7.07
		WI/2	49.51	5.53

a comparative analysis across different galleries, we also performed LS-WI/2 matching. The results of *fingerphoto to live-scan* matching are reported in Tables 9.8 and 9.9.

In this case, LS-WO and LS-WI performs the best. It is because the plain white background makes segmentation easier. Similarly, matching probe NI

TABLE 9.9

EER (%) of the proposed algorithm for *live scan-fingerphoto* matching in comparison to matching based on CompCode and MCC descriptors

	Gallery	Probe	CompCode	MCC descriptors	ScatNet+PCA +RDF
Expt2	LS	WO	14.74	12.92	**7.12**
		NI	10.60	18.05	**10.43**
		NO	11.38	12.76	**10.38**
		WI	14.58	29.92	**7.07**
		WI/2	21.07	31.01	**5.53**

images has the lowest in both *Expt1* and *Expt2*. The close background can be one of the reasons for such performance because a closer background to skin in the fingerphoto makes segmentation challenging.

While performing fingerphoto-fingerphoto matching, it can be observed that WI-NO gives the best matching performance. In the case of live scan-fingerphoto matching, LS-WO performs the best. It is because outdoor illumination has a uniform surrounding illumination compared to focused illumination indoors due to which shadows are formed. The results show that when matching probe NI images with the WI gallery images, the performance is the lowest in both *Expt1* and *Expt2*. The close background can be one of the reasons for such performance because a closer background to skin in the fingerphoto makes segmentation challenging.

9.5 Summary

In this chapter, we have studied two specific challenging cases of uncontrolled fingerprint capture: latent fingerprints and fingerphoto images. We have highlighted the need for using deep representation learning algorithms to extract more robust features for uncontrolled fingerprints. We present a thorough literature study on the use of deep learning in matching controlled and uncontrolled fingerprints. We present a deep GSAE-based latent fingerprint matching and deep ScatNet-based fingerphoto matching algorithms are also presented. Experimental analysis and comparison with existing algorithms demonstrate the promises of deep learning for unconstrained fingerprint matching. Further, research in this domain can be directed toward building a lights-out system using deep neural architectures and building larger data sets. The results of the proposed approach is tabulated and compared with the existing approaches, showing an improvement in performance in the publicly available data sets.

References

1. Report reveals fingerprints dominate global biometrics market. http://www.biometricupdate.com/201703/report-reveals-fingerprints-dominate-global-biometrics-market. [Accessed March 10, 2017].

2. D. Maltoni, D. Maio, A. Jain, and S. Prabhakar. *Handbook of Fingerprint Recognition.* Springer Science & Business Media, London, UK, 2009.

3. CASIA-Fingerprint V5. *Chinese Academy of Sciences Institute of Automation (CASIA) Fingerprint Image Database Version 5.0.*

4. H. C. Lee and R. E. Gaensslen. Methods of latent fingerprint development. *Advances in Fingerprint Technology*, 2:105–176, 2001.

5. P. Komarinski. *Automated Fingerprint Identification Systems (AFIS).* Academic Press, Burlington, MA, 2005.

6. H. C. Lee and R. E. Gaensslen. Methods of latent fingerprint development. In *Advances in Fingerprint Technology*, 2nd ed., pp. 105–175. CRC Press, Boca Raton, FL, 2001.

7. C. Wilson. Fingerprint vendor technology evaluation 2003: Summary of results and analysis report, NISTIR 7123, 2004. http://fpvte.nist.gov/report/ir_7123_analysis.pdf.

8. M. D. Garris and R. M. McCabe. Fingerprint minutiae from latent and matching tenprint images. In *Tenprint Images, National Institute of Standards and Technology*. Citeseer, 2000.

9. A. Sankaran, A. Malhotra, A. Mittal, M. Vatsa, and R. Singh. On smartphone camera based fingerphoto authentication. In *International Conference on Biometrics Theory, Applications and Systems*, pp. 1–7. IEEE, 2015.

10. M. Vatsa, R. Singh, A. Noore, and S. K. Singh. Combining pores and ridges with minutiae for improved fingerprint verification. *Signal Processing*, 89(12):2676 – 2685, 2009. Special Section: Visual Information Analysis for Security.

11. G. Li, B. Yang, M. A. Olsen, and C. Busch. Quality assessment for fingerprints Collected by Smartphone Cameras. In *Proceedings of the IEEE Conference on Computer Vision and Pattern Recognition Workshops*, pp. 146–153, 2013.

12. A. Sankaran, M. Vatsa, and R. Singh. Latent fingerprint matching: A survey. *IEEE Access*, 2(982–1004):1, 2014.

13. M. Puertas, D. Ramos, J. Fierrez, J. Ortega-Garcia, and N. Exposito. Towards a better understanding of the performance of latent fingerprint recognition in realistic forensic conditions. In *International Conference on Pattern Recognition*, pp. 1638–1641. IAPR, Istambul, Turkey, 2010.

14. D. Lee, K. Choi, and J. Kim. A robust fingerprint matching algorithm using local alignment. In *International Conference on Pattern Recognition*, Vol. 3, pp. 803–806. IAPR, 2002.

15. R. Mueller and R. Sanchez-Reillo. An approach to biometric identity management using low cost equipment. In *International Conference on Intelligent Information Hiding and Multimedia Signal Processing*, pp. 1096–1100. IEEE, 2009.

16. R. Cappelli, M. Ferrara, and D. Maltoni. Minutia cylinder-code: A new representation and matching technique for fingerprint recognition. *IEEE Transactions on Pattern Analysis and Machine Intelligence*, 32(12):2128–2141, 2010.

17. M. H. Izadi and A. Drygajlo. Estimation of cylinder quality measures from quality maps for minutia-cylinder code based latent fingerprint matching. In *ENFSI Proceedings of Biometric Technologies in Forensic Science* (No. EPFL-CONF-189764, pp. 6–10), 2013.

18. R. P. Krish, J. Fierrez, D. Ramos, J. Ortega-Garcia, and J. Bigun. Pre-registration for improved latent fingerprint identification. In *International Conference on Pattern Recognition*, pp. 696–701. IAPR, Stockholm, Sweden, 2014.

19. R. P. Krish, J. Fierrez, D. Ramos, J. Ortega-Garcia, and J. Bigun. Pre-registration of latent fingerprints based on orientation field. *IET Biometrics*, 4(2):42–52, 2015.

20. A. A. Paulino, E. Liu, K. Cao, and A. K. Jain. Latent fingerprint indexing: Fusion of level 1 and level 2 features. In *International Conference on Biometrics: Theory, Applications and Systems*, pp. 1–8. IEEE, 2013.

21. A. K. Jain and J. Feng. Latent fingerprint matching. *IEEE Transactions on Pattern Analysis and Machine Intelligence*, 33(1):88–100, 2011.

22. B. Y. Hiew, A. B. J. Teoh, and O. S. Yin. A secure digital camera based fingerprint verification system. *Journal of Visual Communication and Image Representation*, 21(3):219–231, 2010.

23. A. Kumar and Y. Zhou. Contactless fingerprint identification using level zero features. In *Computer Vision and Pattern Recognition Workshops*, pp. 114–119. IEEE, 2011.

24. K. Tiwari and P. Gupta. A touch-less fingerphoto recognition system for mobile hand-held devices. In *International Conference on Biometrics*, pp. 151–156. IEEE, 2015.

25. P. Baldi and Y. Chauvin. Neural networks for fingerprint recognition. *Neural Computation*, 5(3):402–418, 1993.

26. V. K. Sagar and K. J. B. Alex. Hybrid fuzzy logic and neural network model for fingerprint minutiae extraction. In *International Joint Conference on Neural Networks*, Vol. 5, pp. 3255–3259. IEEE, 1999.

27. M. Sahasrabudhe and A. M. Namboodiri. Learning fingerprint orientation fields using continuous restricted boltzmann machines. In *Asian Conference on Pattern Recognition*, pp. 351–355. IAPR, Okinawa, Japan, 2013.

28. M. Sahasrabudhe and A. M. Namboodiri. Fingerprint enhancement using unsupervised hierarchical feature learning. In *Indian Conference on Computer Vision Graphics and Image Processing*, p. 2. ACM, Scottsdale, AZ, 2014.

29. R. Wang, C. Han, Y. Wu, and T. Guo. Fingerprint classification based on depth neural network. *arXiv preprint arXiv:1409.5188*, 2014.

30. S. Minaee and Y. Wang. Fingerprint recognition using translation invariant scattering network. In *Signal Processing in Medicine and Biology Symposium*, pp. 1–6. IEEE, 2015.

31. L. Jiang, T. Zhao, C. Bai, A. Yong, and M. Wu. A direct fingerprint minutiae extraction approach based on convolutional neural networks. In *International Joint Conference on Neural Networks*, pp. 571–578. IEEE, 2016.

32. D. Michelsanti, Y. Guichi, A.-D. Ene, R. Stef, K. Nasrollahi, and T. B. Moeslund. Fast fingerprint classification with deep neural network. In *Visapp-International Conference on Computer Vision Theory and Applications*, Porto, Portugal, 2017.

33. L. Sifre and S. Mallat. Rotation, scaling and deformation invariant scattering for texture discrimination. In *Conference on Computer Vision and Pattern Recognition*, pp. 1233–1240. IEEE, 2013.

34. The Hong Kong Polytechnic University (PolyU) high-resolution-fingerprint (HRF) Database. http://www4.comp.polyu.edu.hk/~biometrics/HRF/HRF_old.htm. Accessed December 8, 2016.

35. V. Nair and G. E. Hinton. Rectified linear units improve restricted boltzmann machines. In *IMLS Proceedings of the Twenty-Seventh International Conference on Machine Learning*, pp. 807–814, 2010.

36. D. Maio, D. Maltoni, R. Cappelli, J. L. Wayman, and A. K. Jain. FVC2002: Second fingerprint verification competition. In *International Conference on Pattern recognition*, Vol. 3, pp. 811–814. IEEE, 2002.

37. D. Maio, D. Maltoni, R. Cappelli, J. L. Wayman, and A. K. Jain. FVC2000: Fingerprint verification competition. *IEEE Transactions on Pattern Analysis and Machine Intelligence*, 24(3):402–412, 2002.

38. A. Ng. Sparse autoencoder. *CS294A Lecture notes*, 72:1–19, 2011.

39. NIST special database 4. https://www.nist.gov/srd/nist-special-database-4. Accessed December 4, 2016.

40. K. Chatfield, K. Simonyan, A. Vedaldi, and A. Zisserman. Return of the devil in the details: Delving deep into convolutional nets. *arXiv preprint arXiv:1405.3531*, 2014.

41. Fingerprint minutiae from latent and matching tenprint images. *NIST Special Database 27*, 2010.

42. A. Sankaran, T. I. Dhamecha, M. Vatsa, and R. Singh. On matching latent to latent fingerprints. In *International Joint Conference on Biometrics*, Washington, DC, 2010.

43. A. Sankaran, M. Vatsa, and R. Singh. Hierarchical fusion for matching simultaneous latent fingerprint. In *International Conference on Biometrics: Theory, Applications and Systems*. IEEE, 2012.

44. A. Sankaran, M. Vatsa, and R. Singh. Multisensor optical and latent fingerprint database. *IEEE Access*, 3:653–665, 2015.

45. A. A. Paulino, J. Feng, and A. K. Jain. Latent fingerprint matching using descriptor-based hough transform. *IEEE Transactions on Information Forensics and Security*, 8(1):31–45, 2013.

46. Evaluation of latent fingerprint technologies. http://www.nist.gov/itl/iad/ig/latent.cfm.

47. K. Cao and A. K. Jain. Latent orientation field estimation via convolutional neural network. In *International Conference on Biometrics*, pp. 349–356. IAPR, Phuket, Thailand, 2015.

48. NIST digital video of live-scan fingerprint database - NIST special database 24. https://www.nist.gov/srd/nistsd24.html. Accessed December 10, 2016.

49. X. Yang, J. Feng, and J. Zhou. Localized dictionaries based orientation field estimation for latent fingerprints. *IEEE Transactions on Pattern Analysis and Machine Intelligence*, 36(5):955–969, 2014.

50. A. Sankaran, P. Pandey, M. Vatsa, and R. Singh. On latent finger-print minutiae extraction using stacked denoising sparse autoencoders. In *International Joint Conference on Biometrics*, pp. 1–7. IEEE, 2014.

51. NIST special database 14. https://www.nist.gov/srd/nist-special-database-14. Accessed December 10, 2016.

52. X. Jia, X. Yang, Y. Zang, N. Zhang, and J. Tian. A cross-device matching fingerprint database from multi-type sensors. In *International Conference on Pattern Recognition*, pp. 3001–3004. IEEE, 2012.

53. J. Ortega-Garcia, J. Fierrez-Aguilar, D. Simon, J. Gonzalez, M. Faundez-Zanuy, V. Espinosa, A. Satue, I. Hernaez, J.-J. Igarza, and C. Vivara-cho. MCYT baseline corpus: A bimodal biometric database. *Proceedings-Vision, Image and Signal Processing*, 150(6):395–401, 2003.

54. Y. Tang, F. Gao, and J. Feng. Latent fingerprint minutia extraction using fully convolutional network. *arXiv preprint arXiv:1609.09850*, 2016.

55. J. Long, E. Shelhamer, and T. Darrell. Fully convolutional networks for semantic segmentation. In *Conference on Computer Vision and Pattern Recognition*, pp. 3431–3440. IEEE, 2015.

56. M. D. Zeiler and R. Fergus. Visualizing and understanding convolutional networks. In *European Conference on Computer Vision*, pp. 818–833. Springer, Germany, 2014.

57. K. Simonyan and A. Zisserman. Very deep convolutional networks for large-scale image recognition. *arXiv preprint arXiv:1409.1556*, 2014.

58. K. He, X. Zhang, S. Ren, and J. Sun. Deep residual learning for image recognition. *arXiv preprint arXiv:1512.03385*, 2015.

59. A. Sankaran, M. Vatsa, R. Singh, and A. Majumdar. Group sparse au-toencoder. *Image and Vision Computing*, 60:64–74, 2017.

60. NIST 8-bit gray scale images of fingerprint image groups (FIGS). *NIST Special Database 14*, 2010.

61. M. Davenport, R. G. Baraniuk, C. D. Scott, et al. Controlling false alarms with support vector machines. In *International Conference on Acoustics, Speech and Signal Processing*, Vol. 5, 2006.

62. NBIS (NIST Biometric Image Software). *Developed by National Institute of Standards and Technology*. http://www.nist.gov/itl/iad/ig/nbis.cfm.

63. VeriFinger. NeuroTechnology. www.neurotechnology.com/verifinger.html. Accessed August 10, 2012.

64. Y. Song, C. Lee, and J. Kim. A new scheme for touchless fingerprint recognition system. In *International Symposium on Intelligent Signal Processing and Communication Systems*, pp. 524–527. IEEE, 2004.

65. D. Lee, K. Choi, H. Choi, and J. Kim. Recognizable-image selection for fingerprint recognition with a mobile-device camera. *IEEE Transactions on Systems, Man, and Cybernetics, Part B: Cybernetics*, 38(1):233–243, 2008.

66. G. Li, B. Yang, R. Raghavendra, and C. Busch. Testing mobile phone camera based fingerprint recognition under real-life scenarios. In *Norwegian Information Security Conference*, 2012.

67. C. Stein, V. Bouatou, and C. Busch. Video-based fingerphoto recognition with anti-spoofing techniques with smartphone cameras. In *International Conference of the Biometrics Special Interest Group*, pp. 1–12, Darmstadt, Germany, 2013.

68. C. Stein, C. Nickel, and C. Busch. Fingerphoto recognition with smartphone cameras. In *IEEE Proceedings of the International Conference of the Biometrics Special Interest Group*, pp. 1–12, 2012.

69. J. Bruna and S. Mallat. Invariant scattering convolution networks. *IEEE Transactions on Pattern Analysis and Machine Intelligence*, 35(8):1872–1886, 2013.

70. T. K. Ho. Random decision forests. In *International Conference on Document Analysis and Recognition*, Vol. 1, pp. 278–282. IEEE, 1995.

71. T. K. Ho. The random subspace method for constructing decision forests. *IEEE Transactions on Pattern Analysis and Machine Intelligence*, 20(8):832–844, 1998.

72. L. Breiman. Random forests. *Machine Learning*, 45(1):5–32, 2001.

73. A. W.-K. Kong and D. Zhang. Competitive coding scheme for palmprint verification. In *International Conference on Pattern Recognition*, Vol. 1, pp. 520–523, 2004.

10

Person Identification Using Handwriting Dynamics and Convolutional Neural Networks

Gustavo H. Rosa, João P. Papa, and Walter J. Scheirer

CONTENTS

10.1	Introduction	227
10.2	Convolutional Neural Networks	228
	10.2.1 Filter bank	229
	10.2.2 Sampling	230
	10.2.3 Normalization	230
10.3	Methodology	231
	10.3.1 SignRec data set	231
	10.3.2 Modeling time series in CNNs	233
	10.3.3 Experimental setup	236
10.4	Experimental Results	239
10.5	Conclusions	242
	Acknowledgments	242
	References	242

10.1 Introduction

Biometrics as a discipline within computer vision has emerged as the need for reliable systems to automatically identify and authenticate people has increased over the past couple of decades. Because passwords can be stolen or even discovered by some brute-force method, using the unique characteristics inherent to each individual has become a safe and convenient alternative. Among the most common and widely used biometric-based information, one shall cite face, fingerprints, and iris, just to name a few. Although each biometric modality has its own positive and negative aspects, there is a consensus that all of them must be able to withstand spoofing attacks, as well as be discriminative enough to distinguish different individuals.

An interesting and noninvasive biometric modality that can be used to identify people is handwriting style. With respect to this modality, one can find two different possibilities: static or dynamic operation. The former approach, also known as offline, usually uses the shape of the letters to identify people, whereas the latter approach make use of more information than just the shape of the letters, including the pressure, slope, and time to write each character down [1,2]. For instance, Sesa-Nogueras and M. Faundez-Zanuy [3] used self-organizing maps to deal with the problem of stroke identification in handwriting dynamics, and Sanchez-Reillo et al. [4] evaluated the strengths and weaknesses of dynamic handwritten-signature recognition against spoofing attacks.

Recently, Zhang et al. [5] presented a review concerning handwritten Chinese character recognition and established new benchmarks for both online and offline. Additionally, Ferrer et al. [6] developed a methodology to generate dynamic information from handwriting recognition, thus providing a unified comprehensive synthesizer for both static and dynamic signature features. Going back further into the literature, Zhu et al. [7] used texture features to identify writers from Chinese handwritten documents, and Ribeiro et al. [8] used deep neural networks for offline handwriting recognition. Last but not least, a group of Brazilian researchers presented interesting advances concerning offline handwritten recognition in a Brazilian data set for writer identification [9].

Because deep learning-driven techniques have been extensively pursued in the last few years, it is quite common to find their usage in a wide variety of application domains. However, to the best of our knowledge, we have not observed any work that explores online recognition from handwriting dynamics by means of deep learning techniques, which is the main contribution of this chapter. In addition to a novel algorithm, we make available a data set composed of 26 individuals that had their signature captured by means of a smartpen, for the further analysis of their handwriting skills. The remainder of this chapter is organized as follows. Section 10.2 briefly reviews the basics of convolutional neural networks (CNNs). Section 10.3 presents the methodology to build the data set and a new CNN-based approach to online handwriting recognition, and Section 10.4 introduces the related experimental results. Finally, Section 10.5 states conclusions and future research directions.

10.2   Convolutional Neural Networks

CNNs can be seen as a representation of a bigger class of models based on the Hubel and Wiesel's architecture, which was presented in a seminal study in 1962 concerning the primary cortex of cats. This research identified,

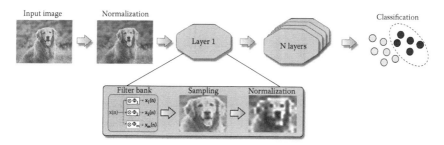

FIGURE 10.1
A simplified, yet typical CNN architecture.

in essence, two kinds of cells: *simple cells*, which possess an analogous duty to the CNN filter bank levels, and *complex cells*, which perform a similar job to the CNN sampling step.

The first model that simulated a computer-based CNN was the well-known "Neocognitron," [10] which implemented an unsupervised training algorithm to establish the filter banks, followed by a supervised training algorithm applied in the last layer. Later on, LeCun et al. [11] simplified this architecture by proposing the use of the back-propagation algorithm to train the network in a supervised way. Thus, several applications that used CNNs emerged in the subsequent decades.

Basically, a CNN can be understood as an N-layered data processing sequence. Thereby, given an input image,* a CNN essentially extracts a high-level representation of it, called a *multispectral image*, whose pixel attributes are concatenated in a feature vector for later application of pattern recognition techniques. Figure 10.1 introduces the naïve architecture of a CNN.

As mentioned, each CNN layer is often composed of three operations: a convolution with a filter bank, followed by a sampling phase, and then by a normalization step. As one can observe in Figure 10.1, there is still a possibility of a normalization operation in the beginning of the whole process. The next sections describe in more detail each of these steps.

10.2.1 Filter bank

Let $\hat{I} = (D_I, \vec{I})$ be a multispectral image such that $D_I \in n \times n$ is the image domain, and $\vec{I} = \{I_1(p), I_2(p), \ldots, I_m(p)\}$ corresponds to a pixel $p = (x_p, y_p) \in D_I$, and m stands for the number of bands. When $\hat{I}$ is a grey-scale image, for instance, we have that $m = 1$ and $\hat{I} = (D_I, I)$.

Let $\phi = (\mathcal{A}, W)$ be a filter with weights $W(q)$ associated with every *pixel* $q \in \mathcal{A}(p)$, where $\mathcal{A}(p)$ denotes a mask of size $L_{\mathcal{A}} \times L_{\mathcal{A}}$, centered

---

*The same procedure can be extended to other nonvisual signal processing–based applications.

at p, and $q \in \mathcal{A}(p)$ if, and only if, $\max\{|x_q - x_p|, |y_q - y_p|\} \leq (L_{\mathcal{A}} - 1)/2$. In case of multispectral filters, their weights can be depicted as vectors $\vec{W}_i(q) = \{w_{i,1}(q), w_{i,2}(q), \ldots, w_{i,m}(q)\}$ for each filter i, and a multispectral filter bank can be then defined as $\phi = \{\phi_1, \phi_2, \ldots, \phi_n\}$, where $\phi_i = (\mathcal{A}, \vec{W}_i)$, $i = \{1, 2, \ldots, n\}$.

Thus, the convolution between an input image $\hat{I}$ and a filter ϕ_i generates the band i of the filtered image $\hat{J} = (D_J, \vec{J})$, where $D_J \in D_I$ and $\vec{J} = \{J_1(p), J_2(p), \ldots, J_n(p)\}$, $\forall p \in D_J$:

$$J_i(p) = \sum_{\forall q \in \mathcal{A}(p)} \vec{I}(q) \otimes \vec{W}_i(q) \tag{10.1}$$

where $\otimes$ denotes the convolution operator. The weights of ϕ_i are usually generated from a uniform distribution, that is, $U(0, 1)$, and afterward normalized with mean zero and unitary norm.

10.2.2 Sampling

This operation is extremely important for a CNN, which provides translational invariance to the extracted features. Let $\mathcal{B}(p)$ be the sampling area of size $L_{\mathcal{B}} \times L_{\mathcal{B}}$ centered at p. Additionally, let $D_K = D_J/s$ be a regular sampling operation every s pixels. Therefore, the resulting sampling operation in the image $\hat{K} = (D_K, \vec{K})$ is defined as follows:

$$K_i(p) = \sqrt[\alpha]{\sum_{\forall q \in \mathcal{B}(p)} J_i(q)^\alpha} \tag{10.2}$$

where:

$p \in D_K$ denotes every pixel of the new image

$i = \{1, 2, \ldots, n^2\}$

α stands for the stride parameter, controlling the downsampling factor of the operation

10.2.3 Normalization

The last operation of a CNN is its normalization, which is a widely employed mechanism to enhance its perfomance [12]. This operation is based on the apparatus found on cortical neurons [13], being also defined under a squared-area $\mathcal{C}(p)$ of size $L_{\mathcal{C}} \times L_{\mathcal{C}}$ centered at pixel p, such as:

$$O_i(p) = \frac{K_i(p)}{\sum_{j=1}^{n} \sum_{\forall q \in \mathcal{C}(p)} K_j(q) K_i(q)} \tag{10.3}$$

Thus, the operation is accomplished for each pixel $p \in D_O \subset D_k$ of the resulting image $\hat{O} = (D_O, \vec{O})$.

10.3   Methodology

In this section, we present the methodology used to design the data set, as well as the proposed approach to analyze the pen-based features (signals) by means of CNNs. In addition, we present the experimental setup as well.

10.3.1   SignRec data set

The SignRec data set,* which stands for signature-recognition data set, aims at characterizing an individual's writing style as a form, such as the one depicted in Figure 10.2, is filled out. The idea of the form is to ask a person to perform some specific tasks so that their writing can be captured by the a set of sensors from a smartpen and recorded for subsequent analysis.

In this work, we used the Biometric Smart Pen (*BiSP®*) [14], which is a multisensory pen system, capable of recording and analyzing handwriting, drawing, and gesture movements, regardless of whether they were made on paper or in the air. The smartpen is intended to monitor hand and finger motor characteristics and is thus equipped with several measuring sensors, which are illustrated in Figure 10.3 and described herein:

- CH 1: Microphone

- CH 2: Fingergrip

- CH 3: Axial Pressure of ink refill

- CH 4: Tilt and Acceleration in the "X direction"

- CH 5: Tilt and Acceleration in the "Y direction"

- CH 6: Tilt and Acceleration in the "Z direction"

Unlike common table-based input devices, the *BiSP®* data are sampled solely by the pen, transferring its outputs to devices like PCs, notebooks, and cell phones, among others. It is able to provide important object-related parameters as well as neuromotor and biometric features from a human being.

The SignRec data set was collected at the Faculty of Sciences, São Paulo State University, Bauru, Brazil. To build this initial data set,† we used signals extracted from handwriting dynamics. The data set consists of 26 individuals; 23 are male and 3 are female. Each person was asked to write down their name using the smartpen starting from left to right. This activity is focused on the

---

*http://www2.fc.unesp.br/~papa/recogna/biometric_recognition.html
†Ongoing research is expanding the data set to incorporate random phrases.

UNIVERSIDADE ESTADUAL PAULISTA
"JÚLIO DE MESQUITA FILHO"
Campus de Bauru

Faculdade de Ciências

Name: _____	**Biometric Identification**
Age: _____	**Universidade Estadual Paulista** Faculdade de Ciências (FC) Bauru
Gender: ( ) F ( ) M	
Dominant hand: ( ) Right ( ) Left	**ID:** _____

1	After sound signal, write full name along the line below, without leaving the current box area. ◀
2	After sound signal, write full name along the line below, without leaving the current box area. ◀
3	After sound signal, write full name along the line below, without leaving the current box area. ◀
4	After sound signal, write full name along the line below, without leaving the current box area. ◀
5	After sound signal, write full name along the line below, without leaving the current box area. ◀
6	After sound signal, write full name along the line below, without leaving the current box area. ◀
7	After sound signal, write full name along the line below, without leaving the current box area. ◀
8	After sound signal, write full name along the line below, without leaving the current box area. ◀
9	After sound signal, write full name along the line below, without leaving the current box area. ◀
10	After sound signal, write full name along the line below, without leaving the current box area. ◀

Departamento de Computação/FC
Av. Engº Luiz Edmundo Carrijo Coube, nº 14-01 - Vargem Limpa Bauru-SP CEP: 17033-360
Fone: (14) 3103-6079/6034   Voip:7979/7739   email: dcogeral@fc.unesp.br   site: http://www.fc.unesp.br/depto_computacao

FIGURE 10.2
Form used to assess the handwritten skills of a given individual during the data collection performed for this work.

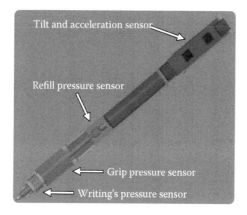

FIGURE 10.3
Biometric smartpen: Sensors are located at four different points points (extracted from Peueker et al. [15]).

analysis of the movement provided by the act of writing a signature, which quantifies different motor activities and different writing patterns for each individual. Figures 10.4a and 10.4b depict the signals extracted from three channels (channels 2, 3, and the average between channels 5 and 6) from two distinct individuals.

Looking at Figure 10.4, the differences between the two individuals can clearly be seen by comparing the curves in the plots. Such a difference can be expressed as a feature space and used by a biometric system to discriminate between different users. We decided to not use channels 1 and 4 (i.e., the microphone and x-axis displacement) because in our experiments we observed that such channels did not play a big role in distinguishing different individuals.

10.3.2 Modeling time series in CNNs

We propose to model the problem of distinguishing individuals as an image recognition task by means of CNNs. Roughly speaking, the signals provided by the smartpen are transformed into pictures. Each acquisition is composed of r rows (acquisition time in milliseconds) and six columns, which are for the aforementioned six signal channels (e.g., sensors). In this chapter, we propose to map the signals obtained by the smartpen to images through the recurrence plot methodology 15, which is a gray-scale image produced for each channel that captures information from the raw data (but not features). Therefore, to compose an RGB image, each acquisition session needs to be rescaled into three channels, and each channel needs to be

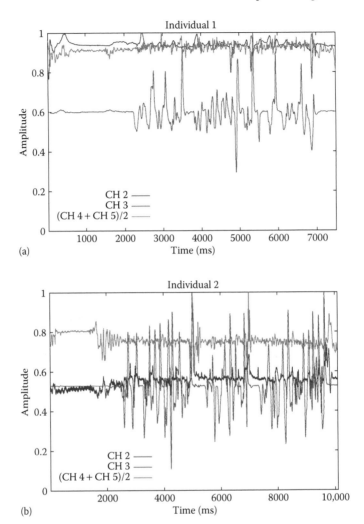

FIGURE 10.4
Signals recorded from (a) individual 1 and (b) individual 2.

normalized between 0 and 1. As we are trying to identify the writing pattern of each person, we discarded both channels 1 and 4, because we observed that the microphone and the displacement in the x-axis are commonly equal for all acquisition sessions. Finally, regarding the last channel, we employed an arithmetic mean between channels 5 and 6. Figure 10.5 illustrates some gray-scale recurrence plot-based images, and Figure 10.6 illustrates their corresponding RGB versions. One can observe distinct patterns between different individuals.

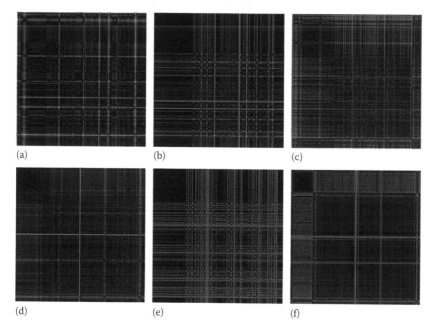

FIGURE 10.5
Signature gray-scale samples from individual 1: (a) R channel, (b) G channel, and (c) B channel, and individual 2: (d) R channel, (e) G channel, and (f) B channel.

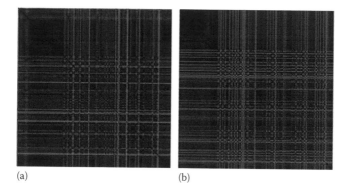

FIGURE 10.6
Signature RGB samples from: (a) individual 1 and (b) individual 2.

10.3.3 Experimental setup

In this work, we classified signature images drawn by different individuals using a CNN-based approach. The data set is composed of 260 images, with 10 images per individual. Note that all images were rescaled to 256×256 pixels for consistency within the experiment. Also, we split our data set into a training set containing 90% of the images and a testing set containing 10% of the images.

In regard to the source code, we used the well-known Caffe library* [17], which is developed under a general purpose GPU platform, thus providing more efficient implementations. Each experiment was evaluated using the CIFAR10_full[†] architecture provided by Caffe with a few small modifications. Because we have larger image resolutions in our data set compared to the images found in the CIFAR10 data set, we multiplied by 4 the number of outputs for all convolution layers and we doubled their pads, kernel sizes, and strides. Also, we are using 1000 training iterations with mini-batches of size 4, 8, and 16, and a smaller learning rate (0.00001) to avoid over-fitting.

To provide a statistical analysis by means of a Wilcoxon signed-rank test with significance of 0.05 [18], we conducted a cross-validation with 20 folds. Figure 10.7 illustrates the proposed architecture. Note that *conv* stands for the convolution layer, *pool* for pooling, *relu* for rectified linear unit activation, *norm* for normalization, *ip* for inner product or fully connected layer, *accuracy* for the final accuracy output, and *loss* for the output of the loss function. Additionally, we employed the standard CaffeNet[‡] architecture for comparison purposes, using the same hyperparameter values from CIFAR10_full for training iterations, batch sizes, and learning rate. Because of the network size, we opted not to employ a figure of its own, however, one can check the full architecture on its default prototxt file. Finally, we also evaluated the standard MNIST architecture[§] again using the same hyperparameters, except for the training iterations, where we used half of the original value (i.e., 5000) to avoid over-fitting. Figure 10.8 illustrates this architecture.

---

*http://caffe.berkeleyvision.org
[†]https://github.com/BVLC/caffe/blob/master/examples/cifar10/cifar10_full_train_test.prototxt
[‡]https://github.com/BVLC/caffe/blob/master/models/bvlc_reference_caffenet/train_val.prototxt
[§]https://github.com/BVLC/caffe/blob/master/examples/mnist/lenet_train_test.prototxt

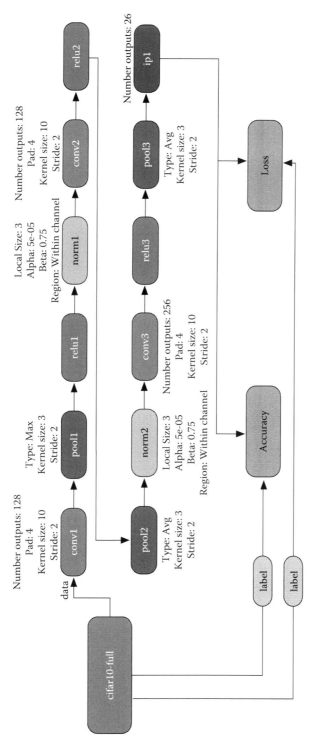

FIGURE 10.7
CIFAR10 full architecture.

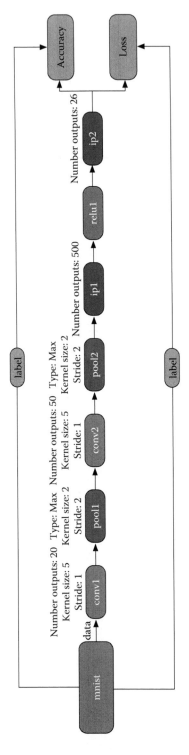

FIGURE 10.8
MNIST architecture.

10.4 Experimental Results

This section aims at presenting the experimental results concerning the CNN-based person identification. The most accurate results, according to a Wilcoxon signed-rank test, are in bold. Notice the overall accuracy is computed using the standard formulation, that, $(1 - \frac{errors}{data\ set\ size}) * 100$. Table 10.1 displays the mean accuracy results in parenthesis, as well as the maximum accuracy obtained along the iterations for each configuration of architecture and batch size.

Because the data set, in its current version, is quite small for a deep learning environment, the most complex architecture (i.e., CaffeNet) obtained the worst results, which is expected as a result of the lack of data. On the other hand, a simple approach did not obtain good results either, as one can observe in the results over the MNIST data set (although they are still better than the results from CaffeNet). The best results were obtained by the CIFAR10_full architecture, which can be seen as a trade-off between the other two architectures. In all cases, results were above chance, and in many cases, much higher. However, we can still observe some over-fitting behavior, as displayed in Figure 10.9.

Figures 10.9a and 10.9b depict the accuracies along the iterations during the learning process, as well as the values of the loss function along the iterations, respectively. Over-fitting is a consideration in these experiments because the accuracy starts to drop after 1800 iterations, and the loss function starts to oscillate after 1400 iterations, before going up. This might be a problem of having a small amount of data because the data set was meant to serve as a pilot for experimentation and not large-scale experiments. As mentioned previously, we are now working on a larger data set, with more individuals and different handwriting forms. However, the confusion matrices show us some interesting results, as one can observe in Figure 10.10. Although our results are preliminary, the confusion matrices showed to be useful to demonstrate

TABLE 10.1
Maximum accuracy (average mean) over the test set

	Batch size		
	4	8	16
CaffeNet	4.04% (7.69%)	4.04% (7.69%)	**5.58% (11.54%)**
CIFAR10_full	**61.54% (50.38%)**	**61.54% (52.12%)**	**61.54% (51.92%)**
MNIST	34.62% (21.54%)	**50.00% (32.88%)**	42.31% (28.85%)

Note: Chance performance is approximately 4% for these experiments.

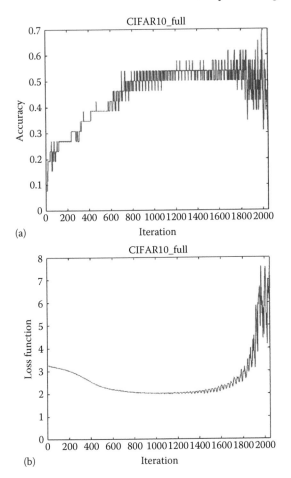

FIGURE 10.9
Convergence process using CIFAR10_full architecture concerning: (a) accuracy values and (b) loss function.

the potential in using features learned from CNNs with raw data obtained by means of a smartpen. Although the proposed approach has mistaken some individuals, a large number of them were correctly recognized, as one can observe by the diagonal line, which is well-defined and with a number of positions colored red.

The batch size did not appear to affect the results considerably, with the exception of the MNIST data set. By using batches of size 8, the results were better, but larger batch sizes (i.e., 16) seemed to slow down the convergence, or even overshoot the local/global minimal with possible oscillatory behaviors.

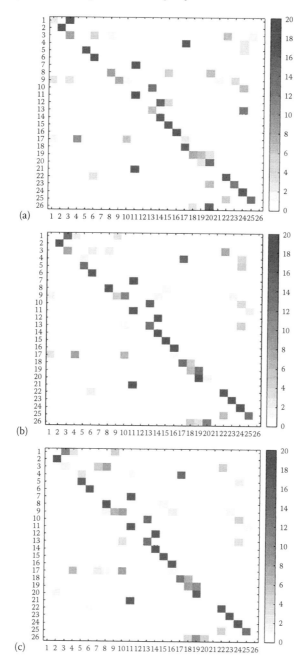

FIGURE 10.10
CIFAR10_full confusion matrices (heat maps) using batch sizes of: (a) 4, (b) 8, and (c) 16.

Even though other batch sizes could be considered, we believe the data set size is a hard constraint on performance. We also tried different approaches to map the raw signals into images, such as a simple concatenation of the signals, but the results were worse than the ones using recurrence plots, which we believe are worth further consideration in future work.

10.5   Conclusions

In this work, we dealt with the problem of biometric identification by means of handwriting dynamics. After designing a data set composed of 26 individuals, their handwritten features were captured by a smartpen with six sensors. Further, these signals were mapped into images by means of recurrence plots and then used as input to CNNs.

This pilot study made two main contributions: the collection and release of a data set to foster new research on handwriting dynamics-based biometrics and the evaluation of both CNNs and recurrence plots in this context. We evaluated three distinct CNN architectures, CIFAR10_full, CaffeNet, and MNIST, with CIFAR10_full coming out on top as the most promising one. As a function of data set size, all of the architectures suffered from over-fitting in some capacity; thus the results still have a ways to go to reach viability in an operational setting. To address this, we intend to make incremental updates to the data set to increase its size, as well as to employ ideas beyond recurrence plots to map signals into CNN parseable images. Nevertheless, this chapter has shown feasibility for a CNN-based approach to this problem for a challenging new data set, which will be an excellent resource to researchers working on handwriting biometrics.

Acknowledgments

The authors are grateful to FAPESP grants #2014/16250-9, #2015/25739-4 and #2016/21243-7, as well as CNPq grant #306166/2014-3.

References

1. R. Plamondon and S. N. Srihari. Online and off-line handwriting recognition: A comprehensive survey. *IEEE Transactions on Pattern Analysis and Machine Intelligence*, 22(1):63–84, 2000.

2. J. Chapran. Biometric writer identification: Feature analysis and classification. *International Journal of Pattern Recognition and Artificial Intelligence*, 20(4):483–503, 2006.

3. E. Sesa-Nogueras and M. Faundez-Zanuy. Biometric recognition using online uppercase handwritten text. *Pattern Recognition*, 45(1):128–144, 2012.

4. R. Sanchez-Reillo, H. C. Quiros-Sandoval, J. Liu-Jimenez, and I. Goicoechea-Telleria. Evaluation of strengths and weaknesses of dynamic handwritten signature recognition against forgeries. In *2015 International Carnahan Conference on Security Technology*, pp. 373–378, IEEE, Piscataway, NJ, 2015.

5. X.-Y. Zhang, Y. Bengio, and C.-L. Liu. Online and offline handwritten chinese character recognition: A comprehensive study and new benchmark. *Pattern Recognition*, 61:348–360, 2017.

6. M. A. Ferrer, M. Diaz, C. Carmona, and A. Morales. A behavioral handwriting model for static and dynamic signature synthesis. *IEEE Transactions on Pattern Analysis and Machine Intelligence*, PP(99):1–1, 2016.

7. Y. Zhu, T. Tan, and Y. Wang. Biometric personal identification based on handwriting. In *Proceedings 15th International Conference on Pattern Recognition*, Vol. 2, pp. 797–800, IEEE, Piscataway, NJ, 2000.

8. B. Ribeiro, I. Gonçalves, S. Santos, and A. Kovacec. *Deep Learning Networks for Off-Line Handwritten Signature Recognition*, pp. 523–532. Springer, Berlin, Germany, 2011.

9. L. G. Hafemann, R. Sabourin, and L. S. Oliveira. Writer-independent feature learning for offline signature verification using deep convolutional neural networks. *CoRR*, abs/1604.00974, 2016.

10. K. Fukushima and S. Miyake. Neocognitron: A new algorithm for pattern recognition tolerant of deformations and shifts in position. *Pattern Recognition*, 15(6):455–469, 1982.

11. Y. LeCun, B. Boser, J. S. Denker, D. Henderson, R. E. Howard, W. Hubbard, and L. D. Jackel. Backpropagation applied to handwritten zip code recognition. *Neural Computation*, 1(4):541–551, 1989.

12. D. Cox and N. Pinto. Beyond simple features: A large-scale feature search approach to unconstrained face recognition. In *Proceedings of the IEEE International Conference on Automatic Face Gesture Recognition and Workshops*, pp. 8–15, 2011.

13. W. S. Geisler and D. G. Albrecht. Cortical neurons: Isolation of contrast gain control. *Vision Research*, 32(8):1409–1410, 1992.

14. C. Hook, J. Kempf, and G. Scharfenberg. New pen device for biometrical 3d pressure analysis of handwritten characters, words and signatures. In *Proceedings of the 2003 ACM SIGMM workshop on Biometrics methods and applications*, pp. 38–44. ACM, New York, 2003.

15. D. Peueker, G. Scharfenberg, and C. Hook. Feature selection for the detection of fine motor movement disorders in parkinsons patients. In *Advanced Research Conference 2011*, Minneapolis, MN, 2011.

16. F. A. Faria, J. Almeida, B. Alberton, L. P. C. Morellato, and R. D. S. Torres. Fusion of time series representations for plant recognition in phenology studies. *Pattern Recognition Letters*, 83, Part 2:205–214, 2016.

17. Y. Jia, E. Shelhamer, J. Donahue, S. Karayev, J. Long, R. Girshick, S. Guadarrama, and T. Darrell. Caffe: Convolutional architecture for fast feature embedding. *arXiv preprint arXiv:1408.5093*, 2014.

18. F. Wilcoxon. Individual comparisons by ranking methods. *Biometrics Bulletin*, 1(6):80–83, 1945.

11

Counteracting Presentation Attacks in Face, Fingerprint, and Iris Recognition

Allan Pinto, Helio Pedrini, Michael Krumdick, Benedict Becker,
Adam Czajka, Kevin W. Bowyer, and Anderson Rocha

CONTENTS

11.1	Introduction	246
11.2	Related Work	249
	11.2.1 Face PAD	249
	11.2.2 Fingerprint PAD	252
	11.2.3 Iris PAD	254
	11.2.4 Unified frameworks to PAD	256
11.3	Methodology	257
	11.3.1 Network architecture	257
	11.3.2 Training and testing	258
	11.3.3 Memory footprint	260
11.4	Metrics and Data Sets	260
	11.4.1 Video-based face-spoofing benchmarks	260
	11.4.1.1 Replay-attack	260
	11.4.1.2 CASIA	262
	11.4.2 Fingerprint-spoofing benchmarks	262
	11.4.2.1 The LivDet2009 benchmark	262
	11.4.2.2 The LivDet2013 benchmark	263
	11.4.3 Iris-spoofing benchmarks	263
	11.4.3.1 AVTS	263
	11.4.3.2 Warsaw LivDet2015	263
	11.4.4 Error metrics	264
11.5	Results	264
	11.5.1 Face	264
	11.5.1.1 Same-dataset results	265
	11.5.1.2 Cross-dataset results	268
	11.5.2 Fingerprints	268
	11.5.2.1 Same-sensor results	268
	11.5.2.2 Cross-sensor results	271

    11.5.3 Iris 271
            11.5.3.1 Same-dataset results 274
            11.5.3.2 Cross-dataset results 274
11.6 Conclusions 283
Acknowledgment 284
References 284

11.1 Introduction

Biometric authentication is a technology designed to recognize humans auto-matically based on their behavior, physical, and chemical traits. Recently, this technology emerged as an important mechanism for access control in many modern applications, in which the traditional methods including the ones based on knowledge (e.g., keywords) or based on tokens (e.g., smart cards) might be ineffective because they are easily shared, lost, stolen, or manipu-lated [1]. Biometric technologies are increasingly used as the main authenti-cating factor for access control and also jointly with traditional authentication mechanisms, as a "step-up authentication" factor in two- or three-factor authentication systems.

In this context, face, iris, and fingerprint are the most commonly used biometric traits. In fact, the choice of the trait to be used takes into account some issues such as universality, easiness to measure the biometric charac-teristics, performance, or difficulty to circumvent the system [1]. However, a common disadvantage of these traits is that an impostor might produce a synthetic replica that can be presented to the biometric sensor to circumvent the authentication process. In the literature, the mechanisms to protect the biometric system against this type of attack are referred to as *spoofing detec-tion*, *liveness detection*, or *presentation attack detection*. Hereinafter, we will use the most generic term, presentation attack detection (PAD), which was initially proposed by SC37 experts in ISO/IEC 30107—Presentation Attack Detection—Framework (Part 1), Data Formats (Part 2), and Testing and Reporting (Part 3).

The idea of spoofing biometric recognition is surprisingly older than bio-metrics itself. A careful reader of the Old Testament can find an imperson-ation attempt described in Genesis, based on presentation of a goat's fur put on Jacob's hand to imitate properties of Esau's skin, so that Jacob would be blessed by Isaac. A fictitious example that is surprisingly realistic is the description of how to copy someone's fingerprint using a wax mold and gelatin presented by Austin Freeman in his crime novel *The Red Thumb Mark*. The novel appeared in 1907, and the technique described is still used almost 100 years later to spoof fingerprint sensors. Note that this description appeared only four years after fingerprints were adopted by Scotland Yard and long before the first fingerprint sensor appeared on the market.

Recent scientific studies and open challenges such as LivDet (www.livdet. org) suggest that presentation attacks are still an open problem in biometrics. Phan [2] and Boulkenafet [3] suggest that face-recognition systems are vulnerable to presentation attacks with an equal error rate (related to distinguishing presentation attacks from genuine samples) reaching as high as 9%. Fingerprint-based recognition systems still face the same problem, with an average classification error rate achieving 2.9% [4]. Iris-based authentication, considered by many to be one of the most reliable biometrics, awaits efficient PAD methodology. Recent proposals in this area still report an average classification error rate around 1% [5].

Besides the laboratory testing of the biometric system's vulnerability to attack, a few real cases also confirm the problem. In the small city of Ferraz de Vasconcelos, in the outskirts of São Paulo, Brazil, a physician of the service of mobile health care and urgency was caught red-handed by the police in a scam that used silicone fingers to bypass an authentication system and confirm the presence of several colleagues at work [6]. A similar case was investigated by the Brazilian Federal Police in 2014, when workers at the Paranaguá Harbor in the Brazilian southern state of Paraná, were suspected of using silicone fingers to circumvent a time attendance biometric system [7]. In Germany, the biometric hacking team in the Chaos Computer Club managed to hack Apple's iPhone Touch ID [8] a few days after its launch, demonstrating that a biometric system without an adequate protection is unsuitable as a reliable access-control method. Other cases of spoofing surveillance systems with three-dimensional masks to change their apparent age or race can also be found in [9,10].

Considering the three aforementioned modalities, when we look at the literature and analyze the algorithms to prevent presentation attacks, we observe that the most promising in terms of errors and minimum effort of implementation or cost often share an interesting feature: they belong to a group of algorithms referred to as data-driven characterization algorithms. According to Pinto et al. [11], methods based on data-driven characterization exploit only the data that come from a standard biometric sensor looking for evidence of artifacts in the already acquired biometric sample. Such approaches are preferable in practice because they are easily integrable with the existing recognition systems because there is no extra requirement in terms of hardware nor is there the need of human interaction to detect attempted attacks.

Although the existing methods following this idea have led to good detection rates, we note that some aspects still need to be taken into account when evaluating a PAD approach (e.g., different types of attack, variety of devices to perform attempted attacks, and attacks directed to different sensors). Another aspect that is normally overlooked is that most detection

methods are custom-tailored to specific types of presentation attacks, in what
we refer to as hand-crafting of the features. With the emergence of deep learning methods and their success in tasks such as image classification, voice recognition, and language translation, in this chapter, we set forth the objective
of exploiting deep learning solutions for detecting presentation attacks using
data-driven solutions. In these cases, the biometric designer is responsible for
choosing an appropriate architecture for PAD and training solely from the
existing data available. We believe that this type of solution is the next step
when designing robust presentation attack detectors and also that they can,
if carefully designed, better deal with the challenging cross-dataset scenario.
The cross-dataset scenario arises when the system is trained with a data set
from one sensor or one scenario, and then later tested on data from a different
sensor or scenario. Figure 11.1 depicts the general pipeline we exploit in this
chapter. We start with pretrained deep neural networks and tune them independently for each modality (face, iris, and fingerprint) with different data sets
before building the final classifiers to distinguish between authentic images of
faces, irises, and fingerprints from their static counterparts.

   We organize the rest of this chapter as follows. Section 11.2 discusses state-
of-the-art methods for PAD considering the three modalities considered in this
chapter (face, iris, and fingerprint). Section 11.3 details the data-driven PAD
solution that we advocate as promising for this problem, and Section 11.5
shows the experiments and validations for different biometric spoofing data
sets. We close the chapter with some final considerations in Section 11.6.

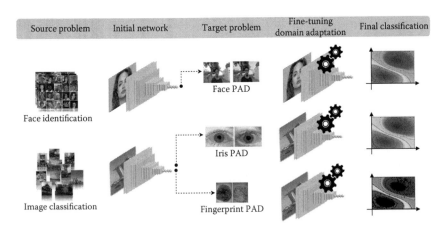

FIGURE 11.1
General pipeline exploited in this work. Initial network architectures,
originally proposed for other problems, are independently fine-tuned with
appropriate PAD examples from different data sets leading to discriminative features. Ultimately, classifiers are trained to separate between authentic
images of faces, irises, and fingerprints from their presentation attack versions.

11.2 Related Work

In this section, we review some of the most important PAD methods published in the literature for iris, face, and fingerprint.

11.2.1 Face PAD

The existing face antispoofing techniques can be categorized into four groups [12]: user behavior modeling [13,14] (e.g., eye blinking, small face movements), methods that require additional hardware [15] (e.g., infrared cameras and depth sensors), methods based on user cooperation (e.g., challenge questions), and finally, data-driven characterization approaches, which is the focus of our work herein.

We start this section reviewing frequency-based approaches, which are methods that rely on analyzing artifacts that are better visible in the frequency domain. Early studies followed this idea [16], and nowadays we have several works that support the effectiveness of this approach in detecting face spoofing. Li et al. [16] proposed a face-spoofing detection that emerged from the observation that the faces in photographs are smaller than the real one, and that the expressions and poses of the faces in photographs are invariant. Based on these observations, the authors devised a threshold-based decision method for detecting photo-based attempted attacks based on the energy rate of the high-frequency components in the two-dimensional Fourier spectrum. The major limitation of the technique proposed by Li et al. is that the high-frequency components are affected by illumination, which makes this frequency band too noisy [16,17]. To reduce that effect, Tan et al. [17] exploited the difference of image variability in the high-middle band. This is done using difference of Gaussian (DoG) bandpass filtering, which keeps as much detail as possible without introducing noisy or aliasing artifacts.

Pinto et al. [18] introduced an idea seeking to overcome the illumination effect when working in the frequency domain. In that work, the authors proposed a face antispoofing method for detecting video-based attempted attacks based on Fourier analysis of the noise signature extracted from videos, instead of using the image pixel values directly. Basically, after isolating the noise signal present in the video frames, the authors transformed that information to the Fourier domain and used the visual rhythm technique to capture the most important frequency components to detect an attempted attack, taking advantage of the spectral and temporal information. In a more recent work [19], the same authors expanded on this technique, taking advantage of the spectral, temporal, and spatial information from the noise signature by using the concept of visual codebooks. According to the authors, the new method enabled them to detect different types of attacks such as print- and mask-based attempted attacks as well.

Lee et al. [20] proposed an antispoofing technique based on the cardiac pulse measurements using video imaging [21]. The authors extended on previous work proposed by Poh et al. [21] by adding a threshold-based decision level based on the entropy measure. It was calculated from the power spectrum obtained from normalized RGB channels after eliminating the cross-channel noise, caused by the environment interference, using the Independent Component Analysis.

Another expressive branch of face antispoofing algorithms reported in the literature consists of texture-based approaches. In general, those algorithms exploit textural cues inserted in the fake biometric samples during its production and presentation to the biometric sensor under attack (e.g., printing defects, aliasing, and blurring effects). Tan et al. [17] proposed a texture-based approach to detect attacks with printed photographs based on the difference of the surface roughness of an attempted attack and a real face. The authors estimate the luminance and reflectance of the image under analysis and classify them using Sparse Low Rank Bilinear Logistic Regression methods. Their work was extended on by Peixoto et al. [22], who incorporated measures for different illumination conditions.

Similar to Tan et al. [17], Kose and Dugelary [23] evaluated a solution based on reflectance to detect attacks performed with printed masks. To decompose the images into components of illumination and reflectance, the Variational Retinex [24] algorithm was applied.

Määttä et al. [25,26] relied on microtextures for face-spoofing detection, inspired by the characterization of printing artifacts and by differences in light reflection when comparing real samples and presentation attack samples. The authors proposed a fusion scheme based on the local binary pattern (LBP) [27], Gabor wavelets [28], and histogram of oriented gradients [29]. Similarly, to find a holistic representation of the face able to reveal an attempted attack, Schwartz et al. [12] proposed a method that employs different attributes of the images (e.g., color, texture, and shape of the face).

Chingovska et al. [30] investigated the use of different variations of the LBP operator used in Määttä et al. [25]. The histograms generated from these descriptors were classified using a χ^2 histogram comparison, linear discriminant analysis (LDA), and a support vector machine (SVM).

Face-spoofing attacks performed with static masks have also been considered in the literature. Erdogmus and Marcel [31] explored a database with six types of attacks using facial information of four subjects. To detect attempted attacks, the authors used two algorithms based on Gabor wavelet [32] with a Gabor-phase based similarity measure [33].

Pereira et al. [34] proposed a score-level fusion strategy for detecting various types of attacks. The authors trained classifiers using different databases and used the Q statistics to evaluate the dependency between classifiers. In a follow-up work, Pereira et al. [35] proposed an antispoofing solution based on the dynamic texture, which is a spatiotemporal version of the original LBP.

Garcia and Queiroz [36] proposed an antispoofing method based on detection of the Moiré patterns, which appear as a result of the overlap of the digital grids. To find these patterns, the authors used a peak-detector algorithm based on maximum-correlation thresholding, in that strong peaks reveal an attempted attack. Similar to that, Patel et al. [37] proposed a PAD technique also based on the Moiré pattern detection, which uses the the multiscale version of the LBP (M-LBP) descriptor.

Tronci et al. [38] exploited the motion information and clues that are extracted from the scene by combining two types of processes, referred to as *static* and *video-based analyses*. The static analysis consists of combining different visual features such as color, edge, and Gabor textures, whereas the video-based analysis combines simple motion-related measures such as eye blink, mouth movement, and facial expression change.

Anjos and Marcel [39] proposed a method for detecting photo-based attacks assuming a stationary face-recognition system. According to the authors, the intensity of the relative motion between the face region and the background can be used as a clue to distinguish valid access of attempted attacks because the motion variations between face and background regions exhibit greater correlation in the case of attempted attacks.

Wen et al. [40] proposed a face spoof-detection algorithm based on image distortion analysis, describing different features such as specular reflection, blurriness, chromatic moment, and color diversity. These features are concatenate to generate feature vectors, which are used to generate an ensemble classifier, each one specialized to detect a type of attempted attack.

Kim et al. [41] proposed a method based on the diffusion speed of a single image to detect attempted attacks. The authors define the local patterns of the diffusion speed, namely local speed patterns via Total Variation (TV) flow [42], which are used as feature vectors to train a linear classifier, using the SVM, to determine whether the given face is fake. In turn, Boulkenafet et al. [43] proposed an antispoofing technique using a color texture analysis. Basically, the authors perform a microtexture analysis considering the color-texture information from the luminance and the chrominance channels by extracting feature descriptions from different color spaces.

Different from the previous methods, which focus on defining a PAD that does not leverage the identity information present in the gallery, Yang et al. [44] proposed a person-specific face antispoofing approach, in which a classifier was built for each person. According to the authors, this strategy minimizes the interferences among subjects.

Virtually all previous methods exploit hand-crafted features to analyze possible clues related to a presentation attack attempt. Whether these features are related to texture, color, gradients, noise, or even reflection, blurriness, and chromatic moment, they always come down to the observation of specific artifacts present in the images and how they can be captured properly. In this regard, LBP stands out as the staple of face-based spoofing research thus far. Departing from this hand-crafted characterization modeling strategy, a recent trend in the literature has been devoted to designing and deploying solutions able to directly learn, from the existing available training data, the intrinsic discriminative features of the classes of interest, the so-called data-driven characterization techniques, probably motivated by the huge success these approaches have been showing in other vision-related problems [45,106]. Out of those, the ones based on deep learning solutions stand out right away as promising for being highly adaptive to different situations.

Menotti et al. [46] aimed at hyperparameter optimization of network architectures [47,48] (architecture optimization) and on learning filter weights via the well-known back-propagation algorithm [49] (filter optimization) to design a face-spoofing detection approach. The first approach consists of learning suitable convolutional network architectures for each domain, whereas the second approach focuses on learning the weights of the network via back-propagation.

Manjani et al. [50] proposed an antispoofing solution based on a deep dictionary learning technique originally proposed in [51] to detect attempted attacks performed using silicone masks. According to the authors, deep dictionary learning combines concepts of two most prominent paradigms for representation learning, deep learning, and dictionary learning, which enabled the authors to achieve a good representation even using a small data for training.

11.2.2   Fingerprint PAD

Fingerprint PAD methods can be categorized into two groups: hardware- and software-based solutions [52]. Methods falling into the first group use information provided from additional sensors to gather artifacts that reveal a spoofing attack that is outside of the fingerprint image. Software-based techniques rely solely on the information acquired by the biometric sensor of the fingerprint authentication system.

Based on several quality measures (e.g., ridge strength or directionality, ridge continuity), Galbally et al. [53,54] proposed a set of features aiming at fingerprint PAD, which were used to feed a linear discriminant analysis classifier.

Gragnaniello et al. [55] proposed an antispoofing solution based on a Weber Local Descriptor (WLD) operating jointly with other texture descriptors such as a Local Phase Quantization (LPQ) and LBP descriptor. The experimental results suggest that WLD and LPQ complement one another, and their joint usage can greatly improve their discriminating ability, even when compared individually or combined with LBP.

Inspired by previous works based on an LBP descriptor, Jia et al. [56] proposed a spoofing-detection scheme based on multiscale block local ternary patterns (MBLTP) [57]. According to the authors, the computation of the LTP descriptor is based on average values of block subregions rather than individual pixels, which makes it less sensitive to noise because the computation is based on a three-value code representation and on average values of block subregions, rather than on individual pixels.

Ghiani et al. [58] proposed the use of binarized statistical image features (BSIF), a textural binary descriptor whose design was inspired by the LBP and LPQ methods. Basically, the BSIF descriptor learns a filter set by using statistics of natural images [59], leading to descriptors better adapted to the problem. The same authors also explored the LPQ descriptor to find a feature space insensitive to blurring effects [60].

Gottschlich [61] proposed another idea based on filter-learning convolution comparison pattern. To detect a fingerprint spoofing, the authors compute the discrete cosine transform from rotation invariant patches, and compute their binary patterns by comparing pairs of discrete cosine transform coefficients. These patterns are gathered in a histogram, which was used to feed a linear SVM classifier.

Rattani et al. [62] introduced a scheme for automatic adaptation of a liveness detector to new spoofing materials in the operational phase. The aim of the proposed approach is to reduce the security risk posed by new spoof materials on an antispoofing system. The authors proposed a novel material detector specialized to detect new spoof materials, pointing out the need for retraining the system with the new material spotted.

Similar to that, Rattani et al. [63] proposed an automatic adaptation anti-spoofing system composed of an open-set fingerprint spoofing detector and by a novel material detector, both based on Weibull-calibrated SVM (W-SVM) [64]. The novel material detector was built with a multiclass W-SVM, composed by an ensemble of pairs of 1-Class and binary SVMs, whereas the open set fingerprint-spoofing detector was trained with features based on textural [60], physiological [65], and anatomical [66] attributes.

Gragnaniello et al. [67] proposed a fingerprint-spoofing detection based on both spatial and frequency information to extract local amplitude contrast and local behavior of the image, which were synthesized by considering the phase of some selected transform coefficients generated by the short-time Fourier transform. This information generates a bidimensional contrast-phase histogram, which was used to train a linear SVM classifier.

Kumpituck et al. [68] exploited an antispoofing schema based on wavelet decomposition and LBP operator. In this work, the authors extract LBP histograms from several wavelet subband images, which were concatenated and used to feed an SVM classifier. The authors also evaluated a more conventional approach that consists of calculating the energy from wavelet subbands instead of the LBP histograms. Experimental results show that wavelet-LBP descriptor achieved a better discrimination than wavelet-energy and LBP descriptors used separately, besides achieving competitive results with the state-of-the-art methods.

Finally, also departing from the traditional modeling, which uses basically texture patterns to characterize fingerprint images, Nogueira et al. [4]

proposed a fingerprint antispoofing technique based on the concept of pre-trained convolutional neural networks (CNNs). Basically, the authors use well-known CNN architectures in the literature such as AlexNet [106] and VGG [105] as their starting point for learning the network weights for fingerprint-spoofing detection.

Marasco et al. [69] investigated two well-known CNN architectures, the GoogLeNet [70], CaffeNet [106], to analyze their robustness in detecting unseen spoof materials and fake samples from new sensors. As mentioned previously, Menotti et al. [46] also proposed hyperparameter optimization of network architectures along with filter optimization techniques for detecting fingerprints presentation attacks.

11.2.3 Iris PAD

Early work on iris-spoofing detection dates back to the 1990s, when Daugman [71] discussed the feasibility of some attacks on iris-recognition systems. In that work, he proposed to detect such attempts using the fast Fourier transform to verify the high-frequency spectral magnitude.

According to Czajka [72], solutions for iris-liveness detection can be categorized into four groups, as Cartesian product of two dimensions: type of measurement (passive or active) and type of model of the object under test (static or dynamic). Passive solutions mean that the object is not stimulated more than it is needed to acquire an iris image for recognition purpose. Hence, it typically means that no extra hardware is required to detect an attempted attack. Active solutions try to stimulate an eye and observe the response to that stimuli. It means that typically some extra hardware elements are required. In turn, the classification between static and dynamic objects means that the algorithm can detect an attempted attack using just one (static) image from the biometric sensor or needs to use a sequence of images to observe selected dynamic features. In this section, we review only passive and static methods, which is the focus of this chapter.

Pacut Czajka [73] introduced three iris liveness-detection algorithms based on the analysis of the image frequency spectrum, controlled light reflection from the cornea, and pupil dynamics. These approaches were evaluated with paper printouts produced with different printers and printout carriers and shown to be able to spoof two commercial iris-recognition systems. A small hole was made in the place of the pupil, and this trick was enough to deceive commercial iris-recognition systems used in their study. The experimental results obtained on the evaluation set composed of 77 pairs of fake and live iris images showed that the controlled light reflections and pupil dynamics achieve zero for both false acceptance rate and false rejection rate. In turn, two commercial cameras were not able to detect 73.1% and 15.6% of iris paper printouts and matched them to biometric references of authentic eyes.

Galbally et al. [74] proposed an approach based on 22 image quality measures (e.g., focus, occlusion, and pupil dilation). The authors use sequential floating feature selection [75] to single out the best features, which were used to feed a quadratic discriminant classifier. To validate the proposed approach, the authors used the BioSec [76,77] benchmark, which contains print-based iris-spoofing attacks. Similarly, Sequeira et al. [78] also exploited image quality measures [74] and three different classification techniques, validating the work on BioSec [76,77] and Clarkson [79] benchmarks and introducing the MobBIO-fake benchmark comprising 800 iris images. Sequeira et al. [80] expanded on previous work using a feature selection step to obtain a better representation to detect an attempted attack. The authors also applied iris segmentation [81] to obtain the iris contour and adapted the feature extraction processes to the resulting noncircular iris regions.

Wei et al. [82] addressed the problem of iris-liveness detection based on three texture measures: iris edge sharpness, iris-texton feature for characterizing the visual primitives of iris texture, and using selected features based on co-occurrence matrix. In particular, they used fake iris-wearing color and textured contact lenses. The experiments showed that the edge sharpness feature achieved comparable results to the state-of-the-art methods at that time, and the iris texture and co-occurrence matrix measures outperformed the state-of-the-art algorithms.

Czajka [83] proposed a solution based on frequency analysis to detect printed irises. The author associated peaks found in the frequency spectrum to regular patterns observed for printed samples. This method, tuned to achieve a close-to-zero false rejection rate (i.e., not introducing additional false alarms to the entire system), was able to detect 95% of printed irises. This paper also introduced the Warsaw LivDet-Iris-2013 data set containing 729 fake images and 1274 images of real eyes.

Texture analysis has also been explored for iris-spoofing detection. In the MobILive [84] iris-spoofing detection competition, the winning team relied on three texture descriptors: LBP [27], LPQ [85], and binary Gabor pattern [86]. Sun et al. [87] recently proposed a general framework for iris image classification based on a hierarchical visual codebook (HVC). The codebook encodes the texture primitives of iris images and is based on two existing bag-of-words models. The method achieved a state-of-the-art performance for iris-spoofing detection, among other tasks related to iris recognition.

Doyle et al. [88] proposed a solution based on a modified LBP [89] descriptor. In this work, the authors show that although it is possible to obtain good classification, results using texture information extracted by the modified LBP descriptor, when lenses produced by different manufacturers are used, the performance of this method drops significantly. They report 83% and 96% of correct classification when measured on two separated data sets, and a significant drop in accuracy when the same method was trained on the one data set and tested on the other data set: 42% and 53%, respectively.

This cross-dataset validation has been shown to be challenging and seems to be recommended in several validation setups for PAD. Yadav et al. [90] expanded on the previous work by analyzing the effect of soft and textured contact lenses on iris recognition.

Raja et al. [91] proposed an antispoofing method based on eulerian video magnification, [92] which was applied to enhance the subtle phase information in the eye region. The authors proposed a decision rule based on cumulative phase information, which was applied by using a sliding window approach on the phase component for detecting the rate of the change in the phase with respect to time.

Raghavendra and Busch [5] proposed a novel spoofing-detection scheme based on a multiscale version of the BSIF and linear SVM. Gupta et al. [93] proposed an antispoofing technique based on local descriptors such as LBP [27], histogram of oriented gradient [29], and GIST [94], which provide a representation space by using attributes of the images such as color, texture, position, spatial frequency, and size of objects present in the image. The authors used the feature vectors produced by the three descriptors to feed a nonlinear classifier and decide whether an image under analysis is fake.

Czajka [95] proposed an iris-spoofing detection based on pupil dynamics. In that work, the author used the pupil dynamics model proposed by Kohn and Clynes [96] to describe its reaction after a positive light stimuli. To decide whether the eye is alive, the author used variants of the SVM to classify feature vectors that contain the pupil dynamic information of a target user. This work has been further extended to a mixture of negative and positive light stimuli [72] and presented close-to-perfect recognition of objects not reacting to light stimuli as expected for a living eye.

Finally, Lovish et al. [97] proposed a cosmetic contact lens detection method based on LPQ and binary Gabor patterns, which combines the benefits of both LBP and Gabor filters [86]. The histograms produced for both descriptors were concatenated and used to build a classification model based on the SVM algorithm.

Similarly to the approaches tackling the presentation attack problem in fingerprint and faces, hand-crafted texture features seem to be the preferred choice in iris-spoofing detection. Methods inspired by LBP, visual codebooks, and quality metrics are the most popular methods so far. In this sense, the works of Menotti et al. [46] and Silva et al. [98], which exploit data-driven solutions for this problem, are sufficiently different from the previous methods and present very promising results.

11.2.4   Unified frameworks to PAD

Galbally et al. [99] proposed a general approach based on 25 image-quality features to detect attempt attacks in face, iris, and fingerprint biometric systems simultaneously. Evaluations performed on popular benchmarks for three modalities show that this approach is highly competitive, considering the state-of-the-art methods dedicated for single modalities.

Gragnaniello et al. [100] evaluated several local descriptors for face-, fingerprint-, and iris-based biometrics in addition to the investigation of promising descriptors using the Bag-of-Visual-Word model [101], Scale-Invariant Feature Transform [102], DAISY [103], and the Shift-Invariant Descriptor [104].

Menotti et al. [46] showed that the combination of architecture optimization and filter optimization provides better comprehension of how these approaches interplay for face, iris, and fingerprint PAD and also outperforms the best known approaches for several benchmarks.

In this chapter, we decided to explore data-driven solutions for spoofing detection in different modalities based on deeper architectures than the one used in [46] and evaluate the effects of such decision. Our objective is to show the potential of this approach, but also highlight its limitations, especially related to cross-dataset experiments.

11.3 Methodology

In this section, we present the CNN that we adopted to PAD for face, fingerprint, and iris. Our objective is simply to show that this new trend in the literature is also relevant for the task of PAD and that research in this direction needs to be considered. At the same time, we also show that even when adopting a powerful image-classification technique such as deep neural networks, we still cannot deal effectively with the challenging cross-dataset problem. As a result, it is clear that the research community now needs to shift its attention to cross-dataset validation setups (or, more general, open-set classification) because they are closer to real-world operational conditions when deploying biometric systems.

11.3.1 Network architecture

For this work, we adopted the VGG network architecture proposed by Simonyan and Zisserman [105]. However, that network was first proposed for object recognition and not PAD. Therefore, for each problem of interest (PAD in face, iris, and fingerprint), we adapt the network's architecture as well as fine-tune its weights to our two-class problem of interest. Training the network from scratch to our problem is also a possibility if enough training samples (normal and presentation attack samples) are available. However, because this is not often the case in this area, it is recommended to start the network weights with a related (source) problem and then adapt these weights with training examples of a target problem.

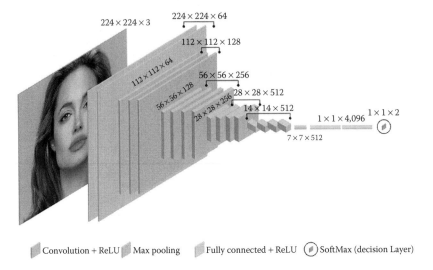

Convolution + ReLU   Max pooling   Fully connected + ReLU   ⊛ SoftMax (decision Layer)

FIGURE 11.2
Adopted network architecture, originally proposed for object recognition by
the Visual Geometry Group and thus referred to as VGG network [105].

Figure 11.2 depicts the network architecture we adopted in this work.
During training, the network's input consists of fixed-size 224×224 RGB
images that go through a stack of convolutional layers comprising filters with
a small receptive field (3×3). In this network, the convolution stride is fixed
to one pixel and the spatial zero-padding for convolutional operation is also of
one pixel. There are five max-pooling layers in this network (carefully placed
after some convolution layers). The max-poolings are performed over a 2×2
pixel window, with stride 2.

The stack of convolutional layers is followed by three fully connected (FC)
layers: the first two have 4096 units each, while the the third layer performs the
two-way spoofing classification problem of our interest (originally this was an
FC layer with 1000 units for the ImageNet 1000-way classification problem).
The final layer is the soft-max layer translating the outputs of two-unit layer
into a posterior probabilities of class membership. Each unit in the hidden
layers has a rectified linear (ReLU) activation function [106]. The depth of
convolution layers or, in other words, their number of channels, starts with
64 and is iteratively doubled after each max-pooling layer to a maximum
of 512.

11.3.2   Training and testing

For training, we start with the network trained to a source problem when-
ever it is possible. To detect presentation attacks with faces, we initialize

the network with the weights learned for face recognition [107]. However, the closest problem we had for iris and fingerprints was general image classification. Therefore, PAD for iris and fingerprints is performed by the network initialized with the weights precomputed for the ImageNet classification problem. The first convolutional layers act mostly as generic feature detectors (such as edges) and are suitable for different computer vision tasks. However, each next convolutional layer is more context-focused and extracts features that are task-related. Hence, using last layers trained for general object recognition in visual-spoofing detection is not optimal, and a large improvement may be achieved by specializing the network. Certainly, a preferable solution is to initialize the weights with those used in networks solving iris- and fingerprint-related tasks, as the network would have been specialized to this type of imagery. However, because training of such networks from scratch requires a lot data and effort, it is still a good move to initialize our own network with image-related weights than just purely at random and tune its weights if there is not enough available training data.

Once a source set of weights to initialize the network is chosen, the fine-tuning follows a standard procedure: selects the training set of the target domain and uses it to perform forward passes and back-propagation in the network. The test for an input image is straightforward. Just resize it to the network's input size and feed it to the network. As the network has been fully adapted to the target problem of interest, it will already produce a two-class output.

More specifically, for the cases of fingerprints, the input images in a data set are center-cropped and resized to 224×224 pixels, which is the standard input size of the VGG network. The centering happens through calculating the average of black pixels in the binary fingerprint image and keeping all the rows/columns with a density of black pixels greater than the global image average plus or minus 1.8 standard deviations of each respective row/column. This is used to eliminate the borders without any useful information. For optimizing the network in the fingerprint case, we use the standard SGD solver implemented in Caffe with the following hyperparameters: base learning rate of 0.0001, step lr policy, step size of 2000, momentum of 0.9, weight decay of 0.0002, gamma of 0.5, and maximum of 2001 iterations.

In the case of faces, we center-cropped the images based on the eye coordinates calculated with the aid of Face++.* On center-cropping, the image is resized to 224×224 pixels. For optimizing the network in the face case, we use the standard SGD solver implemented in Caffe with the following hyperparameters: base learning rate of 0.001, step lr policy, step size of 1000, momentum of 0.9, weight decay of 0.0005, gamma of 0.001 and maximum number of iterations of 4000.

---

*http://www.faceplusplus.com/

For irises, we resize the images to the network's standard input size of 224×224 pixels and employ the same parameters as for the face optimization problem.

11.3.3   Memory footprint

The chosen network has an average size of 140 MB. Most of its parameters (and memory) are in the convolution and FC layers. The first FC layer contains 100 million weights, out of a total of 134 million for the entire adapted network.

11.4   Metrics and Data Sets

In this section, we describe the benchmarks (data sets) and selected accuracy estimators considered in this work. All data sets used in this chapter were freely available to us, and we believe that it is the case for other researchers on request sent directly to their creators. Data sets composing our testing environment are the most commonly used benchmarks to evaluate PAD for face, iris, and fingerprints. Because all the benchmarks have been already divided by their creators into training and testing subsets, we decided to follow these divisions. Each training subset was divided by us into two disjoint subsets multiple times to perform cross-validation–based training to increase generalization capabilities of the winning model and to minimize an overfitting. The results reported further in this chapter are those obtained on testing sets. The next subsections characterize briefly all data sets, and Table 11.1 shows their major features, in particular the number of samples in each benchmark and their assignment to training and testing subsets.

11.4.1   Video-based face-spoofing benchmarks

We use two benchmarks used to evaluate the performance of PAD algorithms for face modality, Replay-Attack [30] and CASIA Face Anti-Spoofing [108] data sets. These data sets contain five types of attempted attacks performed with fake samples presenting different qualities.

11.4.1.1   Replay-attack

This benchmark contains short video recordings of both valid accesses and video-based attacks of 50 different subjects. To generate valid access videos, each person was recorded in two sessions in a controlled and in an adverse environment with a regular webcam. Then, spoofing attempts were generated using three techniques:

- *Print attack*: Hard copies of high-resolution digital photographs were presented to the acquisition sensor; these samples were printed with a Triumph-Adler DCC 2520 color laser printer.

TABLE 11.1
Main features of the benchmarks considered herein

Modality	Benchmark	Color	Dimension $cols \times rows$	# Training			# Testing		
				Live	Fake	Total	Live	Fake	Total
Face	Replay-Attack	Yes	320×240	600	3000	3600	4000	800	4800
	CASIA	Yes	1280×720	120	120	240	180	180	360
Iris	Warsaw LivDet2015	No	640×480	852	815	1667	2002	3890	5892
	AVTS	No	640×480	200	200	400	600	600	1200
Fingerprint	LivDet2009: CrossMatch	No	640×480	500	500	1000	1500	1500	3000
	LivDet2009: Identix	No	720×720	375	375	750	1125	1125	2250
	LivDet2009: Biometrika	No	312×372	500	500	1000	1500	1500	3000
	LivDet2013: Biometrika	No	312×372	1000	1000	2000	1000	1000	2000
	LivDet2013: CrossMatch	No	800×750	1250	1000	2250	1250	1000	2250
	LivDet2013: Italdata	No	640×480	1000	1000	2000	1000	1000	2000
	LivDet2013: Swipe	No	208×1500	1250	1000	2250	1250	1000	2250

- *Mobile attack*: Videos displayed on an iPhone screen were presented to the acquisition sensor; these videos were taken also with the iPhone.

- *High-definition attack*: High-resolution photos and videos taken with an iPad were presented to the acquisition sensor using the iPad screen.

11.4.1.2   CASIA

This benchmark was based on samples acquired from 50 subjects. Genuine images were acquired by three different sensors presenting different acquisition quality (from low to high): "long-time-used USB camera," "newly bought USB camera," and Sony NEX-5. Pixel resolution of images was either 640×480 (both webcams) or 1920×1080 (Sony sensor). Sony images were cropped to 1280×720 by the authors. During the acquisition, subjects were asked to blink. Three kinds of presentation attacks were carried out:

- *Warped photo attack*: High-quality photos were printed on a copper paper and videos were recorded by Sony sensor; the printed images were intentionally warped to imitate face micromovements.

- *Cut photo attack*: Eyes were cut from the paper printouts and an attacker hidden behind an artifact imitated the blinking behavior when acquiring the video by the Sony sensor.

- *Video attack*: High-quality genuine videos were displayed on an iPad screen of 1280×720 pixel resolution.

The data originating from 20 subjects was selected for a training set, while remaining samples (acquired for 30 subjects) formed the testing set.

11.4.2   Fingerprint-spoofing benchmarks

Two data sets used in Liveness Detection Competitions (LivDet, www.livdet. org) were employed in this chapter. LivDet is a series of international competitions that compare presentation attack methodologies for fingerprint and iris using a standardized testing protocol and large quantities of spoof and live samples. All the competitions are open to all academic and industrial institutions that have software-based or system-based biometric liveness detection solutions. For fingerprints, we used data sets released in 2009 and 2013.

11.4.2.1   The LivDet2009 benchmark

This benchmark consists of three subsets of samples acquired by Biometrics FX2000, CrossMatch Verifier 300 LC, and Identix DFR2100. Both the spatial scanning resolution and pixel resolution vary across subsets, from 500 DPI to 686 DPI, and from 312×372 to 720×720 pixels, respectively. Three different materials were used to prepare spoofs: Play-Doh, gelatin, and silicone.

11.4.2.2 The LivDet2013 benchmark

This benchmark contains four subsets of real and fake fingerprint samples acquired by four sensors: Biometrika FX2000, Italdata ET10, Crossmatch L Scan Guardian, and Swipe. Inclusion of samples from the Swipe sensor is especially interesting, because it requires, as the name suggests, swiping a finger over the small sensor. This makes the quality of spoofs relatively different when compared to the regular, flat sensors requiring only touching the sensor by the finger. For a more realistic scenario, fake samples acquired by Biometrika and Italdata were generated without user cooperation, whereas fake samples acquired by Crossmatch and Swipe were generated with user cooperation. Several materials for creating the artificial fingerprints were used, including gelatin, silicone, latex, among others. The spatial scanning resolution varies from a small 96 DPI (the Swipe sensor) to 569 (the Biometrika sensor). The pixel resolution is also heterogeneous: from relatively nonstandard 208×1500 to pretty large 800×750. This makes the cross-subset evaluation quite challenging.

11.4.3 Iris-spoofing benchmarks

To evaluate our proposed method in detecting iris presentation attack, we used two benchmarks: AVTS [77] and a new data set Warsaw LivDet2015, which is an extension of Warsaw LivDet2013 [83]. These data sets contain attempted attacks performed with printed iris images, which were produced using different printers and paper types.

11.4.3.1 AVTS

This benchmark was based on live samples collected from 50 volunteers under the European project BioSec (Biometrics and Security). To create spoofing attempts, the authors tested two printers (HP Deskjet 970cxi and HP LaserJet 4200L), various paper types (e.g., cardboard as well as white, recycle, photo, high resolution, and butter papers), and a number of preprocessing operations. The combination that gave the highest probability of image acquisition by the LG IrisAccess EOU3000 sensor used in the study was selected for a final data set collection. The authors printed their samples with the inkjet printer (HP Deskjet 970cxi) on a high-resolution paper and applied an Open-TopHat preprocessing to each image prior to printing. The pixel resolution of each image was 640×480, which is recommended by ISO/IEC as a standard resolution for iris recognition samples.

11.4.3.2 Warsaw LivDet2015

This data set is an extension of the LivDet-Iris 2013 Warsaw Subset [83] and was used in the 2015 edition of the LivDet-Iris competition (www.livdet.org). It gathers 2854 images of authentic eyes and 4705 images of the paper printouts prepared for almost 400 distinct eyes. The photographed paper printouts were used to successfully forge an example commercial iris-recognition system (i.e., samples used in real and successful presentation attacks). Two printers were used to generate spoofs: HP LaserJet 1320 and Lexmark C534DN. Both real and fake images were captured by an IrisGuard AD100 biometric device with liveness-detection functionality intentionally switched off. (To get a free copy of this data set, follow the instructions given at Warsaw's lab webpage http://zbum.ia.pw.edu.pl/EN/node/46).

11.4.4   Error metrics

In this chapter, we use the error metrics that are specific to PAD and partially considered by ISO/IEC in their PAD-related standards [110].

Attack presentation classification error rate (APCER): Proportion of *attack presentations* incorrectly classified as *bona-fide (genuine) presentations* at the PAD subsystem in a specific scenario. This error metric is analogous to false match rate (FMR) in biometric matching, that is related to a false match of samples belonging to two different subjects. As FMR, the APCER is a function of a decision threshold τ.

Bona-fide presentation classification error rate (BPCER): Proportion of *bona-fide (genuine) presentations* incorrectly classified as *presentation attacks* at the PAD subsystem in a specific scenario. This error metric is analogous to false nonmatch rate (FNMR) in biometric matching, that is related to false nonmatch of samples belonging to the same subject. Again, the BPCER is a function of a decision threshold τ.

Half total error rate (HTER): Combination of APCER and BPCER in a single error rate with a decision threshold as an argument:

$$\text{HTER}(\tau) = \frac{\text{APCER}(\tau) + \text{BPCER}(\tau)}{2} \qquad (11.1)$$

11.5   Results

In this section, we present and discuss the experimental results of the proposed method. Sections 11.5.1 through 11.5.3 show the performance results and the experimental protocols employed to validate the performance of the proposed methodology.

11.5.1   Face

In this section, we present the results of our proposed PAD for face modality. The experiments are conducted considering the original protocol of the data sets used in this chapter (cf., Section 11.4), as well cross-dataset protocol, hereafter referred to as same-dataset and cross-dataset protocols, respectively. In general, a prime requirement of most machine-learning algorithms is that both training and testing sets are independent and identically distributed. But unfortunately, it does not always happen in practice; subsets can be identically distributed (e.g., captured using the same sensor and in the same environment conditions), but totally dependent because of adding of bias in the data (e.g., some dirt in the biometric sensor used to capture both subsets, identities present in two subsets, artifacts added during the attack simulations, etc.). In addition, the effects of the closed-world assumption [64] may mislead us to believe that a given approach is perfect when in fact its performance can be disastrous when deployed in practice for unknown presentation attacks. In this context, both same-dataset and cross-dataset are key experimental protocols in determining more accurate detection rates of an antispoofing system when operating in less-controlled scenarios with different kinds of attacks and sensors.

11.5.1.1 Same-dataset results

Table 11.2 shows the results for Replay-Attack and CASIA data sets, considering that training and testing is performed on the same data set. The VGG network was able to detect all kinds of attempted attacks present in the Replay-Attack data set, and also to detect two methods of attempted attacks (hand-based and fixed-support attacks), which were confirmed by the perfect classification result (HTER of 0.0%). Considering the CASIA data set, the proposed method obtained an HTER of 6.67%. The performance achieved by the proposed method on this data set can be explained by the high degree of variability present in the CASIA data set (e.g., different kinds of attack and resolution) that makes this data set more challenging. In both data sets, we use the k-fold cross-validation technique ($k = 10$) to build a classification model using the training set and also the development set whether it is available. Figures 11.3 and 11.4 present empirical distributions of the difference between two CNN output nodes and the corresponding ROC curves.

TABLE 11.2

Performance results obtained in the **same-dataset** evaluations of the face **PAD**

	APCER (%)	BPCER (%)	HTER (%)	ROC and ePDF
Replay-Attack	0.00	0.00	0.00	Figure 11.3
CASIA	0.00	13.33	6.67	Figure 11.4

Note: Pointers to plots presenting receiver operating characteristics (ROC) and empirical probability distribution functions (ePDF) are added in the last column.

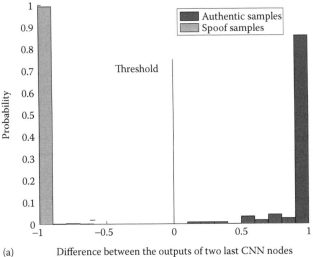

(a)   Difference between the outputs of two last CNN nodes

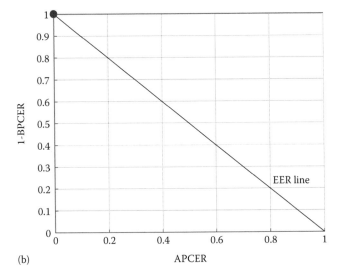

(b)   APCER

FIGURE 11.3
(a) Empirical probability distributions (ePDF) of the difference between two
CNN output nodes (after softmax) obtained separately for authentic and spoof
face samples. (b) ROC curve. Variant: training on **Replay-Attack**, testing
on **Replay-Attack**. The threshold shown in blue color on the left plot and
the blue dot on the ROC plot correspond to the approach when the predicted
label is determined by the node with the larger output.

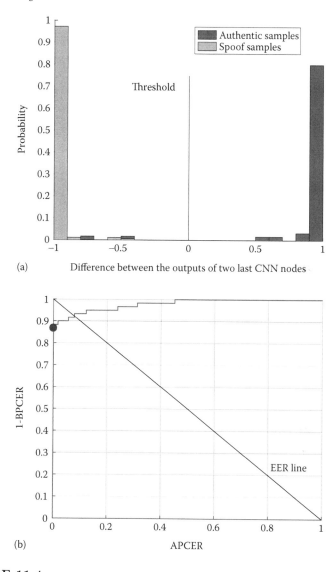

(a)

(b)

FIGURE 11.4

Same as Figure 11.3 except the variant: training on **CASIA**, testing on **CASIA**.

TABLE 11.3
Performance results obtained with the **cross-dataset** evaluations of **face PAD** and using the overall testing set of each data set

Training	Test	APCER (%)	BPCER (%)	HTER (%)	ROC and ePDF
Replay-Attack	CASIA	42.67	51.67	47.16	Figure 11.5
CASIA	Replay-Attack	89.44	10.0	49.72	Figure 11.6

11.5.1.2 Cross-dataset results

Table 11.3 and Figures 11.5 and 11.6 show the results obtained in cross-dataset evaluation protocol. We can clearly see a dramatic drop in the performance when we train and test on different data sets. Several sources of variability between the data sets may contribute to this result. The first one is that the data sets contain different kinds of attack. The Replay-Attack data set contains three kinds of attacks (high definition-based, mobile-based, and video-based attacks), whereas the CASIA data set includes an additional two kinds of attack (warp-based and cut-based photo attacks). Another source is the fact that data come from different sensors, which potentially produce samples with different resolutions, color distributions, backgrounds, and so on. The VGG architecture finds specific features and even when it is tuned to the specific problem, it does not generalize well to be agnostic to specific properties of data acquisition process.

11.5.2 Fingerprints

This section presents how our VGG-based approaches perform in detection of fingerprint attack presentation. As for experiments with face benchmarks, we used the training subsets (as defined by data set creators) to make a cross-validation–based training and separate testing subsets in final performance evaluation. Fingerprint benchmarks are composed of subsets gathering mixed attacks (e.g., glue, silicone, or gelatin artifacts) and acquired by different sensors (Table 11.1).

11.5.2.1 Same-sensor results

In this scenario, samples acquired by different sensors are not mixed together. That is, if the classifier is trained with samples acquired by sensor X, only sensor X samples are used in both the validation and final testing. As in previous experiments, 10 statistically independent estimation-validation pairs of nonoverlapping subsets were created, and the solution presenting the lowest HTER over 10 validations was selected for testing. Table 11.4 as well as Figures 11.7 and 11.8 show the same-sensor testing results averaged over all sensors (used to build a given data set) and presented for each benchmark

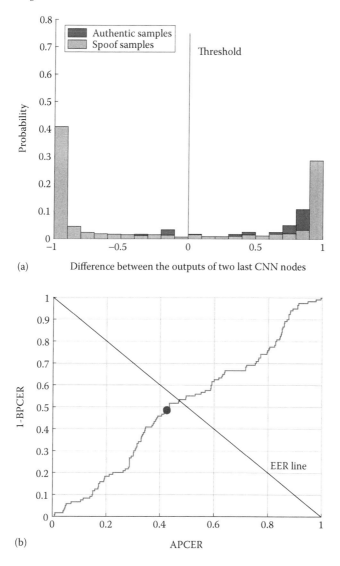

(a) Difference between the outputs of two last CNN nodes

(b) APCER

FIGURE 11.5

Same as Figure 11.3 except the variant: training on **Replay-Attack**, testing on **CASIA** (cross-dataset testing).

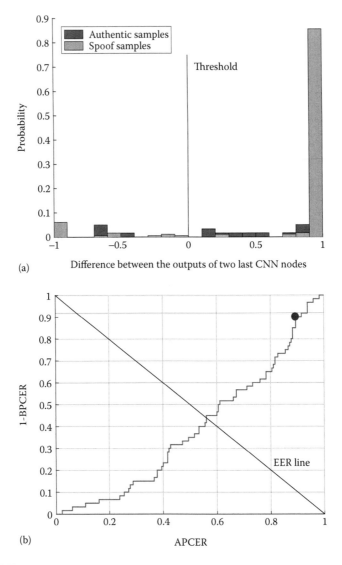

FIGURE 11.6
Same as Figure 11.3 except the variant: training on **CASIA**, testing on **Replay-Attack**.

TABLE 11.4

Performance results obtained in **same-dataset** evaluations of **fingerprint PAD** using a part of testing samples acquired by the same sensor as in the training procedure

Training	Testing	APCER (%)	BPCER (%)	HTER (%)	ROC and ePDF
LivDet2009	LivDet2009	19.37	3.45	11.4	Figure 11.7
LivDet2013	LivDet2013	6.8	2.79	4.795	Figure 11.8

Notes: Results are averaged over all subsets representing different sensors.

separately. These results suggest that the older benchmark (LivDet2009) is relatively difficult because almost 20% of spoofing samples were falsely accepted in a solution that falsely rejects only 3.45% of authentic examples.

11.5.2.2 Cross-sensor results

For cross-sensor analysis, the newer benchmark (LivDet2013) was selected. Each subset (estimation, validation, and testing) was divided into two disjoint subsets of samples: acquired by ItalData and Swipe sensors and acquired by Biometrika and CrossMatch sensors. Table 11.5 shows that, as with the other modalities, we can observe serious problems with recognition of both artifacts or genuine samples (two first rows of Table 11.5). Figures 11.9 and 11.10, illustrating these results, suggest that a better balance between APCER and BPCER can be found if there is a possibility to adjust the acceptance threshold.

For completeness, same-sensor results are also presented on this data set in two last rows of Table 11.5 and in Figures 11.11 and 11.12. As expected, a solution based on deep network achieves much better accuracy when the type of sensor in known.

11.5.3 Iris

This last section presents the results of iris presentation attacks detection. Two iris PAD benchmarks were used (as described in Section 11.4), and both same-dataset and cross-dataset experiments were carried out. Each data set (Warsaw LivDet2015 and AVTS) is already split by their creators into training and testing subsets. We followed this split and used the testing subset only in final performance evaluation. The training subset, used in method development, was randomly divided 10 times into estimation and validation disjoint subsets used in cross-validation when training the classifiers.

The average HTER's over 10 splits calculated for validation subsets were approx. 0.0001 and 0.0 for Warsaw and AVTS data sets, respectively. HTER = 0.0 for 5 out of 10 splits of Warsaw training data set. This means that the VGG-based feature extractor followed by a classification layer trained on our

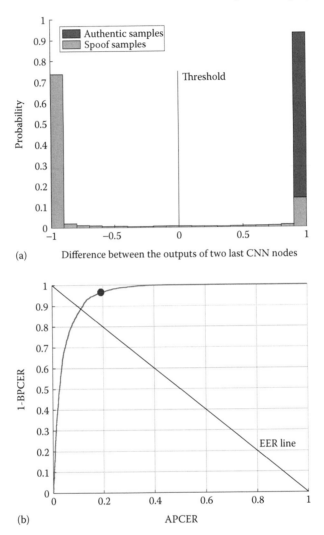

FIGURE 11.7
(a) Empirical distributions of the difference between two CNN output nodes (after softmax) obtained separately for authentic and spoof **fingerprint** samples. (b) ROC curve. Variant: training on **LivDet2009**, testing on **LivDet2009**. As in previous plots, the threshold shown in blue color on the left plot and the blue dot on the ROC plot correspond to the approach when the predicted label is determined by the node with the larger output.

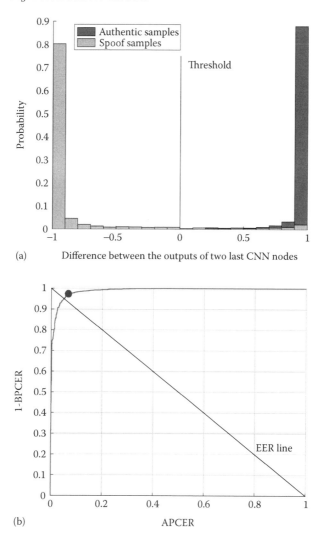

(a) Difference between the outputs of two last CNN nodes

(b) APCER

FIGURE 11.8
Same as Figure 11.7 except the variant: training on **LivDet2013**, testing on **LivDet2013**.

TABLE 11.5

Performance results obtained in **cross-dataset** evaluations of **fingerprint PAD** using a part of testing samples acquired by different sensor as in the training procedure

Training	Testing	APCER (%)	BPCER (%)	HTER (%)	ROC and ePDF
IS	BC	24.9	4.01	14.1	Figure 11.9
BC	IS	2.8	75.6	39.18	Figure 11.10
IS	IS	3.4	2.37	2.88	Figure 11.11
BC	BC	2.65	13.1	7.87	Figure 11.12

Note: All data comes for LivDet2013 fingerprint benchmark. BC, Biometrika+ CrossMatch; IS, Italdata+Swipe.

data was perfect on the AVTS data set, and also it was perfect on half of the splits of the Warsaw benchmark. Because there is no "best split" for either of two data sets, we picked one trained solution presenting perfect performance on the training subsets to evaluate them on the test sets.

11.5.3.1 Same-dataset results

Table 11.6 presents the testing results obtained in the scenario when both training and testing sets come from the same benchmark. APCER and BPCER refer to classification task, that is each sample belonging to the testing set was classified to one of two classes (authentic or presentation attack) based on posteriori probabilities of class membership estimated by the softmax layer of the trained network. Hence, single APCER and BPCER (point estimators) are presented because this protocol is equivalent to a single acceptance threshold. The results obtained in this scenario are astonishing: the classifiers trained on disjoint subsets of samples originating from the same data set are either perfect (ATVS benchmark) or close to perfect (a perfect recognition of spoofing samples of Warsaw benchmark with only 0.15% of authentic samples falsely rejected). Figures 11.13 and 11.14 present empirical distributions of the difference between two CNN output nodes and the corresponding ROC curves. The distributions are well separated for both benchmarks, suggesting high performance of the VGG-based solution applied for known spoofing samples.

11.5.3.2 Cross-dataset results

Table 11.7 shows how catastrophically bad this method may be if tested on **cross-dataset** samples. ATVS and Warsaw samples differ significantly in terms of image properties such as contrast and visibility of iris texture. Especially, all the printouts used to produce Warsaw fake samples were able to spoof an example commercial iris-recognition system, which is not the case in the ATVS benchmark. Hence, because of nonaccidental quality of Warsaw

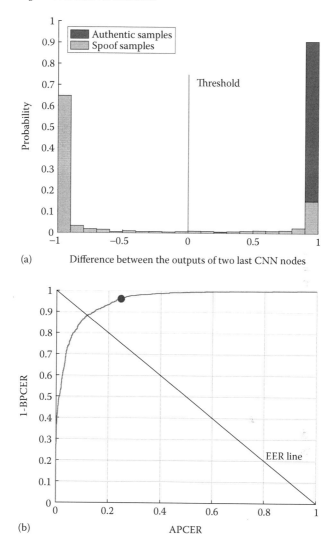

(a) Difference between the outputs of two last CNN nodes

(b) APCER

FIGURE 11.9

Same as Figure 11.7 except the variant: training on **Italdata+Swipe**, testing on **Biometrika+CrossMatch**.

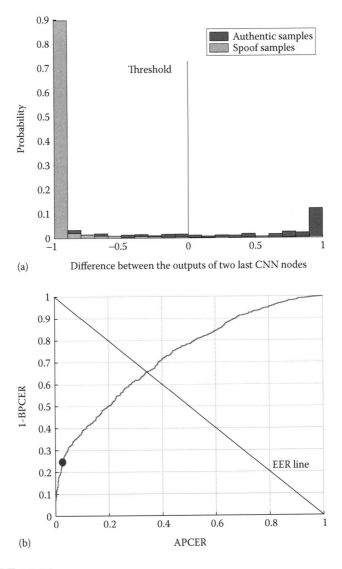

(a)

(b)

FIGURE 11.10
Same as Figure 11.7 except the cross-sensor that training is realized on samples
composed **Biometrika+CrossMatch**, testing on **Italdata+Swipe**.

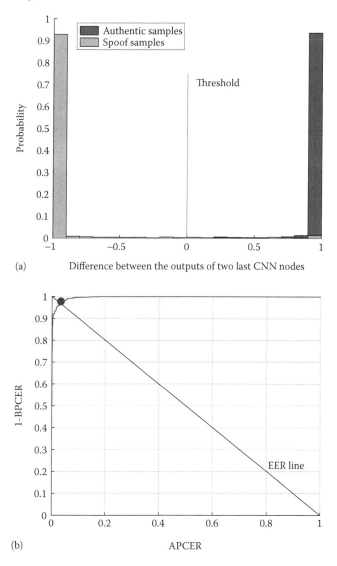

(a) Difference between the outputs of two last CNN nodes

(b) APCER

FIGURE 11.11

Same as Figure 11.7 except the variant: training on **Italdata+Swipe**, testing on **Italdata+Swipe**.

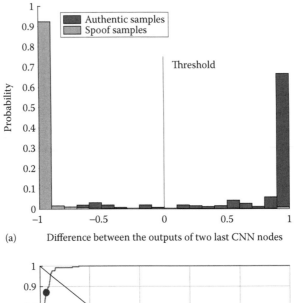

(a) Difference between the outputs of two last CNN nodes

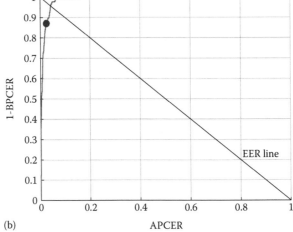

(b) APCER

FIGURE 11.12
Same as Figure 11.7 except the variant: training on **Biometrika+ CrossMatch**.

TABLE 11.6
Performance results obtained in **same-dataset** evaluations of **iris PAD** using the overall testing set of each data set

Training	Testing	APCER (%)	BPCER (%)	HTER (%)	ROC and ePDF
Warsaw	Warsaw	0.0	0.15	0.075	Figure 11.13
ATVS	ATVS	0.0	0.0	0.0	Figure 11.14

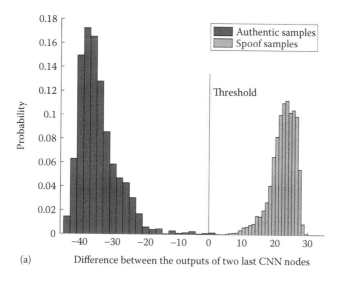

(a) Difference between the outputs of two last CNN nodes

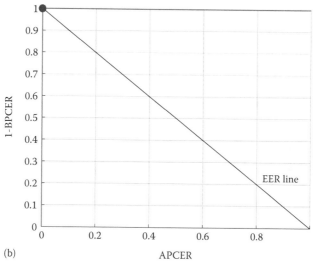

(b) APCER

FIGURE 11.13

(a) Empirical distributions of the difference between two CNN output nodes (before softmax) obtained separately for authentic and spoof **iris** samples. (b) ROC curve. Variant: training on **Warsaw LivDet2015**, testing on **Warsaw LivDet2015**. The threshold shown in blue color on the left plot and the blue dot on the ROC plot correspond to the approach when the predicted label is determined by the node with the larger output.

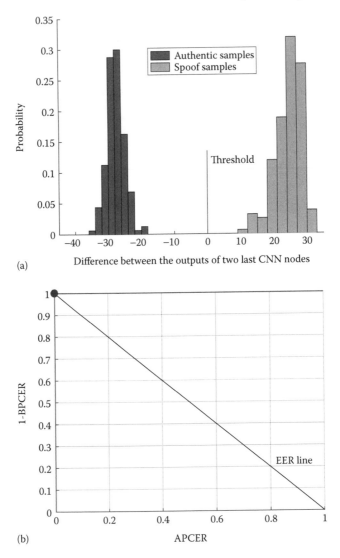

FIGURE 11.14
Same as Figure 11.13 except the variant: training on **ATVS**, testing on **ATVS**.

samples, this database seems to be more realistic and more difficult to process than the ATVS. Indeed, training on Warsaw (the "difficult" benchmark) and testing on ATVS (the "easier" benchmark) yields good results. Figure 11.15 presents well-separated empirical distributions of the difference between the output nodes of the network obtained for authentic samples and spoofs.

TABLE 11.7

Performance results obtained in **cross-dataset** evaluations of **iris PAD** using the overall testing set of each data set

Training	Testing	APCER (%)	BPCER (%)	HTER (%)	ROC and ePDF
Warsaw	ATVS	0.0	0.625	0.312	Figure 11.15
ATVS	Warsaw	99.9	0.0	49.99	Figure 11.16

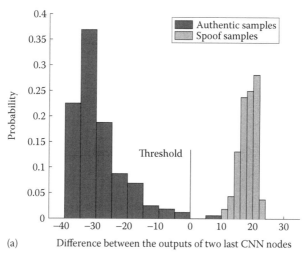

(a) Difference between the outputs of two last CNN nodes

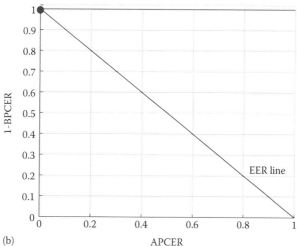

(b) APCER

FIGURE 11.15

Same as Figure 11.13 except the variant: training on **Warsaw LivDet2015**, testing on **ATVS**.

However, training on ATVS and testing on Warsaw yields almost null abilities to detect spoofs (APCER = 99.9%). This may suggest that exchanging a single layer put on top of the VGG-based feature extraction (trained for a different problem than spoofing detection) is not enough to model various qualities of iris printouts prepared independently by different teams and using different acquisition hardware. Figure 11.16 confirms that almost all

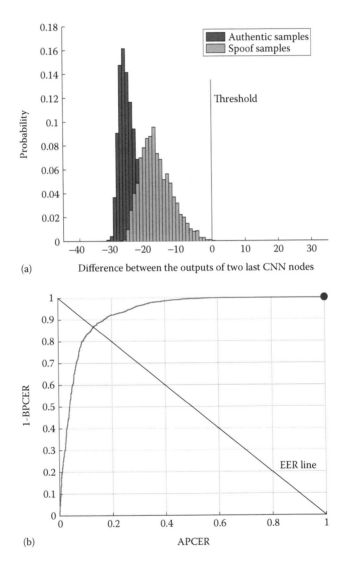

FIGURE 11.16
Same as Figure 11.13 except the variant: training on **ATVS**, testing on **Warsaw LivDet2015**.

scores obtained for spoofing samples are on the same side of the threshold as for authentic samples. Certainly, if the threshold can be adapted (which is not typically done in the tests), one can find other proportion between APCER and BPCER, for instance a threshold shifted from 0 to -21.9 results in the EER 13.2%.

11.6   Conclusions

In this chapter, we proposed a PAD solution for three modalities widely employed for designing biometric systems (i.e., face, iris, and fingerprint) based on VGG network architecture, a deep network architecture originally proposed for object recognition. We showed a methodology to adapt the VGG network to the two-class spoofing classification problem, which was evaluated using six benchmarks available for scientific purposes. The experiments were conducted taking into account the main challenges existing in this research field such as classification across different types of attempted attacks, biometric sensors, and qualities of samples used during attack. In this section, we discuss two main takeaways observed after the analysis presented in this chapter.

The first conclusion is that deep learning is an astonishingly powerful approach to detect image-based presentation attacks in three considered modalities. Note that the final solution is a subtle modification of the VGG network, trained for a different task, not related to PAD. In the case of iris and fingerprints, the starting network is not even related to the same object-recognition task. The results showed that we can use deep learning to detect spoofing attacks in some cases (AVTS iris benchmark) even perfectly. In this simple approach, we have changed only the last layer, connected strictly to the classification task performed by the VGG network. However, one can consider replacing two or all FC layers and use the output of the convolutional part of the network more efficiently.

The second takeaway comes from the cross-dataset and cross-sensor experiments. These exceptionally poor results seem to be related to the flexibility that characterizes convolutional networks. The flexibility allows them to "decide" which discovered properties of the input data they use in the classification task. But if they are not trained on data that contain a reasonable sampling of the situations present during testing, then they fail terribly because most of the features no longer correspond to the new data.

This, however, is not a surprising result and simply calls for solutions that take prior knowledge about the modeled phenomenon into account. Apparently the current fascination with deep learning has brought back an old debate: should we use models that are based on our understanding of the problem, which is neither full nor accurate (called *feature engineering* or *hand-crafted solutions*) or rather flexible models that learn everything from the data (called *feature learning* or *data-driven solutions*)? It seems that a reasonable mixture of both approaches should present the best reliability. We firmly believe the solution to this problem is in taking the *best of both worlds*.

Acknowledgment

We thank Brazilian Coordination for the Improvement of Higher Education Personnel (CAPES) through the DeepEyes project, the São Paulo Research Foundation (FAPESP) through the DéjàVu project (Grants #2015/19222-9 and #2017/12646-3), and Microsoft Research for the financial support.

References

1. A. Jain, A. Ross, and K. Nandakumar. Introduction. In *Introduction to Biometrics*, edited by Anil Jain, Arun A. Ross, and Karthik Nandakumar, pp. 1–49. Springer US, Boston, MA, 2011.

2. Q. T. Phan, D. T. Dang-Nguyen, G. Boato, and F. G. B. D. Natale. Face spoofing detection using ldp-top. In *IEEE International Conference on Image Processing (ICIP)*, pp. 404–408, September 2016.

3. Z. Boulkenafet, J. Komulainen, X. Feng, and A. Hadid. Scale space texture analysis for face anti-spoofing. In *IAPR International Conference on Biometrics (ICB)*, pp. 1–6, June 2016.

4. R. F. Nogueira, R. de Alencar Lotufo, and R. C. Machado. Fingerprint liveness detection using convolutional neural networks. *IEEE Transactions on Information Forensics and Security (TIFS)*, 11(6):1206–1213, 2016.

5. R. Raghavendra and C. Busch. Robust scheme for iris presentation attack detection using multiscale binarized statistical image features. *IEEE Transactions on Information Forensics and Security (TIFS)*, 10(4): 703–715, 2015.

6. F. Lourenço and D. Pires. Video shows Samu's medical using silicone fingers, in Ferraz, March 2013. http://tinyurl.com/akzcrgw. Accessed February 20, 2016.

7. E. Carazzai. Paranaguá Harbor employees used silicone fingers to circumvent biometric system in Paraná, February 2014. http://tinyurl.com/hkoj2jg. Accessed February 20, 2016.

8. C. Arthur. iPhone 5S fingerprint sensor hacked by germany's chaos computer club, September 2013. http://tinyurl.com/pkz59rg. Accessed February 20, 2016.

9. D. M. Reporter. Face off: Man arrested after boarding plane as an old man—only to land as youthful refugee, November 2010. http://tinyurl.com/33l9laz. Accessed February 20, 2016.

10. D. M. Reporter. The white robber who carried out six raids disguised as a black man (and very nearly got away with it), December 2010. http://tinyurl.com/2cvuq59. Accessed February 20, 2016.

11. A. Pinto, H. Pedrini, W. Schwartz, and A. Rocha. Face spoofing detection through visual codebooks of spectral temporal cubes. *IEEE Transactions on Image Processing (TIP)*, 24(12):4726–4740, 2015.

12. W. Schwartz, A. Rocha, and H. Pedrini. Face spoofing detection through partial least squares and low-level descriptors. In *IEEE International Joint Conference on Biometrics (IJCB)*, pp. 1–8, October 2011.

13. G. Pan, L. Sun, Z. Wu, and S. Lao. Eyeblink-based anti-spoofing in face recognition from a generic webcamera. In *IEEE International Conference on Computer Vision (ICCV)*, pp. 1–8, October 2007.

14. C. Xu, Y. Zheng, and Z. Wang. Eye states detection by boosting local binary pattern histogram features. In *IEEE International Conference on Image Processing (ICIP)*, pp. 1480–1483, October 2008.

15. N. Erdogmus and S. Marcel. Spoofing in 2D face recognition with 3D masks and anti-spoofing with kinect. In *IEEE International Conference on Biometrics: Theory Applications and Systems (BTAS)*, pp. 1–6, September 2013.

16. J. Li, Y. Wang, T. Tan, and A. Jain. Live face detection based on the analysis of fourier spectra. In *Biometric Technology for Human Identification (BTHI)*, Vol. 5404, pp. 296–303. Proc. SPIE, 2004.

17. X. Tan, Y. Li, J. Liu, and L. Jiang. Face liveness detection from a single image with sparse low rank bilinear discriminative model. In *European Conference on Computer Vision (ECCV)*, pp. 504–517, 2010.

18. A. Pinto, H. Pedrini, W. Schwartz, and A. Rocha. Video-based face spoofing detection through visual rhythm analysis. In *Conference on Graphics, Patterns and Images (SIBGRAPI)*, pp. 221–228, August 2012.

19. A. Pinto, W. Schwartz, H. Pedrini, and A. Rocha. Using visual rhythms for detecting video-based facial spoof attacks. *IEEE Transactions on Information Forensics and Security (TIFS)*, 10(5):1025–1038, 2015.

20. T.-W. Lee, G.-H. Ju, H.-S. Liu, and Y.-S. Wu. Liveness detection using frequency entropy of image sequences. In *IEEE International Conference on Acoustics, Speech, and Signal Processing (ICASSP)*, pp. 2367–2370, 2013.

21. M.-Z. Poh, D. J. McDuff, and R. W. Picard. Non-contact, automated cardiac pulse measurements using video imaging and blind source separation. *Optics Express*, 18(10):10762–10774, 2010.

22. B. Peixoto, C. Michelassi, and A. Rocha. Face liveness detection under bad illumination conditions. In *IEEE International Conference on Image Processing (ICIP)*, pp. 3557–3560, September 2011.

23. N. Kose and J.-L. Dugelay. Reflectance analysis based countermeasure technique to detect face mask attacks. In *International Conference on Digital Signal Processing (ICDSP)*, pp. 1–6, 2013.

24. N. Almoussa. Variational retinex and shadow removal. Technical report, University of California, Department of Mathematics, 2009.

25. J. Määttä, A. Hadid, and M. Pietikäinen. Face spoofing detection from single images using micro-texture analysis. In *IEEE International Joint Conference on Biometrics (IJCB)*, pp. 1–7, October 2011.

26. J. Määttä, A. Hadid, and M. Pietikäinen. Face spoofing detection from single images using texture and local shape analysis. *IET Biometrics*, 1(1):3–10, 2012.

27. T. Ojala, M. Pietikainen, and T. Maenpaa. Multiresolution gray-scale and rotation invariant texture classification with local binary patterns. *IEEE Transactions on Pattern Analysis and Machine Intelligence (TPAMI)*, 24(7):971–987, 2002.

28. J. G. Daugman. Uncertainty relation for resolution in space, spatial frequency, and orientation optimized by two-dimensional visual cortical filters. *Journal of the Optical Society of America*, 2(7):1160–1169, 1985.

29. N. Dalal and B. Triggs. Histograms of oriented gradients for human detection. In *IEEE International Conference on Computer Vision and Pattern Recognition (CVPR)*, Vol. 1, pp. 886–893, June 2005.

30. I. Chingovska, A. Anjos, and S. Marcel. On the effectiveness of local binary patterns in face anti-spoofing. In *International Conference of the Biometrics Special Interest Group (BIOSIG)*, pp. 1–7, September 2012.

31. N. Erdogmus and S. Marcel. Spoofing 2D face recognition systems with 3D masks. In *International Conference of the Biometrics Special Interest Group (BIOSIG)*, pp. 1–8, 2013.

32. T. S. Lee. Image representation using 2d gabor wavelets. *IEEE Transactions on Pattern Analysis and Machine Intelligence (TPAMI)*, 18(10):959–971, 1996.

33. M. Günther, D. Haufe, and R. Würtz. Face recognition with disparity corrected gabor phase differences. In *International Conference on Artificial Neural Networks and Machine Learning (ICANN)*, pp. 411–418, 2012.

34. T. Pereira, A. Anjos, J. de Martino, and S. Marcel. Can face anti-spoofing countermeasures work in a real world scenario? In *IAPR International Conference on Biometrics (ICB)*, pp. 1–8, 2013.

35. T. Pereira, J. Komulainen, A. Anjos, J. de Martino, A. Hadid, M. Pietikäinen, and S. Marcel. Face liveness detection using dynamic texture. *EURASIP Journal on Image and Video Processing (JIVP)*, 2014(1):2, 2014.

36. D. Garcia and R. de Queiroz. Face-spoofing 2D-detection based on moiré-pattern analysis. *IEEE Transactions on Information Forensics and Security (TIFS)*, 10(4):778–786, 2015.

37. K. Patel, H. Han, A. Jain, and G. Ott. Live face video vs. spoof face video: Use of moiré; patterns to detect replay video attacks. In *IAPR International Conference on Biometrics (ICB)*, pp. 98–105, May 2015.

38. R. Tronci, D. Muntoni, G. Fadda, M. Pili, N. Sirena, G. Murgia, M. Ristori, and F. Roli. Fusion of multiple clues for photo-attack detection in face recognition systems. In *IEEE International Joint Conference on Biometrics (IJCB)*, pp. 1–6, October 2011.

39. A. Anjos and S. Marcel. Counter-measures to photo attacks in face recognition: A public database and a baseline. In *IEEE International Joint Conference on Biometrics (IJCB)*, pp. 1–7, October 2011.

40. D. Wen, H. Han, and A. Jain. Face spoof detection with image distortion analysis. *IEEE Transactions on Information Forensics and Security (TIFS)*, 10(4):746–761, 2015.

41. W. Kim, S. Suh, and J.-J. Han. Face liveness detection from a single image via diffusion speed model. *IEEE Transactions on Image Processing (TIP)*, 24(8):2456–2465, 2015.

42. M. Rousson, T. Brox, and R. Deriche. Active unsupervised texture segmentation on a diffusion based feature space. In *IEEE International Conference on Computer Vision and Pattern Recognition (CVPR)*, Vol. 2, pp. 699–704, June 2003.

43. Z. Boulkenafet, J. Komulainen, and A. Hadid. Face spoofing detection using colour texture analysis. *IEEE Transactions on Information Forensics and Security (TIFS)*, 11(8):1818–1830, 2016.

44. J. Yang, Z. Lei, D. Yi, and S. Li. Person-specific face antispoofing with subject domain adaptation. *IEEE Transactions on Information Forensics and Security (TIFS)*, 10(4):797–809, 2015.

45. Y. Taigman, M. Yang, M. Ranzato, and L. Wolf. Deepface: Closing the gap to human-level performance in face verification. In *IEEE International Conference on Computer Vision and Pattern Recognition (CVPR)*, pp. 1701–1708, June 2014.

46. D. Menotti, G. Chiachia, A. Pinto, W. Schwartz, H. Pedrini, A. Falcao, and A. Rocha. Deep representations for iris, face, and fingerprint spoofing detection. *IEEE Transactions on Information Forensics and Security (TIFS)*, 10(4):864–879, 2015.

47. J. Bergstra and Y. Bengio. Random search for hyper-parameter optimization. *Journal of Machine Learning Research (JMLR)*, 13:281–305, 2012.

48. N. Pinto, D. Doukhan, J. DiCarlo, and D. Cox. A high-throughput screening approach to discovering good forms of biologically-inspired visual representation. *PLoS ONE*, 5(11):e1000579, 2009.

49. Y. LeCun, L. Bottou, Y. Bengio, and P. Haffner. Gradient-based learning applied to document recognition. *Proceedings of the IEEE*, 86(11):2278–2324, 1998.

50. I. Manjani, S. Tariyal, M. Vatsa, R. Singh, and A. Majumdar. Detecting silicone mask based presentation attack via deep dictionary learning. *IEEE Transactions on Information Forensics and Security (TIFS)*, PP(99):1–1, 2017.

51. S. Tariyal, A. Majumdar, R. Singh, and M. Vatsa. Deep dictionary learning. *IEEE Access*, 4:10096–10109, 2016.

52. L. Ghiani, D. Yambay, V. Mura, S. Tocco, G. Marcialis, F. Roli, and S. Schuckcrs. LivDet 2013—fingerprint liveness detection competition. In *IAPR International Conference on Biometrics (ICB)*, pp. 1–6, 2013.

53. J. Galbally, F. Alonso-Fernandez, J. Fierrez, and J. Ortega-Garcia. Fingerprint liveness detection based on quality measures. In *International Conference on Biometrics, Identity and Security (BIdS)*, pp. 1–8, September 2009.

54. J. Galbally, F. Alonso-Fernandez, J. Fierrez, and J. Ortega-Garcia. A high performance fingerprint liveness detection method based on quality related features. *Future Generation Computer Systems (FGCS)*, 28(1):311–321, 2012.

55. D. Gragnaniello, G. Poggi, C. Sansone, and L. Verdoliva. Fingerprint liveness detection based on weber local image descriptor. In *IEEE Workshop on Biometric Measurements and Systems for Security and Medical Applications (BIOMS)*, pp. 46–50, September 2013.

56. X. Jia, X. Yang, Y. Zang, N. Zhang, R. Dai, J. Tian, and J. Zhao. Multi-scale block local ternary patterns for fingerprints vitality detection. In *IAPR International Conference on Biometrics (ICB)*, pp. 1–6, June 2013.

57. X. Tan and B. Triggs. Enhanced local texture feature sets for face recognition under difficult lighting conditions. *IEEE Transactions on Image Processing (TIP)*, 19(6):1635–1650, 2010.

58. L. Ghiani, A. Hadid, G. Marcialis, and F. Roli. Fingerprint liveness detection using binarized statistical image features. In *IEEE International Conference on Biometrics: Theory Applications and Systems (BTAS)*, pp. 1–6, September 2013.

59. A. Hyvrinen, J. Hurri, and P. Hoyer. *Natural Image Statistics: A Probabilistic Approach to Early Computational Vision.* Springer Publishing Company, 1st ed., 2009.

60. L. Ghiani, G. Marcialis, and F. Roli. Fingerprint liveness detection by local phase quantization. In *International Conference on Pattern Recognition (ICPR)*, pp. 537–540, November 2012.

61. C. Gottschlich. Convolution comparison pattern: An efficient local image descriptor for fingerprint liveness detection. *PLoS ONE*, 11(2):12, 2016.

62. A. Rattani and A. Ross. Automatic adaptation of fingerprint liveness detector to new spoof materials. In *IEEE International Joint Conference on Biometrics (IJCB)*, pp. 1–8, September 2014.

63. A. Rattani, W. Scheirer, and A. Ross. Open set fingerprint spoof detection across novel fabrication materials. *IEEE Transactions on Information Forensics and Security (TIFS)*, 10(11):2447–2460, 2015.

64. W. Scheirer, A. Rocha, A. Sapkota, and T. Boult. Toward open set recognition. *IEEE Transactions on Pattern Analysis and Machine Intelligence (TPAMI)*, 35(7):1757–1772, 2013.

65. E. Marasco and C. Sansone. Combining perspiration- and morphology-based static features for fingerprint liveness detection. *Pattern Recognition Letters (PRL)*, 33(9):1148–1156, 2012.

66. B. Tan and S. Schuckers. Spoofing protection for fingerprint scanner by fusing ridge signal and valley noise. *Pattern Recognition (PR)*, 43(8):2845–2857, 2010.

67. D. Gragnaniello, G. Poggi, C. Sansone, and L. Verdoliva. Local contrast phase descriptor for fingerprint liveness detection. *Pattern Recognition (PR)*, 48(4):1050–1058, 2015.

68. S. Kumpituck, D. Li, H. Kunieda, and T. Isshiki. Fingerprint spoof detection using wavelet based local binary pattern. In *International Conference on Graphic and Image Processing (ICGIP)*, Vol. 10225, pp. 102251C–102251C–8, 2017.

69. E. Marasco, P. Wild, and B. Cukic. Robust and interoperable fingerprint spoof detection via convolutional neural networks. In *IEEE Symposium on Technologies for Homeland Security (HST)*, pp. 1–6, May 2016.

70. C. Szegedy, W. Liu, Y. Jia, P. Sermanet, S. Reed, D. Anguelov, D. Erhan, V. Vanhoucke, and A. Rabinovich. Going deeper with convolutions. In *IEEE International Conference on Computer Vision and Pattern Recognition (CVPR)*, pp. 1–9, June 2015.

71. J. Daugman. High confidence visual recognition of persons by a test of statistical independence. *IEEE Transactions on Pattern Analysis and Machine Intellgence*, 15(11):1148–1161, 1993.

72. A. Czajka. Iris liveness detection by modeling dynamic pupil features. In K. W. Bowyer and M. J. Burge (Eds.), *Handbook of Iris Recognition*, pp. 439–467. Springer, London, UK, 2016.

73. A. Pacut and A. Czajka. Aliveness detection for iris biometrics. In *IEEE International Carnahan Conferences Security Technology (ICCST)*, pp. 122–129, October 2006.

74. J. Galbally, J. Ortiz-Lopez, J. Fierrez, and J. Ortega-Garcia. Iris liveness detection based on quality related features. In *IAPR International Conference on Biometrics (ICB)*, pp. 271–276, 2012.

75. P. Pudil, J. Novovičová, and J. Kittler. Floating search methods in feature selection. *Pattern Recognition Letters (PRL)*, 15(11):1119–1125, 1994.

76. J. Fierrez-Aguilar, J. Ortega-garcia, D. Torre-toledano, and J. Gonzalez-rodriguez. Biosec baseline corpus: A multimodal biometric database. *Pattern Recognition (PR)*, 40:1389–1392, 2007.

77. V. Ruiz-Albacete, P. Tome-Gonzalez, F. Alonso-Fernandez, J. Galbally, J. Fierrez, and J. Ortega-Garcia. Direct attacks using fake images in iris verification. In *First European Workshop on Biometrics and Identity Management (BioID)*, Vol. 5372 of *Lecture Notes in Computer Science*, pp. 181–190. Springer, 2008.

78. A. Sequeira, J. Murari, and J. Cardoso. Iris liveness detection methods in mobile applications. In *International Conference on Computer Vision Theory and Applications (VISAPP)*, Vol. 3, pp. 22–33, January 2014.

79. S. Schuckers, K. Bowyer, A.C., and D. Yambay. LivDet 2013—liveness detection iris competition, 2013.

80. A. Sequeira, J. Murari, and J. Cardoso. Iris liveness detection methods in the mobile biometrics scenario. In *International Joint Conference on Neural Network (IJCNN)*, pp. 3002–3008, July 2014.

81. J. Monteiro, A. Sequeira, H. Oliveira, and J. Cardoso. Robust iris localisation in challenging scenarios. In *Communications in Computer and Information Science (CCIS)*. Springer-Verlag, 2004.

82. Z. Wei, X. Qiu, Z. Sun, and T. Tan. Counterfeit iris detection based on texture analysis. In *International Conference on Pattern Recognition (ICPR)*, pp. 1–4, 2008.

83. A. Czajka. Database of iris printouts and its application: Development of liveness detection method for iris recognition. In *International Conference on Methods and Models in Automation and Robotics (ICMMAR)*, pp. 28–33, August 2013.

84. A. Sequeira, H. Oliveira, J. Monteiro, J. Monteiro, and J. Cardoso. MobILive 2014—mobile iris liveness detection competition. In *IEEE International Joint Conference on Biometrics (IJCB)*, pp. 1–6, September 2014.

85. V. Ojansivu and J. Heikkilä. Blur insensitive texture classification using local phase quantization. In *Image and Signal Processing (ISP)*, edited by Abderrahim Elmoataz, Olivier Lezoray, Fathallah Nouboud, and Driss Mammass, pp. 236–243. Springer, Berlin, Germany, 2008.

86. L. Zhang, Z. Zhou, and H. Li. Binary Gabor pattern: An efficient and robust descriptor for texture classification. In *IEEE International Conference on Image Processing (ICIP)*, pp. 81–84, September 2012.

87. Z. Sun, H. Zhang, T. Tan, and J. Wang. Iris image classification based on hierarchical visual codebook. *IEEE Transactions on Pattern Analysis and Machine Intelligence (TPAMI)*, 36(6):1120–1133, 2014.

88. J. S. Doyle, K. W. Bowyer, and P. J. Flynn. Variation in accuracy of textured contact lens detection based on sensor and lens pattern. In *IEEE International Conference on Biometrics: Theory Applications and Systems (BTAS)*, pp. 1–7, September 2013.

89. T. Ojala, M. Pietikäinen, and D. Harwood. A comparative study of texture measures with classification based on featured distributions. *Pattern Recognition*, 29(1):51–59, 1996.

90. D. Yadav, N. Kohli, J. Doyle, R. Singh, M. Vatsa, and K. Bowyer. Unraveling the effect of textured contact lenses on iris recognition. *IEEE Transactions on Information Forensics and Security (TIFS)*, 9(5):851–862, 2014.

91. K. Raja, R. Raghavendra, and C. Busch. Video presentation attack detection in visible spectrum iris recognition using magnified phase information. *IEEE Transactions on Information Forensics and Security (TIFS)*, 10(10):2048–2056, 2015.

92. S. Bharadwaj, T. Dhamecha, M. Vatsa, and R. Singh. Computationally efficient face spoofing detection with motion magnification. In *IEEE Computer Society Conference on Computer Vision and Pattern Recognition Workshops (CVPRW)*, pp. 105–110, June 2013.

93. P. Gupta, S. Behera, M. Vatsa, and R. Singh. On iris spoofing using print attack. In *International Conference on Pattern Recognition (ICPR)*, pp. 1681–1686, August 2014.

94. A. Oliva and A. Torralba. Modeling the shape of the scene: A holistic representation of the spatial envelope. *International Journal of Computer Vision (IJCV)*, 42(3):145–175, 2001.

95. A. Czajka. Pupil dynamics for iris liveness detection. *IEEE Transactions on Information Forensics and Security (TIFS)*, 10(4):726–735, 2015.

96. M. Kohn and M. Clynes. Color dynamics of the pupil. *Annals of the New York Academy of Sciences*, 156(2):931–950, 1969.

97. Lovish, A. Nigam, B. Kumar, and P. Gupta. Robust contact lens detection using local phase quantization and binary gabor pattern. In G. Azzopardi and N. Petkov (Eds.), *Computer Analysis of Images and Patterns (CAIP)*, pp. 702–714. Springer International Publishing, 2015.

98. P. Silva, E. Luz, R. Baeta, H. Pedrini, A. X. Falcao, and D. Menotti. An approach to iris contact lens detection based on deep image representations. In *Conference on Graphics, Patterns and Images (SIBGRAPI)*, pp. 157–164. IEEE, 2015.

99. J. Galbally, S. Marcel, and J. Fierrez. Image quality assessment for fake biometric detection: Application to iris, fingerprint, and face recognition. *IEEE Transactions on Image Processing (TIP)*, 23(2):710–724, 2014.

100. D. Gragnaniello, G. Poggi, C. Sansone, and L. Verdoliva. An investigation of local descriptors for biometric spoofing detection. *IEEE Transactions on Information Forensics and Security (TIFS)*, 10(4):849–863, 2015.

101. J. Sivic and A. Zisserman. Video Google: A text retrieval approach to object matching in videos. In *IEEE International Conference on Computer Vision (ICCV)*, pp. 1470–1477, October 2003.

102. D. Lowe. Distinctive image features from scale-invariant keypoints. *International Journal of Computer Vision (IJCV)*, 60(2):91–110, 2004.

103. E. Tola, V. Lepetit, and P. Fua. Daisy: An efficient dense descriptor applied to wide-baseline stereo. *IEEE Transactions on Pattern Analysis and Machine Intelligence (TPAMI)*, 32(5):815–830, 2010.

104. I. Kokkinos and A. Yuille. Scale invariance without scale selection. In *IEEE International Conference on Computer Vision and Pattern Recognition (CVPR)*, pp. 1–8, June 2008.

105. K. Simonyan and A. Zisserman. Very deep convolutional networks for large-scale image recognition. *arXiv preprint arXiv:1409.1556*, 2014.

106. A. Krizhevsky, I. Sutskever, and G. E. Hinton. Imagenet classification with deep convolutional neural networks. In *Advances in Neural Information Processing Systems*, pp. 1097–1105, 2012.

107. O. M. Parkhi, A. Vedaldi, and A. Zisserman. Deep face recognition. In *British Machine Vision Conference*, number 3 in 1, p. 6, 2015.

108. Z. Zhang, J. Yan, S. Liu, Z. Lei, D. Yi, and S. Li. A face antispoofing database with diverse attacks. In *IAPR International Conference on Biometrics (ICB)*, pp. 26–31, April 2012.

109. G. Marcialis, A. Lewicke, B. Tan, P. Coli, D. Grimberg, A. Congiu, A. Tidu, F. Roli, and S. Schuckers. First international fingerprint liveness detection competition—LivDet 2009. In *Image Analysis and Processing (IAP)*, edited by Pasquale Foggia, Carlo Sansone, and Mario Vento, pp. 12–23. Springer, Berlin, Germany, 2009.

110. ISO/IEC. DIS (Draft International Standard) 30107-3, Information technology—Biometric presentation attack detection—Part 3: Testing and reporting, 2016.

12

Fingervein Presentation Attack Detection Using Transferable Features from Deep Convolution Neural Networks

Raghavendra Ramachandra, Kiran B. Raja, Sushma Venkatesh, and Christoph Busch

CONTENTS

12.1   Introduction 295
12.2   Transferable Features for Fingervein PAD 298
12.3   Experiments and Results 300
    12.3.1   Fingervein video presentation attack database
          (Database-1) 300
    12.3.2   Fingervein image presentation attack database
          (Database-2) 300
    12.3.3   Performance evaluation protocol and metrics 300
    12.3.4   Results on fingervein video presentation attack
          database (Database-1) 301
    12.3.5   Results on fingervein image presentation attack
          database (Database-2) 302
12.4   Discussion and Summary 304
    12.4.1   Future research avenues 304
References 304

12.1 Introduction

The ease of authenticating any subject without requiring them to carry special kinds of tokens or additional devices has resulted in the wide use of biometrics. Along with the advantage of not carrying a token device, biometrics also offers a high level of security. Further, various biometric characteristics ranging from face, iris, fingerprint, and fingervein provide a differing security level. Based on the high level of security provided by fingervein recognition, many commercial applications such as banks in many parts of Asia have employed fingervein as the primary authentication mode for interacting with automated teller machines (ATMs) [1]. Fingervein is difficult to spoof through the creation of artifacts because the vein pattern is beneath the dermal layer and masked by tissues making it less vulnerable to attacks such as lifted fingerprint attacks.

Thus, it is challenging to collect a fingervein sample without cooperation of the enrolled data subject. Although the creation of fingervein artifacts is difficult, limited works in this direction have shown that fingervein-capture devices themselves can be easily attacked with some engineering efforts such as presentation of printed fingervein samples with various quality, presenting the fingervein samples using smartphones, or by using additional illumination to outcast the sensor illumination source such that the artifact is captured as a real sample [2]. Such kinds of attacks at the sensor level through the aid of artifacts are popularly called presentation attacks (or *spoofing attacks*) [2]. A previous work in this direction demonstrated the vulnerability of fingervein-capture devices toward presentation attack by a simple mechanism of printed fingervein sample using a conventional laser printer [3]. Figure 12.1 shows the example of the fingervein bona fide and the artifact samples that indicates the similar visual quality to that of bona-fide image.

In their work, a detailed analysis of vulnerability was presented using 100 unique fingervein images captured from 50 different subjects which indicated an Impostor Attack Presentation Match Rate (IAPMR) or Spoof False Acceptance Rate (SFAR) of 86%. An alternative to such image-capture–based fingervein systems can be video-capture–based systems. Another work studied fingervein systems operating with video-based acquisition, and it was demonstrated that even those systems are vulnerable to attacks [2]. Specifically, a detailed analysis of vulnerability was carried on a relatively large fingervein database of 300 unique fingervein videos [2]. In that study, each unique fingervein biometric instance was captured twice with 25 frames of video and one of them was used to generate different kinds of artifacts such as printed samples using both laserjet and inkjet printers. The analysis indicated the vulnerability of video-based systems, which resulted in an IAPMR of 90.62% for the inkjet printed artifact species and 91.87% for the laserjet printed artifact species [2]. A recent work also analyzed the vulnerability toward electronically presented artifacts through smartphone display, which indicated an IAPMR of 100%.

To address these flaws toward presentation attacks on fingervein sensors, a number of presentation attack detection (PAD) schemes have been

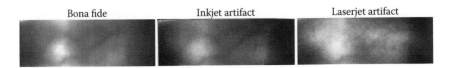

FIGURE 12.1
Example of fingervein artifact samples.

proposed that range from exploring quality-based features and texture features to motion features. A fingervein PAD algorithm based on spatial-frequency and time-frequency analysis using Fourier and Wavelet transform was proposed in Nguyen et al. [4]. Preliminary experiments were carried out on a small database of seven subjects that indicated the applicability of frequency information to detect fingervein artifacts. Along similar lines, a number of approaches were provided in the first competition on fingervein PAD [5]. The techniques submitted to this competition were evaluated on the publicly available fingervein artifact database collected from 50 subjects [3]. Four different texture-based approaches were explored by learning a support vector machine (SVM) to classify the artifacts against the normal or bona-fide presentations. The evaluation of approaches such as binarized statistical image features, Riesz transform, local binary pattern, extensions of local binary patterns, and local phase quantization iterated the employability of texture-based features [3,6]. Following the paradigm of texture-based approaches, Steerable Pyramids were explored for detecting the fingervein presentation attacks [7]. Steerable Pyramids were shown to detect three different species of presentation attacks instruments (PAIs) that included printed and electronic screen attacks.

Another work in this direction employed the motion characteristics in vein pattern by collecting the video of fingerveins [2]. The flow of blood in the fingervein was magnified and analyzed with a simple decision module to distinguish the real presentations against artifact presentations. The evaluation of the blood-flow–based method indicated superiority in detecting artifacts on a video database of 300 unique fingervein samples [2]. A key outcome of this work indicated generalizability to unseen attacks on fingervein biometric systems unlike most of the texture-based approaches, which fail to detect unseen attacks [2]. It has to be noted that robustness of biometric systems is largely based on good performance while addressing any unknown attacks.

Recent advances in machine learning have resulted in more sophisticated algorithms based on convolutional neural networks (CNNs) that have demonstrated an excellent classification accuracy on various applications [8]. The effectiveness of CNNs strongly depends on the availability of the large-scale training data with significant variability. However, by employing the analogy of transfer learning that involves using the existing fully trained networks and then fine-tuning the network to adapt to the precise application is also explored in the machine-learning domain. The transferable CNNs can be fine-tuned using a small set of data that can be made applicable for the applications that lack the large-scale data sets. Thus, it is reasonable to explore the transfer learning approach using CNNs for the precise application of fingervein PAD (or spoof detection). In Raghavendra et al. [9], the transferable features from AlexNet were explored to detect the presentation attacks on the fingervein biometric system. To this extent, we have explored AlexNet [8], which is further augmented with five additional layers with consecutive fully connected and dropout layers to lower the over-fitting of network. Note that AlexNet was originally trained with 15 million labeled high-resolution images belonging to roughly 22,000 categories. The images were collected from the web and were as such not similar in their properties to fingervein samples. Finally, the classification is carried out using the softmax classifier to detect the presentation attack on the fingervein samples.

In this chapter, we further extend our study on the transferable features from deep CNNs by exploring the decision-level fusion. To this extent, we explore the majority voting for combining the decision from the individual classifiers. We follow the similar fine-tuning steps as indicated in Raghavendra et al. [9] to obtain the fully connected features. We then propose to use three different classifiers such as: softmax, linear discriminant analysis (LDA), and SVM in which the decision from individual classifiers are combined using majority voting. Extensive experiments are carried out using two different databases comprised of 300 unique fingervein samples from the fingervein video presentation attack database [2] and fingervein image PAD database [7]. We also present a comparative performance of the proposed approach with six different existing methods on the fingervein video database and 12 different existing methods on the fingervein image database.

The rest of the chapter is organized as follows: Section 12.2 presents the transferable features based on deep-CNN features for fingervein PAD, and Section 12.3 presents the experimental results and discussion. Finally, Section 12.4 draws the summary of this work.

12.2   Transferable Features for Fingervein PAD

Figure 12.2 shows the block diagram of our proposed scheme based on the transferable features using deep CNN for finger PAD. The proposed scheme employs the well-known, pretrained AlexNet [8] and then extends it further with five continuous layers with fully connected layers and dropout layers. The main motivation for these additional layers, especially, the dropout

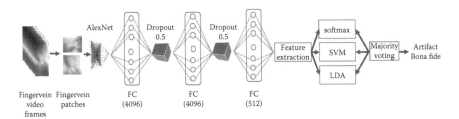

FIGURE 12.2
Block diagram of our fingervein PAD scheme based on transferable deep-CNN features.

layers- is that they are expected to reduce the over-fitting of the network and thus can learn a robust feature set to detect the fingervein presentation attack prominently. We then extract the features from the last fully connected layer of dimension 1×512 and use that vector to perform the classification. We have explored three different classifiers such as softmax, LDA, and the SVM classifier that are trained independently on the deep-CNN features. Given the frame from the probe video, the classification is carried out using all three classifiers, and the final decision is made by considering the majority voting.

The crucial part of the proposed method is fine-tuning the network, mainly to control the learning rates of the newly added layers. We follow the conventional method on adjusting the learning rate that will modify the learning rate quickly and learn the weights of the newer layers. Thus, we have used the *weight learning rate factor* as 10 and *bias learning rate factor* as 20. We employ the fingervein video PAD database [2] to fine-tune the network. We have used the training set comprised of 50 unique fingervein instances of bona-fide samples, laser artifact samples, and inkjet artifact samples. For each subject, two bona-fide videos are recorded at a rate of 15 frames/second for the duration of 3 seconds in two different sessions. Thus, each video has 45 frames that will result in a total of 50 *unique fingervein instances* × 2 *sessions* × 1 *videos* = 100 *videos*, which equals $100 \times 45 = 4500$ frames. Each frame is of the dimension 300×100 *pixels*. To effectively fine-tune the deep-CNN architecture, we carried out the data augmentation by dividing each frame into 100 nonoverlapping image patches of size 64×64 *pixels*. This will result in a total of 4500 *frames* × $100 = 450,000$ bona-fide image patches used for fine-tuning the network. Because the same number of the artifact (or spoof) samples are available for both laser and inkjet print artifact, we have used 450,000 laser and inkjet print artifact image patches to pretrain the network separately.

We train two independent models separately for two different kinds of artifact species such that *Model-1*: is trained using only laser print artifacts and the bona-fide fingervein samples. *Model-2*: is trained using only inkjet print artifact and the bona-fide fingervein samples. Because the proposed approach is based on developing a classification scheme using deep-CNN features, we extract the features from the last fully connected layer and train three different classifiers corresponding to *Model-1* and *Model-2*. Given the fingervein probe video, we first decompose it to get 45 frames. Then each fingervein sample is first divided into 100 different image patches of size 64×64 pixels. The classification of the fingervein video as either bona fide or artifact is carried out in two steps: (1) For each of the image patches, we first obtain the classification results for all three classifiers. Then, majority voting on the classifier decision is computed to label the given patch as either artifact or bona fide. This step is repeated for all 100 image patches to obtain the label for each patch. (2) The majority voting of the decision made on each patch is employed to make the final decision on the given fingervein as bona fide or artifact.

12.3   Experiments and Results

In this section, we discuss the experimental results of the proposed method on two different fingervein artifact databases namely; fingervein video presentation attack database [2] and fingervein image presentation attack database [7]. These two databases are captured using the same fingervein sensor, but the resolution of the captured data differs significantly from the video-based database to the image-based database. In the following section, we present the brief information on both of these databases employed in this work.

12.3.1   Fingervein video presentation attack database (Database-1)

This database is collected using the custom fingervein sensor available in Raghavendra et al. [10] and comprised of the video recording of 300 unique fingervein instances with two different artifact species such as laser print artifact and inkjet print artifact. We have used the same protocol described in Raghavendra et al. [2] with 50 unique fingervein instances with two videos each for training and 250 unique fingervein instances with two videos each for testing.

12.3.2   Fingervein image presentation attack database (Database-2)

This database is also collected using the custom sensor, which is the same as the one used with the Database-1. This database is comprised of 300 unique fingervein image instances. Each unique instance has two bona-fide captures; one artifact sample generated using the laser printer, one artifact sample created using the inkjet printer, and one artifact sample made using a display attack. In this work, we have used this database only to test the performance of the proposed scheme, which is trained only on the Database-1. In this work; we choose only two artifact species corresponding to laser and inkjet print method. Figure 12.3 shows the example of the sample images from the database.

12.3.3   Performance evaluation protocol and metrics

The pretraining of the proposed network is carried out using the training set of the Database-1. The testing set from both Database-1 and Database-2 is used to report the results to benchmark the performance of the proposed scheme with existing techniques.

The quantitative performance of the PAD algorithms are presented according to the metric developed in ISO/IEC 30107-3 [11] in terms of: (1) Attack Presentation Classification Error Rate (APCER), which is defined as

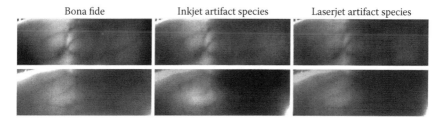

Bona fide   Inkjet artifact species   Laserjet artifact species

FIGURE 12.3
Example images from fingervein image presentation attack database (Database-2) corresponding to two subjects [7].

a proportion of attack presentation incorrectly classified as bona-fide (or real) presentation; (2) Bona-fide Presentation Classification Error Rate (BPCER), which is defined as a proportion of bona-fide presentation incorrectly classified as attack presentation.

12.3.4 Results on fingervein video presentation attack database (Database-1)

Tables 12.1 and 12.2 indicate the performance of the proposed scheme on the fingervein video presentation attack database (Database-1) [2]. The performance of the proposed method is benchmarked against five different state-of-the-art fingervein PAD schemes that are evaluated using the proposed scheme. Table 12.1 shows the performance of the proposed scheme for the inkjet artifact species that has demonstrated the best performance and the proposed method has indicated the best performance with the APCER of 1.82% and BPCER of 0%.

Table 12.2 shows the performance of the proposed scheme on the inkjet print artifact species. Here as well, it can be observed that the proposed system

TABLE 12.1
Performance of the proposed scheme on inkjet printed artifact species

Method	APCER (%)	BPCER (%)
Riesz transform-SVM[a] [5]	9.20	84.40
LPQ+WLD-SVM[a] [5]	22.80	0.40
LBP-SVM[a] [5]	34.40	2.40
M-BSIF-SVM[a] [5]	20.00	5.60
Liveness Measure [2]	2.40	2.00
Transferable D-CNN [9]	3.48	0
Proposed scheme	**1.82**	**0**

[a]Reimplemented in MATLAB®

TABLE 12.2

Performance of the proposed scheme on laserjet printed
artifact species

Method	APCER (%)	BPCER (%)
Riesz transform-SVM[a] [5]	7.20	79.60
LPQ+WLD-SVM[a] [5]	13.20	1.60
LBP-SVM[a] [5]	10.00	6.00
M-BSIF-SVM[a] [5]	8.00	14.00
Liveness Measure [2]	5.20	2.00
Transferable D-CNN [9]	0	0
Proposed scheme	**0**	**0**

[a]Reimplemented in MATLAB

has demonstrated an outstanding performance with APCER and BPCER of
0%. Analyzing the obtained results, the proposed method has emerged as
the best performing scheme on both laser and inkjet artifact species on the
fingervein video presentation attack database.

12.3.5   Results on fingervein image presentation attack database (Database-2)

This section presents the results of the proposed scheme on the fingervein
image presentation attack database [7]. Figure 12.3 shows the example of the
bona-fide and artifact samples from the fingervein image presentation attack
database [7]. Table 12.3 shows the quantitative results of the proposed method
on inkjet printed artifact species and a benchmark with 11 different state-of-
the-art schemes. Based on the obtained results, the proposed method has
indicated best results with the lowest APCER of 1.2% and BPCER of 0%.
The obtained results have shown robustness of the proposed approach when
compared to 11 different state-of-the-art fingervein PAD algorithm.

Table 12.4 indicates the quantitative results of the proposed scheme on the
laserjet artifact species. The performance of the proposed scheme is compared
with 11 different state-of-the-art systems evaluated on the same database following the same protocol. The obtained results with the proposed method
have shown the best performance with lowest error rates of APCER = 0%
and BPCER = 0%.

Based on the obtained performance of the proposed scheme on two different
fingervein presentation attack databases, it can be noted that:

- The proposed method has indicated the best performance on both laser
and inkjet print artifact species from both fingervein presentation attack
databases.
- The proposed method has consistently shown a BPCER of 0% on both
artifact species from both fingervein presentation attack databases.

TABLE 12.3

Performance results for the proposed PAD algorithm on the inkjet printed artifact species

Type	Techniques	APCER	BPCER
Texture-based schemes	Ensemble of BSIF - SVM	13.2	8.4
	GLCM - SVM	5.6	40
	$LBP^{u2}_{(3X3)}$ - SVM	22.4	6
	LPQ - SVM	7.6	6
	LBP Variance - SVM	21.6	36.4
	Steerable Pyramids - SVM	4.4	2.8
Frequency-based techniques	2D FFT - SVM	47.2	9.6
	2D Cepstrum - SVM	9.6	43.2
Quality-based techniques	Local Entropy Map - SVM	12.4	7.6
	Block-wise Sharpness - SVM	10.8	19.2
	Block-wise standard deviation - SVM	13.2	9.6
Transferable Deep-CNN [9]	Deep-CNN with softmax	3.2	0
Proposed method	Deep-CNN with classifier fusion	**1.2**	**0**

TABLE 12.4

Performance results for the proposed PAD algorithm on the laserjet artifact species

Type	Techniques	APCER	BPCER
Texture-based schemes	Ensemble of BSIF - SVM	15.6	11.2
	GLCM - SVM	18.4	36
	$LBP^{u2}_{(3X3)}$ - SVM	1.2	20.8
	LPQ - SVM	0.4	20.4
	LBP Variance - SVM	14	47.2
	Steerable Pyramids - SVM	0.4	5.66
Frequency-based techniques	2D FFT - SVM	14	82
	2D Cepstrum - SVM	3.6	27.6
Quality-based techniques	Local Entropy Map SVM	6.8	52
	Block-wise Sharpness - SVM	3.2	41.2
	Block-wise standard deviation - SVM	4.4	42.8
Transferable Deep-CNN [9]	Deep-CNN with softmax	0.4	0
Proposed method	Deep-CNN with classifier fusion	**0**	**0**

12.4 Discussion and Summary

The evolution of Deep CNNs and the ability to adopt a pretrained network to other applications has indicated their applicability to several real-life applications. In this chapter, we have explored transferable features from the pretrained deep CNN (AlexNet) and trained this network to adapt to our precise problem of fingervein PAD. To facilitate the learning of robust features by reducing the over-fitting of the network, we augmented the pretrained network with five additional layers that are arranged with consecutive fully connected and dropout layers. We then use the final fully connected layer to obtain a feature vector and train three different classifiers with softmax, LDA, and the SVM classifier. Given the probe video/image of the fingervein, the final decision is made by combing the decisions from all three classifiers using majority voting. Extensive evaluation of the proposed scheme is carried out using two different databases such as the fingervein video presentation attack database (Database 1) and fingervein image presentation attack database (Database 2). The quantitative results obtained using the proposed method has indicated the best performance on both fingervein video presentation attack database when compared with five different state-of-the-art schemes and fingervein image presentation attack database when compared with 11 different state-of-the-art systems. The obtained results have clearly shown the superiority of the proposed scheme for fingervein PAD. The future work involves in exploring the proposed scheme for unseen attacks and cross database evaluation.

12.4.1 Future research avenues

It is well demonstrated that the use of the transferable features from the prelearned deep-CNN networks are robust to detect the presentation attacks on the fingervein sensor. The future work should focus on both generating a presentation attack and also on detecting the same. New kinds of presentation attack instruments need to be developed that can simulate the artificial blood flow that can be used to bypass the liveness measure techniques. It is well known that the use of the learning-based schemes can show the better performance, but lacks generalizability when it comes to the unseen fingervein artifacts. Thus, it is necessary to develop newer PAD schemes that can be generalized on the unseen PAI.

References

1. VeinID. http://www.hitachi.eu/veinid/atmbanking.html. Accessed February 1, 2017.

2. R. Raghavendra, M. Avinash, C. Busch, and S. Marcel. Finger vein liveness detection using motion magnification. In *IEEE International Conference on Biometrics: Theory, Applications and Systems*, September 2015.

3. P. Tome, M. Vanoni, and S. Marcel. On the vulnerability of finger vein recognition to spoofing. In *International Conference of the Biometrics Special Interest Group (BIOSIG)*, pp. 1–10, IEEE, Darmstadt, Germany, September 2014.

4. D. T. Nguyen, Y. H. Park, K. Y. Shin, S. Y. Kwon, H. C. Lee, and K. R. Park. Fake finger-vein image detection based on fourier and wavelet transforms. *Digital Signal Processing*, 23(5):1401–1413, 2013.

5. P. Tome, R. Raghavendra, C. Busch, S. Tirunagari, N. Poh, B. H. Shekar, D. Gragnaniello, C. Sansone, L. Verdoliva, and S. Marcel. The 1st competition on counter measures to finger vein spoofing attacks. In *The 8th IAPR International Conference on Biometrics (ICB)*, Phuket,Thailand, May 2015.

6. D. Kocher, S. Schwarz, and A. Uhl. Empirical evaluation of lbp-extension features for finger vein spoofing detection. In *Proceedings of the International Conference of the Biometrics Special Interest Group (BIOSIG16)*, pp. 1–8, 2016.

7. R. Raghavendra and C. Busch. Presentation attack detection algorithms for finger vein biometrics: A comprehensive study. In *11th International Conference on Signal-Image Technology Internet-Based Systems (SITIS-2015)*, pp. 628–632, IEEE, Bangkok, Thailand, November 2015.

8. A. Krizhevsky, I. Sutskever, and G. E. Hinton. Imagenet classification with deep convolutional neural networks. In F. Pereira, C. J. C. Burges, L. Bottou, and K. Q. Weinberger (Eds.), *Advances in Neural Information Processing Systems 25*, pp. 1097–1105. Curran Associates, Red Hook, NY, 2012.

9. R. Raghavendra, K. B. Raja, S. Venkatesh, and C. Busch. Transferable deep convolutional neural network features for fingervein presentation attack detection. In *International Workshop on Biometrics and Forensics (IWBF-2017)*, pp. 1–6, IEEE, Warwick, UK, April 2017.

10. R. Raghavendra, K. B. Raja, J. Surbiryala, and C. Busch. A low-cost multimodal biometric sensor to capture finger vein and fingerprint. In *International Joint Conference on Biometrics (IJCB)*, pp. 1–7, IEEE, Florida, September 2014.

11. ISO/IEC JTC1 SC37 Biometrics. *ISO/IEC DIS 30107-3:2016 Information Technology—Presentation Attack Detection—Part 3: Testing and Reporting and Classification of Attacks*. International Organization for Standardization, Geneva, Switzerland, 2016.

Index

Note: Page numbers followed by f and t refer to figures and tables respectively.

2D score maps, 87–89
2v-SVM, 210–211

A
Aadhaar, 2
Activation map, 14
AEs. *See* Autoencoders (AEs)
AFFACT (Alignment Free Facial
    Attribute Classification
    Technique), 157
AFIS (automated fingerprint
    identification system), 199
AFLW
    in HyperFace, 41
    in proposed deep cascaded
        regression, 97–99
AFLW-Full database, 97, 99,
    101, 104f
AFW
    database, 97
    data set, 48–49
Age-invariant face verification, 145–146
AlexNet, 16, 19, 34, 297
Alignment Free Facial Attribute
    Classification Technique
    (AFFACT), 157
ANet (attribute classification
    network), 156
Arch orientation, 207
Attack presentation classification
    error rate (APCER), 264,
    300–301
Attribute classification network
    (ANet), 156
Attribute prediction experiments,
    159–161, 160f

Autoencoders (AEs), 9–11
    bimodal deep, 13
    class representative, 13
    contractive, 13
    CsAE, 11–12
    deeply coupled, 13–14
    linear, 10
    multimodal deep, 13
    single-layer, 10, 10f
    stacked, 10–11
        denoising, 12–13
    supervised, 11
        transfer learning-based, 12
Automated biometric system, 1
Automated DCNN-based face
        identification/verification
        system, 38, 39f
    deep convolutional networks,
        44–46
    HyperFace, 40–41, 40f
        face detection, 41
        gender recognition, 43–44
        landmark localization, 42
        pose estimation, 42
        visibility factor, 42
    issues, 56
    joint Bayesian metric learning,
        46–47
Automated face-recognition
        system, 34
    face detection, 35–36
    facial landmark detection,
        36–37
    feature representation, 37–38
    metric learning, 38

Automated fingerprint identification
system (AFIS), 199
Automatic adaptation antispoofing
system, 253
AVTS, 263, 280f, 282f

B
Back-propagation technique, 15–16
Bimodal deep AE, 13
Binarized statistical image features
(BSIF), 253
Binomial deviance, 113
Biometric(s), 1
    authentication, 246
        system, 1–2, 1f
    feature extraction, 2–3
    hand-crafted features, 3–4
    RBM for, 8–9
Biometric Smart Pen (*BiSP*®),
231, 233f
Boltzmann machine, 5, 5f
Bona-fide presentation classification
error rate (BPCER),
264, 301
BSIF (binarized statistical image
features), 253

C
Caffe library, 236
Cascade-based face detectors, 35
CASIA, 262, 267f, 270f
CASIA-WebFace, 73
    dataset for face recognition, 45
CCA (Correct Classification
Accuracy), 212
CDBN (convolutional deep belief
network), 8, 206
CIFAR10 full architecture, 236, 237f
    confusion matrices, 239, 241f
    convergence process, 239, 240f
Class representative AE, 13
Class Sparsity Signature-based RBM
(cssRBM), 7
CMC. *See* Cumulative match
characteristic (CMC)

CNNs. *See* Convolutional neural
networks (CNNs)
Combined Refinement (CR) method,
184, 187, 188f
Constrained fingerprint recognition,
203–207, 204t
Constrained metric embedding,
118–119
Contractive AE, 13
Contrastive AE, 11–12
Contrastive loss function, 70–71
ConvNet-RBM, 8–9
Convolutional deep belief network
(CDBN), 8, 206
Convolutional neural networks
(CNNs), 14, 110,
228–229, 297
    architectures, 15, 229, 229f
        AlexNet, 16, 19
        back-propagation technique,
        15–16
        Deep Dense Face Detector, 19
        DeepFace, 18
        FaceNet, 19
        GoogLeNet, 18
        Le-Net5, 16
        network-in-network, 17
        R-CNN, 19
        ResNet, 18–19
        VGGNet, 17–18
        ZFNet, 17
    complex cells, 229
    filter bank, 229–230
    Inception in, 18
    normalization, 230
    sampling, 230
    simple cells, 229
    traditional, 14, 16f
        convolutional layer, 14
        fully connected layer, 15
        pooling layer, 15
        ReLU, 14–15
Cornell kinship databases, 138,
138f, 142

Correct Classification Accuracy (CCA), 212
Correlation filter (CF), 175
CR (Combined Refinement) method, 184, 187, 188f
CsAE (contrastive AE), 11–12
cssRBM (Class Sparsity Signature-based RBM), 7
CUHK01 data set, 119
CUHK03 data set, 119
CUHK Campus data set, 114
Cumulative match characteristic (CMC)
    EDM, 120, 121f
    scores, 47
Cut photo attack, 262

D
Databases
    kinship verification
      Cornell kinship, 138, 138f, 142
      KinFaceW-I, 138f, 139, 142–143
      KinFaceW-II, 138f, 139, 142–143
      UB KinFace, 138f, 139, 143, 146
      WVU kinship, 138f, 139, 143–145
    Look-alike face, 73
    proposed deep cascaded regression on
      AFLW-Full database, 97, 99, 101, 104f
      AFW, 97
      and baselines, 97–99
      challenging databases performance, 101–104
      PIFA-AFLW, 99–100, 101t
      UHDB31 database, 97, 99
Data-driven characterization algorithms, 247, 252
DBM. *See* Deep Boltzmann machine (DBM)

DBN. *See* Deep belief network (DBN)
DCNNs. *See* Deep convolutional neural networks (DCNNs)
DeCAF (Deep Convolutional Activation Feature), 157
DeconvNet, 17, 87f, 93, 95
    HGN, 99
Deep belief network (DBN), 6, 133, 135
    fc-DBN, 9
    multiresolution, 8
    sparse, 7–8
Deep Boltzmann machine (DBM), 6–7
    joint, 9
    multimodal, 8–9
Deep cascaded regression, 86, 88–89
    3D pose-aware score map, 90
    DSL, 90–92, 92f
    GoDP model, 93, 94f–95f, 100–101
      decision pathway, score maps in, 93, 95
      information pathway features, 93
      network structure, 95
    optimized progressive refinement, 89–90
    proposed, experiments of, 96
      architecture analysis and ablation study, 99–101
      challenging databases performance, 101–104
      databases and baselines, 97–99
      PIFA-AFLW database, 99–100, 101t
      robustness evaluation, 101–104
Deep Convolutional Activation Feature (DeCAF), 157
Deep convolutional neural networks (DCNNs), 34, 37
    for face identification, 44t
    feature-activation maps, 45, 45f

Deep Dense Face Detector, 19
Deep Dictionary, 21
DeepFace, 18
Deep learning, 4, 21–22
    algorithms, kinship verification
        DBN, 135
        stacked denoising
            autoencoder, 134–135
    architectures, 4–5, 20–21
        AEs, 9–14, 10f
        CNN, 14–19
        RBM, 5–9
    challenges
        small sample size, 21
        theoretical bounds, lack of, 22
Deep Metric Learning (DML)
        method, 110
    cross-dataset evaluation,
        115, 115t
    metric and cost function,
        112–114
    neural network architecture,
        111–112
    performance, 114–115
Deep neural networks, 153
Deformable part model (DPM),
        35–36
Diagonalizations, LeNet, 170, 170f
Digit-classification errors, 164, 165f
Discriminative multimetric learning
        (DMML), 140
Discriminative restricted Boltzmann
        machine (DRBM), 7
Distance-aware softmax function
        (DSL), 88, 90–92, 92f
Distillation, 165
DML method. *See* Deep Metric
        Learning (DML) method
DMML (discriminative multimetric
        learning), 140
DNA testing, kinship
        verification, 129
DPM (deformable part model),
        35–36
Dropout, 137

DSL (distance-aware softmax
        function), 88, 90–92, 92f
Dynamic Refinement (DYN)
        method, 185, 191f

E
Embedding Deep Metric (EDM), 110
    CMC curves and rank-1
        accuracy, 120, 121f
    constrained metric embedding,
        118–119
    framework, 118, 118f
    metric-learning layers, 118
    moderate positive mining,
        116–118, 116f
    performance, 119–121
Equal error rates (EERs), 72, 186
Eulerian video magnification, 255
Extended MACH (EMACH), 176

F
Face
    antispoofing techniques, 249–250
    biometric systems, 156
    detection, 35–36
        on FDDB data set, 48, 48f
        in HyperFace, 41
    identification and verification, 34
        IJB-A and JANUS CS2 for,
            49–50, 50f
        performance evaluations,
            51–53
    networks analysis, 158–159
        attribute prediction
            experiments, 159–161
        pose-prediction experiments,
            161–163, 162t
    recognition, 34
        DCNN for, 45
        intraclass variations, 3, 3f
        pipeline, 2
Face-frontalization technique, 74
FaceNet, 19
Facial cues, kinship verification
        via, 129

Facial key-point localization, 85–88
  2D score maps in, 89
  softmax loss function in, 91
Facial landmark detection, 36–37
  on IJB-A, 48–49
  landmark localization,
    HyperFace, 42
False Acceptance Rate (FAR), 72
Familial traits, 129
fc-DBN. *See* Filtered contractive (fc)
  DBN
FDDB dataset, face detection on,
  47–48, 48f
Feature extraction, 2–3
  person reidentification, 109
FG-Net database, 146
Filtered contractive (fc) DBN, 9
Filter map, 14
Fingerphotos, 198–199, 215
  using deep ScatNet,
    215–217, 216f
  experimental protocol,
    217–218
  fingerphoto to fingerphoto
    matching, 218, 218f,
    219f, 219t
  fingerphoto to live-scan
    matching, 218–219,
    219t, 220t
Fingerprint
  antispoofing technique, 253–254
  recognition, 197–203
    challenges, 199, 200f
    constrained, 203–207, 204t
    controlled/uncontrolled
      capture mechanisms,
      198, 198f
    features, 199, 201f
    unconstrained, 207–209
    VeriFinger 6.0 SDK, 201, 202f
Fingervein PAD, 295–298
  artifact samples, 296, 296f
  experiments
    image, 300, 301f, 302
    video, 300–302

independent models, 299
performance evaluation, 300
transferable features,
  298–299, 298f
Fisher criterion, 113

G
Gender recognition, HyperFace,
  43–44
Generative adversarial network
  (GAN), 21
Genuine pair, 67, 72
Globally optimized dual-pathway
  (GoDP), 87f
  architecture, 87–88
  in deep cascaded regression, 93,
    94f–95f, 100–101
  decision pathway, score maps
    in, 93, 95
  information pathway
    features, 93
  network structure, 95
GoogLeNet, 18
Greedy layer-by-layer
  optimization, 11
Group sparse autoencoder
  (GSAE), 209

H
Half total error rate (HTER), 264
Hand-crafted features, 3–4
Harmonium, 5
HeadHunter, 36
Heat maps, LeNet feature spaces,
  168, 169f
Hierarchical visual codebook
  (HVC), 255
High-definition attack
  techniques, 262
Homo sapiens, 22
Hourglass network (HGN), 99
HTER (half total error rate), 264
HyperFace, 48
  face detection, 41
  on FDDB data set, 48, 48f

HyperFace (*Continued*)
  gender recognition, 43–44
  landmark localization, 42
  pose estimation, 42
  visibility factor, 42

I
Identical twins
  identification, 68–69
  recognition, 66f
  restricted year, 76
  verification, 76
Identity-correlated attributes, 159
Identity-dependent attributes, 159
Identity-independent attributes, 161
IJB-A. *See* Intelligence Advanced
      Research Projects Activity
      (IARPA) Janus Benchmark
      A (IJB-A) data set
Impostor pair, 72
Individual Refinement (IR) method,
      184–185
Inked fingerprints, 198
Intelligence Advanced Research
      Projects Activity (IARPA)
      Janus Benchmark A
      (IJB-A) data set, 34–35
  for face identification/
      verification, 49–50, 50f
  performance evaluations,
      51–53
  facial landmark detection on,
      48–49
  vs. JANUS CS2, 50
Intraclass
  variations, 3f, 11, 116
  YaleB+ face database, images
      from, 186, 186f
Invariant representation, 163
  analysis and visualization,
      167–170
  angle prediction, 164, 164f
  architectures for enhancement
      of, 165
  μ-LeNet, 165–166

PAIR-LeNet, 166–167
  digit-classification errors,
      164, 165f
  MNIST, preliminary evaluation
      on, 163–165
Iris-liveness detection, 254–255
IR (Individual Refinement) method,
      184–185

J
Janus Challenging set 2 (JANUS
      CS2) data set, 47
  for face identification/
      verification, 49–50, 50f
  performance evaluations,
      51–53
  vs. IJB-A, 50
Joint Bayesian metric learning, 46–47
Joint DBM, 9
JudgeNet, 205

K
KinFaceW-I databases, 138f, 139,
      142–143
KinFaceW-II databases, 138f, 139,
      142–143
Kinship model index, 130
Kinship verification, 128, 128f
  accuracy results, 140, 141t, 142t
  age-invariant face verification,
      145–146
  characteristics, 138, 138f
  databases, performance results
    Cornell kinship data set, 142
    KinFaceW-I and
      KinFaceW-II, 142–143
    UB KinFace, 143, 146
    WVU kinship, 143–145
  data set, 138–139
  deep learning algorithms, 133–134
    DBN, 135
    stacked denoising
      autoencoder, 134–135
  DNA testing, 129
  KVRL framework, 135–138
    deep learning, 139–140

overview, 131, 131t
pairwise, 133
protocol, 140
self-kinship, 133
via facial cues, 129–130
Kinship Verification via
   Representation Learning
   (KVRL) framework, 133,
   135–138
deep learning, 139–140
Kullback–Leibler divergence metric,
   134
KVRL-DBN, 137
KVRL-SDAE, 136

L
Labeled Faces in the Wild
   (LFW), 156
data set, 35, 50, 53–54, 55t, 73
Large Margin Nearest Neighbor
   metric, 38
Latent fingerprints, 198–199, 200f
characteristics of, 207, 208f
minutiae extraction using
   GSAE, 209–212, 210f
architecture, 210, 211f
experimental protocol,
   212–213, 212t
results and analysis, 213–215,
   213t, 214t
LBP. *See* Local binary pattern
   (LBP)
Le-Net5, 16
Level zero ScatNet
   representation, 216
LFW. *See* Labeled Faces in the
   Wild (LFW)
Lineal kin relationships, 129, 129f
Linear AE, 10
LivDet (Liveness Detection
   Competitions), 262
LivDet2009 benchmark, 262,
   268, 272f
LivDet2013 benchmark, 263,
   268, 273f

Liveness detection, 246
Liveness Detection Competitions
   (LivDet), 262
Live-scan fingerprints, 198
LNets (localization networks), 156
Local binary pattern (LBP), 250
descriptor, 252, 255
histograms, 253
modified, 255
spatiotemporal version, 250
Localization networks (LNets), 156
Local Phase Quantization
   (LPQ), 252
descriptor, 253
LocateNet, 205
Log likelihood ratio, 46
Long short-term memory
   (LSTM), 20
Look-alike data set, 80
Look-alike face database, 73
Loss function
binary cross entropy, 41
CsAE, 12
cssRBM, 7
DRBM, 7
RBM, 6
supervised AE, 11
LSTM (long short-term memory), 20

M
Machine-learning algorithms, 265
Mahalanobis distance, 118
Maternal Perinatal Association
   (MPA), 130
Maximum average correlation height
   (MACH), 176
Mean Activation Vector (MAV), 165
Metric learning, 132
automated face-recognition
   system, 38
joint Bayesian, 46–47
Minutia Cylinder Code (MCC), 202
Mixed Objective Optimization
   Network (MOON), 157
Mlpconv, 17

MNIST architecture, 236, 238f
MNRML (multiview neighborhood
     repulsed metric
     learning), 140
Mobile attack techniques, 262
MobILive, 255
Moderate positive mining,
     116–118, 116f
MOON (Mixed Objective
     Optimization Network), 157
μ-LeNet, 165–166
Multilabel conditional RBM, 8
Multimodal DBM model, 8–9
Multimodal deep AE, 13
Multiresolution DBN, 8
Multispectral image, 229
Multiview neighborhood repulsed
     metric learning
     (MNRML), 140

N
Neocognitron, 229
Network-in-network architecture,
     CNN, 17
Normalization constant, 6
Normalized mean error (NME), 99

P
PAD. *See* Presentation attack
     detection (PAD)
PAIR (Perturbations to Advance
     Invariant Representations),
     166, 166f
PAIR-LeNet, 166–167
PAIs (presentation attacks
     instruments), 297
Partition function, 6
Patch-based matching
     algorithm, 110
Peak correlation energy (PCE), 183
Perceptron, 4, 4f
Person reidentification, 109
     DML method, 110–115
     EDM, 110, 116–121

feature extraction, 109
hard positive cases, 116, 116f
learning deep metrics, 113
patch-based matching
     algorithm, 110
similarity learning, 109
Perturbations to Advance Invariant
     Representations (PAIR),
     166, 166f
Pose-prediction experiments,
     161–163, 162t
Presentation attack detection
     (PAD), 246, 248, 248f, 296
in CNN
     memory footprint, 260
     network architecture, 257–258
     training and testing, 258–260
error metrics, 264
face, 249–252, 264–265
     cross-dataset results,
          268, 268t
     same-dataset results,
          265, 265t
     static analysis process, 251
     video-based analysis
          process, 251
fingerprint, 252–254, 268
     cross-sensor results, 271, 274t,
          275f, 276f
     same-sensor results, 268,
          271, 271t
fingerprint-spoofing
     benchmarks, 262
     LivDet2009, 262
     LivDet2013, 263
iris, 254–256, 271, 274
     cross-dataset results, 274, 280,
          281t, 282–283
     same-dataset results,
          274, 278t
iris-spoofing benchmarks, 263
     AVTS, 263
     Warsaw LivDet2015, 263–264
metrics and data sets, 260–264,
     261t

unified frameworks, 256–257
video-based face-spoofing
  benchmarks, 260
  CASIA, 262
  replay-attack, 260, 262
Presentation attacks, 296
Presentation attacks instruments
  (PAIs), 297
PRID data set, 114
Print attack techniques, 260
Pupil dynamics model, 256

R
Random decision forest
  (RDF), 217
Rate-coded RBM, 9
RBM. *See* Restricted Boltzmann
  machine (RBM)
R-CNN, 19
R-CNN Fiducial, 48–49
Receiver operating characteristic
  (ROC) curve, 47, 72, 142,
  142f, 143t, 144f
Rectified linear unit (ReLU),
  14–15, 183
Recurrent neural networks (RNNs),
  20, 20f
The Red Thumb Mark (novel), 246
Replay-attack, face PAD, 260, 262,
  266f, 269f
ResNet, 18–19
Restricted Boltzmann machine
  (RBM), 5–7, 5f, 135
  advances in, 7–8
  for biometrics, 8–9
  cssRBM, 7
  DRBM, 7
  energy function, 5–6
  loss function, 6
  multilabel conditional, 8
  rate-coded, 9
  supervision in, 7
ROC curve. *See* Receiver
  operating characteristic
  (ROC) curve

S
Scattering Network (ScatNet), 205
  fingerphoto matching using,
    215–217, 216f
  experimental protocol,
    217–218
  fingerphoto to fingerphoto,
    218, 218f, 219f, 219t
  fingerphoto to live-scan,
    218–219, 219t, 220t
SCFs. *See* Stacked correlation filters
  (SCFs)
SCNNs (Siamese Convolutional
  Neural Networks), 69, 71, 71f
Score maps
  2D, 89
  3D pose-aware, 90
  in decision pathway of GoDP,
    93, 95
Screen diagrams, LeNet feature
  spaces, 167, 168f
SDAEs (Stacked Denoising
  Autoencoders), 12–13,
  133–135
SDSAE (stacked denoising sparse
  autoencoder), 208
Self-kinship problem, 145–146
Shape-indexed features, 86, 90
Siamese architecture, 68–69
  in twin Siamese discriminative
    model, 69–70, 71f
Siamese Convolutional Neural
  Networks (SCNNs), 69,
  71, 71f
Siamese long short-term memory
  architecture, 110
Siamese network, 18
  DML, 110–111, 111f
  identical twins verification, 67
SignRec data set, 231–233, 232f
Similarity learning, person
  reidentification, 109
Single-layer AE, 10, 10f

Softmax, 16
   loss function, 91
SoftmaxWithLoss, 75
Sparse autoencoder, 206
Spatial Pyramid Learning-based
   feature descriptor
   (SPLE), 132
Spoofing detection, 246
Stacked AE, 10–11
Stacked correlation filters (SCFs),
   176, 178–180
   correlation output refinement,
   183–185, 184f
   experiments and analysis,
   185–191
   initial correlation output,
   180–181, 180f
   overview, 179f
   stacked layers, 182–183
   testing algorithm, 185
   training algorithm, 185
Stacked Denoising Autoencoders
   (SDAEs), 12–13, 133–135
Stacked denoising sparse
   autoencoder (SDSAE), 208
Stacking autoencoders, 136
Steerable Pyramids, 297
Supervised AEs, 11
   transfer learning-based, 12
Support vector machines (SVMs),
   157, 205

T
Twin Days Festival Collection
   (TDFC), 68, 73
Twin Siamese discriminative model,
   69, 71f
   contrastive loss, 70–71
   data sets, 73
   evaluation and verification
   metric, 72
   look-alike data set, face
   verification with, 80
   model pretraining, 74–76
   preprocessing, 74

SCNNs architecture, 71
Siamese architecture, 69–70, 71f
   verification, 76
   with age progression and
   unrestricted year,
   79–80, 79t
   architecture in, 77t
   with restricted year,
   76–78, 78t
   VGG-16, 76
Two-stage KVRL with three-level
   SDAE approach, 135, 136f

U
UB KinFace databases, 138f, 139,
   143, 146
UHDB31 database, 97, 99
Unconstrained face detection, 36
Unconstrained fingerprint
   recognition, 207–209
Unique Identification Authority of
   India, 2

V
Validation Rate (VR), 72
Verification rates (VRs), 186
VeriFinger 6.0 SDK, 201, 202f
VGG-16, 76
VGGNet, 17–18, 75
Video attack, 262
VIPeR data set, 114
Visibility factor, 42
Visual Geometry Group (VGG), 156
   network, 257–258, 258f

W
Warped photo
   attack, 262
Warsaw LivDet2015, 263–264,
   279f, 281f
WVU kinship databases, 138f, 139,
   143–145

Z
ZFNet, 17

Printed and bound by CPI Group (UK) Ltd, Croydon, CR0 4YY

17/10/2024

01775682-0014